The Dynamics of Fluidized Particles

Recent years have seen major progress in the development of equations to describe the motion of fluid–particle mixtures and their application to a limited range of problems. With rapid advances in numerical methods and computing power we are now presented with new opportunities to use direct integration of these equations in the solution of complex practical problems. However, results so obtained are only as good as the equations on which they are based, so it is essential to have a clear understanding of the underlying physics and the extent to which it is reflected properly in these equations.

In *The Dynamics of Fluidized Particles*, the author formulates these equations carefully and then to describe some important existing applications that serve to test their ability to predict salient phenomena. This account will be of value to both novices and established researchers in the field, and also to people interested in applying the equations to practical engineering problems.

Roy Jackson is a Professor of Engineering & Applied Science Emeritus at Princeton University. He has received many academic honours, including the School of Engineering Distinguished Teaching Award from Princeton University, and is a fellow of the Royal Society. The American Chemical Society has also recently published a "festschrift" in honour of his many research contributions.

CAMBRIDGE MONOGRAPHS ON MECHANICS

FOUNDING EDITOR

G. K. Batchelor

GENERAL EDITORS

S. Davis

Walter P. Murphy Professor
Applied Mathematics and Mechanical Engineering
Northwestern University

L. B. Freund

Henry Ledyard Goddard University Professor
Division of Engineering
Brown University

S. Leibovich

Sibley School of Mechanical & Aerospace Engineering
Cornell University

V. Tvergaard

Department of Solid Mechanics
The Technical University of Denmark

The Dynamics of Fluidized Particles

ROY JACKSON

Princeton University

CAMBRIDGE
UNIVERSITY PRESS

PUBLISHED BY THE PRESS SYNDICATE OF THE UNIVERSITY OF CAMBRIDGE
The Pitt Building, Trumpington Street, Cambridge, United Kingdom

CAMBRIDGE UNIVERSITY PRESS
The Edinburgh Building, Cambridge CB2 2RU, UK   http://www.cup.cam.ac.uk
40 West 20th Street, New York, NY 10011-4211, USA   http://www.cup.org
10 Stamford Road, Oakleigh, Melbourne 3166, Australia
Ruiz de Alarcón 13, 28014 Madrid, Spain

First published 2000

Printed in the United States of America

Typeface Times Roman 10/13 pt.   *System* LATEX 2_ε [TB]

A catalog record for this book is available from the British Library.

Library of Congress Cataloging in Publication Data
Jackson, R. (Roy), 1931–
The dynamics of fluidized particles / Roy Jackson.
p.   cm. – (Cambridge monographs on mechanics)
Includes bibliographical references.
ISBN 0-521-78122-1 (hb)
1. Fluidization.   2. Fluid dynamics.   I. Title.   II. Series.
TP156.F65 J33   2000
660′.284292 – dc21   99-086301

ISBN 0 521 78122 1 (hb)

In memory of Susan

Contents

Preface

This book addresses the motion of systems of solid particles immersed in a fluid that may be a liquid or a gas. The focus is on the range of particle concentrations of greatest interest in the operation of process plants, that is, solids volume fractions anywhere from a few percent to random close packing. As typical process applications we might mention hoppers and bunkers, dense fluidized beds, pneumatic transport lines, circulating fluidized beds, standpipes, cyclones, riser reactors, and slurry pipelines, but the same ideas can be used in nonprocess applications such as sediment transport, landslides, and avalanches. The book is intended as an introduction to this field for graduate students and others entering it for the first time but, by drawing together widely scattered material, it is hoped that it may also serve as a useful overview for more experienced workers. Most of the material is covered somewhere in the existing literature, to which the reader's attention is directed, but some appears here for the first time, for example, parts of Chapters 3 and 4.

Many of the figures are taken from other publications and my thanks are due to the copyright holders for permission to reproduce this material. In certain cases these permissions are acknowledged in the captions of the figures in question but, in addition, I am indebted to the following organisations and individuals: Academic Press for Figures 5.1, 5.6, 5.7, and 6.32; Birkhäuser Verlag for Figure 5.26; The Institution of Chemical Engineers for Figures 5.4, 5.5, 5.8, 5.9, 5.10, 5.14, 5.15, and 5.16; T. B. Anderson for Figures 4.8, 4.10, and 4.11; Y-M. Chen for Figures 7.4 and 7.5; B. Glasser for Figure 5.44; G. D. Cody for Figure 3.20; and T. J. Mountziaris for Figure 7.12.

I would not have undertaken the task of writing this book without urging by George Batchelor, whose influence has been a guiding light since my student days. A serious start on the work was made during a half year spent as a Visiting

Fellow Commoner at Trinity College, Cambridge, in 1994 and I am grateful to the College for providing this opportunity for uninterrupted thought.

Such understanding of the subject as I have owes much to informal interactions over the years with my graduate students, and also with colleagues in academia and industry, among whom I should mention Sankaran Sundaresan, John Davidson, John Hinch, Jennifer Sinclair, Bud Homsy, Don Koch, John Gwyn, George Cody, and the late Yuri Buyevich. In particular, my collaboration with Sankaran Sundaresan over the past decade has been a source of special pleasure. At a more general level the intellectual atmosphere of the Chemical Engineering Department at Princeton and stimulating discussions with Dudley Saville, Ioannis Kevrekidis, Pablo Debenedetti, Bill Russel, and Sandra Troian have served to sharpen my fluid-mechanical wits in many ways. In addition Pablo Debenedetti, in his role as department chairman, has been most supportive of my literary efforts. Finally, I must acknowledge the invaluable help provided by Patti Weiss, who has taken care of many of those time-consuming details under whose weight the project might otherwise have foundered.

1

The mathematical modelling of fluidized suspensions

1.1 Introduction

At the high concentrations of interest in this book interactions between particles play a major role. These are of two kinds: interactions via motions induced in the interstitial fluid and interactions by direct contact between solid particles. The former dominate when the interstitial fluid is a liquid and the particle concentration is not too large, while the latter dominate when the fluid is a gas, though they may also be important with a liquid medium if the particles are present at very high concentration.

As a result of fluid–particle and particle–particle interactions the behaviour of these systems is very complicated. The distribution of the particles in space is usually far from uniform; the coexistence of regions of strongly contrasting concentration is apparently an intrinsic feature of fluid–particle systems in motion. For many years it was therefore assumed that the properties of these systems could be predicted only by the use of empirical correlations. Many of these were developed and they have since formed the basis for most engineering design, as reflected in the earlier textbooks on the subject; see, for example, Othmer (1956), Leva (1959), and Zenz & Othmer (1960).

Interest in the mechanisms responsible for the observed heterogeneities, and how simple pictures of their associated flow fields might be integrated into design calculations, was sparked by the monograph of Davidson & Harrison (1963) and it dominated the literature of the following decade (Kunii & Levenspiel, 1969). The same period also marked the beginnings of interest in a more fundamental approach, based on equations of motion for the inter-acting fluid and particles. This was regarded primarily as a means of gaining better understanding of the mechanisms responsible for the complexities of the observed behaviour, rather than as a basis for practical design calculations, and

1

indeed there was considerable progress of this sort. However, the 1980s saw the dawning of a realization that it might eventually be possible to base quantitative design calculations on differential equations of continuity and momentum balance. Indeed, rapid improvements in the speed and memory capacity of digital computers, coupled with better methods for addressing the difficulties attending numerical solution of the equations of motion, tempted a few pioneers to attempt to predict bubble formation in dense fluidized beds by direct integration of these equations (Pritchett et al., 1978; Gidaspow & Ettehadieh, 1983; Gidaspow et al., 1986).

Though empirical correlations remain, quite properly, a central feature of practical design methods, there is now a rapidly increasing interest in the use of methods that can loosely be described as "computational fluid dynamics", or CFD. These make it possible to answer many questions that cannot be addressed by using conventional correlations. For example, correlations may be available that are quite successful in predicting the performance of a riser reactor with certain standard arrangements for introducing particles and gas at its foot, but they can say nothing useful about the effect of a proposed change in the detailed geometry of these arrangements, such as replacement of a single entry point for the particles by two or more entry points, or a change in the angle at which a standpipe meets the bottom of the riser. In the case of a dense fluidized bed one might be interested in the effects of proposed internal baffling of specified geometry, in exploring the most effective disposition of immersed heat transfer tubes, or in comparing different designs for the gas distributor. Questions of this sort could, in principle, be addressed if efficient computational codes were available to solve the equations describing the dynamics of the system.

These equations can be formulated at different levels of detail. At the most fundamental level the motion of the whole system is determined by the Newtonian equations of motion for the translation and rotation of each particle and the Navier–Stokes and continuity equations, to be satisfied at every point of the interstitial fluid. These are linked by the no-slip condition between the solid and the fluid on each particle boundary, and the fluid must also satisfy no-slip conditions everywhere on the walls bounding the entire system of interest. Calculations at this level of detail have been performed successfully, but only for quite small numbers of particles. It is not presently conceivable that they could be extended to systems containing the very large number of particles present in commercial units such as fluidized beds.

A second description, at a less detailed level, can be obtained by replacing the fluid velocity at each point by its average, taken over a spatial domain large enough to contain many particles but still small compared to the whole region occupied by the flowing mixture. The force exerted by the fluid on each particle is then related to the particle's velocity relative to this *locally averaged* fluid velocity, and to the local concentration of the particle assembly, using one of a number of empirical correlations. The Newtonian equations of motion are then solved for each particle separately, taking into account direct collisions between particles when this is appropriate. This procedure, sometimes referred to as "discrete particle modelling", is much less demanding computationally than a complete solution at the first level of detail. Results from this approach have begun to appear in the literature in the past few years (see, for example, Tsuji et al. (1993) and Hoomans et al. (1996)), and their number can be expected to increase.

At a third level of detail both the fluid velocity and the particle velocity are averaged over the local spatial domains introduced above. There are then two local-averaged velocity fields, **u** and **v**, for the fluid and the particles respectively. Each of these is defined at all points of space, so that the resulting equations *look like* the equations of motion one would write for two imaginary fluids, capable of interpenetrating so that every point is occupied simultaneously by both fluids. Consequently, a description at this level of detail is often referred to as a "two fluid model". As we shall see in Chapter 2 the formal process of averaging that leads to these equations leaves behind a number of terms whose form is not determined, and to close the equations they must be related to the fields of **u**, **v**, and particle concentration. This type of model then takes the form of coupled partial differential equations that usually must be solved numerically, subject only to boundary conditions at the boundaries of the system as a whole. This might be expected to be less demanding computationally than the solution of models at the first and second levels of detail and, as we shall see, there are certain important problems for which approximate, or even exact, analytical solutions can be found. Compared with discrete particle models these two-fluid models suffer from the disadvantage that closures must be formulated for certain important terms left undetermined in the averaged momentum equation for the particle phase. However, it is by no means clear that the physical effects corresponding to these terms are represented properly even by the discrete particle models, since these replace the fluid flow field by its locally averaged form.

This book will deal primarily with the third level of description, that is, the so-called two-fluid models, and what light they have been able to throw on a number of important questions concerning the motion of fluidized particles. This focus is in no way intended to reflect adversely on the enormous and continuing importance of more empirically based approaches, nor on the promise of more detailed models at the first or second levels described above. The former have already been covered well in a number of texts, whereas the latter are still at an early stage of development where it would not yet be appropriate to summarize them in book form. The specific problems discussed in Chapters 3 to 7, while having intrinsic importance and serving to illustrate the sort of results obtainable using equations of the two-fluid type, also frankly reflect the author's own interests over the past three and a half decades.

It should be emphasized that the point of departure for our averaged equations of motion is the equations at the first level of detail, referred to above, which are well established. The averaging process is then entirely formal and it leaves behind terms that, though not expressed in terms of the average variables themselves, are nonetheless explicitly related to details of the motion at the "microscopic" scale of the individual particles. We resort to empirical closures for these terms only because we do not know, at present, how to evaluate them exactly, except in the simplest cases. This should be contrasted to an alternative approach that, from the outset, "models" the fluid–particle system as a pair of interpenetrating continuous fluids, then formulates their equations of motion using intuitive ideas, constrained by general principles of continuum mechanics. Though both approaches may lead to similar equations there is a clear distinction between their philosophies.

1.2 A Simple Application of Equations of the Two-Fluid Type

Obviously, the equations of motion for a fluid–particle mixture, in local-averaged form, are more complicated than the equations of motion of a single-phase fluid. To begin with, they are larger in number; there are two scalar equations of continuity and two vector equations of momentum balance, with the latter coupled through the forces of interaction between the two phases. Since, even for a single-phase fluid, exact solutions are available only in quite simple situations one might naturally conclude that the scope for making useful deductions from the two-fluid equations is very limited. However, paradoxically, this is not entirely the case. There are situations in which simplifying assumptions, so radical that they would lose the features of physical interest for single-phase flow, still succeed in accounting for important aspects of the behaviour of the two-phase system. We shall now illustrate this by a simple, but practically important example.

Consider fully developed flow in a vertical pipe of circular cross section. For a single-phase incompressible fluid there is then the exact Poiseuille solution that relates the gravity force, the pressure gradient, and the velocity profile. If the fluid is permitted to slip freely in contact with the pipe wall there is a simpler, one-dimensional solution. The velocity is then the same at all points of the cross section, and it is found that the pressure gradient does not depend on the flow rate but is always equal to the hydrostatic gradient induced by gravity.

However, this solution is of little value since the result of most practical interest, namely the relation between flow rate and pressure gradient, is lost because of the simplification.

For a fluid–particle suspension, in contrast, a comparably radical simplification by no means destroys the usefulness of the solution. A complicated and largely realistic pattern of behaviour can still be predicted, as we shall now demonstrate by means of an example that is both simple and of some practical importance. We shall focus attention on a gas–particle mixture moving up a vertical pipe. Then, permitting free slip of both phases at the pipe wall, we shall seek a solution in which the local average velocity of each phase and the concentration of the particles are all independent of position and time. This means that the flow is steady and fully developed, and both the particle concentration and the velocities of gas and particles are independent of radial position. Then, if ϕ denotes the fraction of the total volume occupied by the particles, and if $\bar{V}_f$ and $\bar{V}_s$ denote the volume flow rates of fluid and solid material per unit cross-sectional area, the local average axial velocities of fluid and particles, u and v, are given by

$$u = \frac{\bar{V}_f}{1 - \phi}, \quad v = \frac{\bar{V}_s}{\phi}. \tag{1.1}$$

For small enough particles the drag force f, per unit total volume, exerted by the fluid on the particles would be expected to be proportional to $u - v$, with a factor of proportionality β that is an increasing function of ϕ:

$$f = \beta(\phi) \left[\frac{\bar{V}_f}{1 - \phi} - \frac{\bar{V}_s}{\phi} \right]. \tag{1.2}$$

We shall assume that β is given by a well-known empirical expression due to Richardson & Zaki (1954), namely

$$\beta(\phi) = \frac{\rho_s \phi g}{v_t (1 - \phi)^n}, \tag{1.3}$$

where ρ_s is the density of the solid material, v_t is the terminal velocity of fall of an isolated particle in an infinite body of the fluid, and n is a number whose value depends on the Reynolds number for a particle moving through the fluid with speed v_t.

As attention is limited to fully developed flow inertial terms vanish and the momentum equation for each phase reduces to a force balance. Since the suspending fluid is a gas, buoyancy forces can also be neglected and, making use of (1.2) and (1.3), a force balance on the particles in unit volume of the

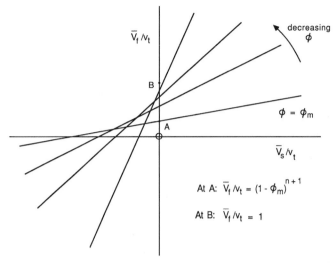

Figure 1.1. Contours of constant ϕ (or pressure gradient) in the plane of $\bar{V}_s/v_t$ and $\bar{V}_f/v_t$. (After Jackson, 1993.)

suspension gives

$$\frac{\bar{V}_f/v_t}{1-\phi} - \frac{\bar{V}_s/v_t}{\phi} = (1-\phi)^n. \tag{1.4}$$

A corresponding force balance on the gas shows that its pressure gradient must support the weight of the suspended particles, so

$$\frac{dp}{dz} = -\rho_s\phi g, \tag{1.5}$$

where z is a coordinate measured vertically upward. This indicates that the pressure gradient or ϕ may be invoked interchangeably, since they are proportional to each other.

One way of presenting these relations graphically is as contours of constant dp/dz (or equivalently ϕ) in the $(\bar{V}_s, \bar{V}_f)$-plane. From (1.4) we see that these are a set of straight lines of slopes $(1-\phi)/\phi$, as shown in Figure 1.1. The line for $\phi = 0$ coincides with the ordinate axis, while the line of smallest slope corresponds to $\phi = \phi_m$, the volume fraction for random close packing. The contours extend into the first, second, and third quadrants, where they represent cocurrent upflow, countercurrent flow, and cocurrent downflow, respectively. Points in the fourth quadrant are, of course, physically impossible since they would represent upward flow of particles with a downward flow of gas. In the second quadrant the set of contours has an envelope, and all points enclosed between this envelope, the positive $\bar{V}_f$ axis, and the negative $\bar{V}_s$ axis represent possible

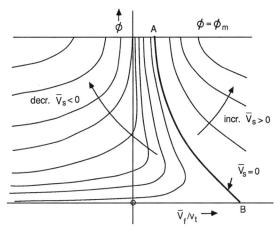

Figure 1.2. Zenz diagram (curves of constant $\bar{V}_s$ in the plane of $\bar{V}_f/v_t$ and ϕ) corresponding to Figure 1.1. (After Jackson, 1993.)

conditions for countercurrent flow. This region is of bounded extent because countercurrent flow is not possible for arbitrarily large values of either the gas flow or the particle phase flow. Points on the envelope correspond to conditions usually referred to as "flooding".

A second method of graphical presentation is the so-called Zenz diagram, which shows a set of curves relating the pressure gradient (or equivalently ϕ) to the gas flow rate, for various fixed values of the particle flow rate. It is easy to translate Figure 1.1 into this alternative form and the result is sketched as Figure 1.2, where ϕ is used as the ordinate. The curve labelled AB, which corresponds to $\bar{V}_s = 0$, represents situations in which the particles are suspended at rest in the flowing gas – in other words, fluidized beds. The corresponding part of Figure 1.1 is the interval AB of the ordinate axis. The part of Figure 1.2 to the right of AB represents cocurrent upflow, the part between AB and the axis $\bar{V}_f = 0$ represents countercurrent flow, and the whole region $\bar{V}_f < 0$ represents cocurrent downflow.

We have already noted that this very simple theoretical model predicts flooding in countercurrent flow. It can also be used to illustrate another striking aspect of suspension flow, namely the importance of interactions between the pipe itself and the devices used to supply the gas and the particles. Let us focus on the characteristics of the gas supply device, assuming that some provision has been made to feed the particles at the constant flow rate represented by $\bar{V}_s$. We envisage a system in which gas starting from atmospheric pressure is compressed and introduced at the bottom of the vertical pipe, where it is joined by the particles, while at the top of the pipe the suspension is discharged into a

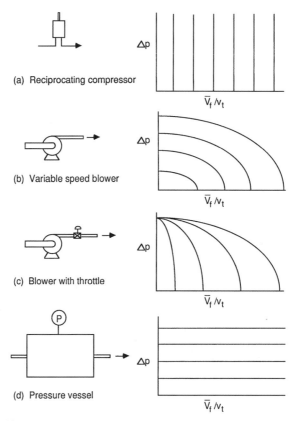

Figure 1.3. Characteristics of different types of gas compression device. (After Jackson, 1993.)

second region, also at atmospheric pressure. Then the device used to compress the gas must raise its pressure by an amount that exactly balances the drop in gas pressure along the pipe. Denote the magnitude of each of these pressure changes by Δp.

In general the pressure rise across the gas compression device is related to the flow rate, and a curve showing this relation is called a *characteristic* of the device. Figure 1.3 shows sketches of these characteristics for various devices.

Panel (a) represents an idealized reciprocating compressor, for which the flow rate is almost independent of Δp. Panel (b) is a centrifugal blower whose speed can be adjusted. There is then a separate characteristic curve for each value of the speed, and on each curve Δp decreases as the delivered flow increases, falling to zero at some finite value of the flow. Panel (c) again represents a centrifugal blower, but in this case the speed is constant and a throttle valve

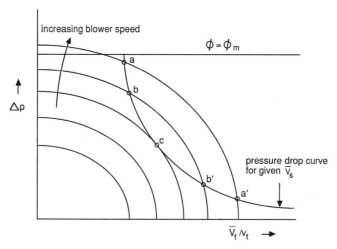

increasing blower speed

$\phi = \phi_m$

a

b

c

pressure drop curve
for given $\bar{V}_s$

b'

a'

Δp

$\bar{V}_f / \bar{V}_t \longrightarrow$

Figure 1.4. Determination of operating conditions by superimposing the pressure drop curve for the pipe and the characteristic curves of a variable speed blower. (After Jackson, 1993.)

in the delivery line is used for control. Then there is a separate characteristic curve for each setting of this valve, as indicated. Finally, panel (d) shows a large pressurized buffer vessel as a source of the gas. The pressure to which this is charged then determines Δp, and the characteristics are horizontal lines, one for each value of this pressure.

The operating conditions for the combination of the gas compression device and the pipe are determined by the intersection of the characteristic curve of the compression device with the curve from the Zenz diagram relating Δp to $\bar{V}_f$ for the pipe. This latter curve is obtained from Figure 1.2 by scaling the curve for the appropriate value of $\bar{V}_s$ by a multiplier $\rho_s g L$, where L is the total length of the pipe. For compression device characteristics of the types shown in panels (a) and (d) of Figure 1.3 the intersection in question is clearly unique, but this is not so for the blower characteristics of panels (b) and (c). Figure 1.4 superimposes the pressure drop curve for the pipe, for the specified value of $\bar{V}_s$, and several blower characteristics from panel (b) of Figure 1.3; we see that the number of intersections is either two or zero, depending on the blower speed. If the blower speed is high there are two, such as the points denoted by a and a', and as the blower speed is progressively decreased these move closer together, as seen from the pair bb'. Finally a critical value of the blower speed is reached at which the intersection points coincide (point c), and for lower speeds there is no intersection, indicating that the particles can no longer ascend the pipe as a suspension in the gas. This represents a condition called "choking"; when the speed falls below this value there is a discontinuous increase in ϕ to the value ϕ_m and the particles continue to ascend the pipe at the specified flow rate only if they can be forced up as a packed bed by the particles injected below them.

When the blower speed is high enough to avoid choking the question remains as to which of the two intersection points represents the actual operating condition of the system. This raises the issue of stability, which can be addressed either intuitively or more formally. At any point in Figure 1.4 representing an operating state the pressure rise in the gas feed device balances the pressure drop in the pipe. If, as a consequence of a small increase in the gas flow rate, the pressure rise in the compression device should become larger than the pressure drop in the pipe, intuition would suggest that the original operating point was unstable. Conversely, if a small increase in the gas flow rate causes the pressure rise in the compression device to become smaller than the pressure drop in the pipe, one would anticipate stability. A more formal analysis confirms these intuitive ideas (Matsumoto, 1986); so we conclude that points a and b in Figure 1.4 represent unstable conditions of operation, whereas points such as a′ and b′, lying to the right of point c, represent conditions that are stable and could therefore be observed in practice. The same criterion applied to other gas feed devices indicates that the device of panel (a) in Figure 1.3 gives unique and stable operating points in all cases, whereas the device of panel (d) would always lead to unstable operation. Consequently a feed device whose characteristics approximate sufficiently closely those of panel (a) will exhibit no choking phenomenon as the gas flow is decreased; instead the particle concentration will increase in a continuous way, eventually forming a dense, moving suspension known as a "fast fluidized bed".

Of course, the fact that the operating conditions are determined by an interaction between the pipe and its feed device comes as no surprise; the same is true for the flow of a single-phase fluid. The interesting features of the two-phase flow are the marked qualitative changes, such as flooding and choking, and the influence of the feed device on the whole pattern of behaviour as the flow rates are changed. These have no analogues in the case of single-phase flow.

In single-phase flow the tangential forces exerted on the fluid at the walls of the pipe give rise to the part of the pressure drop that increases with the flow rate. Though these forces have been omitted in the above discussion, they are also present in the suspension flow, and it is not difficult to see how they affect the outcome. For a given value of the particle flow rate we expect that they will make a contribution to the pressure drop in the pipe that increases with increasing gas flow rate. At low flow rates this contribution will be proportional to $\bar{V}_f$, and at high flow rates to $\bar{V}_f^2$, so the curves of Figure 1.2 are modified as in Figure 1.5, which shows only those corresponding to cocurrent upflow. In

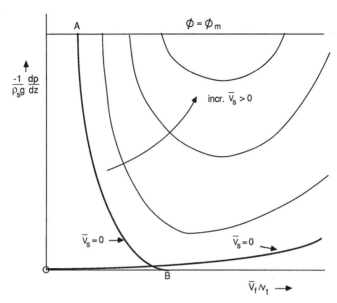

Figure 1.5. Form of the Zenz diagram when wall stresses are taken into account. Compare with Figure 1.2. (After Jackson, 1993.)

contrast to Figure 1.2, where ϕ is a monotone decreasing function of $\bar{V}_f$, each curve in Figure 1.5 (for $\bar{V}_s > 0$) passes through a minimum, then increases without bound as $\bar{V}_f$ is increased. To the left of the minimum the pressure drop is dominated by the hydrostatic contribution representing the weight of the suspended particles, which decreases as the gas flow increases and the suspension becomes more dilute. To the right of the minimum the frictional pressure drop due to the wall stress dominates, and this pressure drop increases with increasing gas flow rate. The curve representing fluidized beds, with $\bar{V}_s = 0$, is still present in Figure 1.5, but now there is a second branch, also with $\bar{V}_s = 0$, that starts from the origin and increases monotonically. This branch represents flow of gas in the absence of any particles and, since we neglect the density of the gas, the pressure drop is entirely the result of wall stress. A corresponding branch is actually present also in Figure 1.2 but, because the effect of wall stress is neglected there, it coincides with the axis $\phi = 0$ and is thus hidden.

This modification of the Zenz diagram by the inclusion of wall stress has some qualitative consequences. Thus, for a pipe with a gas compression device consisting of a pressure vessel (panel (d) of Figure 1.3) the number of possible operating conditions, for given values of the vessel pressure and $\bar{V}_s$, is now two or zero. If the pressure in the feed vessel is high enough the characteristic of this vessel, which is a horizontal line, has two intersections with the pressure

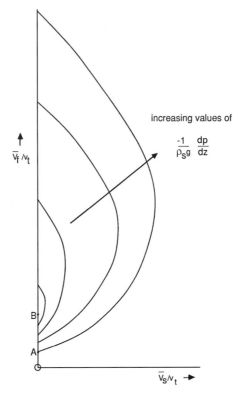

Figure 1.6. Replot of Figure 1.5 to show contours of constant pressure gradient in the plane of $\bar{V}_s/v_t$ and $\bar{V}_f/v_t$. Compare with Figure 1.1. (After Jackson, 1993.)

drop curve for the pipe. That at the lower value of $\bar{V}_f$ is the analogue of the single intersection found using Figure 1.2 and, as in that case, it is unstable. The intersection at the higher value of $\bar{V}_f$, however, represents a stable condition of operation. As the pressure in the feed vessel is lowered the two intersections converge and eventually coincide. For values of the feed pressure lower than this no intersection occurs so the system now exhibits the phenomenon of choking.

The information contained in the Zenz diagram of Figure 1.5 can also be presented in a form analogous to Figure 1.1, as a map of contours of constant pressure gradient in the plane of the two flow rates. This is sketched as Figure 1.6. Like the lines in Figure 1.1, each contour leaves the axis $\bar{V}_s = 0$ with positive slope; indeed, each contour in Figure 1.6 is tangential to the corresponding line of Figure 1.1 at this point. However, the contours now increase in slope on moving away from the ordinate axis, eventually bending around to meet it again at a higher value of $\bar{V}_f$. The lower of the two intersections of a given contour curve with the ordinate axis represents a fluidized bed, while the upper one represents a flow of gas up the pipe in the absence of any particles. Such a flow exists, of course, for every point on the positive ordinate axis.

This simple example has been discussed at some length because it illustrates an important feature of the dynamics of fluidized suspensions. At first sight it might appear that there is little prospect for useful progress in this field by attempting to formulate and solve the dynamical equations of continuity and momentum balance for each phase. First, it is not certain that a closed set of equations exists at the level of local averaged variables. Even if it does, one faces enormous difficulties in deriving the necessary closures by any rigourous reasoning, starting from the point equations of motion of the fluid and the equations of Newtonian mechanics for each particle. Indeed, except in certain special cases of quite restricted applicability, it seems unlikely that we shall ever know the "correct" form of these equations. Second, even if the equations could be established with complete confidence, their solution clearly would present great difficulties. There are twice as many equations as in the case of a single-phase fluid, and each of the momentum balances would be expected to be more complicated than the Navier–Stokes equation.

The reason why progress has, nevertheless, been possible is brought out very well by the above example. Rather than the "correct" form of the equations of motion, whatever that may be, all that we actually need are equations "good enough" to describe the phenomena of interest to us. Fortunately, it turns out that equations modelling only a limited number of rather simple physical mechanisms are adequate to give surprisingly good descriptions of many important aspects of the behaviour of fluidized systems. Phenomena of compelling importance in practice, such as the flooding and choking phenomena described above, or the generation of local inhomogeneities such as clusters of particles, or the bubbles seen in dense gas fluidized beds, turn out to be surprisingly robust, in the sense that their main features are captured correctly by gratifyingly simple forms for the equations of motion. Thus, esoteric though these phenomena may appear, it seems that the physics responsible for them is not very subtle.

The example of the vertical pipe also underlines the importance of interactions between different sections of devices handling flows of fluidized suspensions. This will be brought out with startling emphasis by the example of a standpipe system, treated in Chapter 7.

1.3 Scope of This Text

As noted in Section 1.1 above, this book focusses on an approach to the motion of fluidized suspensions through differential equations of motion, at the level of locally averaged variables, and their associated boundary value problems. For these two-phase systems this is the closest analogue of the familiar methods of single-phase fluid mechanics.

Chapter 2 addresses the problem of formulating suitable equations of motion. Though long and rather dense it provides the basis for the applications that follow in subsequent chapters. In Section 2.1 the formalism of local spatial averaging is developed and it is used to derive averaged forms of the point equations of motion for the fluid and the particles. Although the point equations are known with confidence, the averaging procedure introduces terms that are not related explicitly to the locally averaged velocities and the particle concentration, so the averaged equations are not closed in terms of these variables. Sections 2.2 and 2.3 are concerned with more or less formal derivations of closures for these undetermined terms. These formally derivable closures are applicable only in certain limiting cases, but to cover the conditions of most interest in process applications it is necessary to resort to alternative, empirical, but reasonable closure relations, which are described in Sections 2.4 and 2.5. Chapter 2 then concludes with some comments on the relation between different approaches to the process of averaging.

Perhaps the most basic problem in the field of fluidization is that of providing a description of the processes of fluidizing and defluidizing a bed of particles by alternately increasing and decreasing an upward flow of fluid. Chapter 3 shows that this is not quite so straightforward as might be assumed but that examination of bed expansion and pressure drop curves, for both fluidization and defluidization, should provide valuable information about the nature of stresses appearing in the particle phase momentum equation.

In Chapter 4 we turn to one of the earliest problems studied from the point of view of averaged equations of motion, namely the stability of a uniform fluidized bed against small perturbations. The well-known fact that gas fluidized beds are usually not uniform, but contain bubble-like structures that rise through them, led to speculation that the hypothetical uniform fluidized state must be unstable. Since the early 1960s, when this question was first studied, an extensive literature has grown up on the application of stability theory to the uniformly fluidized state. The results have proved controversial. Initially it was hoped that a stability criterion would emerge to distinguish between those beds that bubble (most gas fluidized beds) and those that do not (most liquid fluidized beds). The present consensus seems to be that this is not so; bubbling is a large-amplitude phenomenon, but experiments reveal a lower amplitude instability to be present even in the nonbubbling beds. Hence the distinction between the two lies beyond the scope of small-amplitude perturbation theory.

This problem is revisited in Chapter 5, which deals with fully developed, large-amplitude structures that can propagate in fluidized beds. In dense fluidized beds these take the form of the bubbles that are such a familiar observation and have been the subject of many experimental studies. In less dense beds structures can also propagate, but now they take the form of streamers or clusters of enhanced particle concentration. As we shall see, recent work along the lines described in Chapter 5 shows how bubbles can grow from small disturbances and gives a new insight into the nature of the distinction between bubbling and nonbubbling behaviour.

Chapters 6 and 7 differ from their predecessors in that they are concerned with the behaviour of practical devices in which fluid and particles flow together through ducts. In Chapter 6 the duct in question is a vertical pipe, or riser, and the direction of flow is upward. The objective is to describe the velocity and concentration fields and hence to relate the flow rates of the two phases and the pressure gradient in the fluid. In practice it is usually observed that the particles concentrate, on average, within a layer adjacent to the wall of the pipe, and one of the challenges facing the theory is to account for this and to predict its extent quantitatively. This is of more than academic interest; the riser is the central element in a circulating fluidized bed system for catalytic reactions, and the most prominent reaction carried out in this way is the catalytic cracking of heavier hydrocarbon stocks to yield gasoline. In terms of tonnage processed this is the world's largest chemical reaction process. Chapter 7 looks at the other half of the circulating fluidized bed loop, namely the standpipe that returns the particles from the top of the riser, where the gas pressure is low, to the bottom where it is higher. Though standpipes have attracted less attention than risers their performance is vital if the circulation of the particles is to be maintained. Experience in practice indicates that they can malfunction seriously for reasons that have not been well understood. However, an analysis based on a theory not much more sophisticated than that of Section 1.2 proves to be capable of shedding a good deal of light on these idiosyncrasies and demonstrates, in a striking way, the importance of interactions among different sections of the system.

References

Davidson, J. F. & Harrison, D. 1963. *Fluidized Particles.* Cambridge University Press.
Gidaspow, D. & Ettehadieh, B. 1983. Fluidization in two-dimensional beds with a jet, Part II: hydrodynamic modeling. *Ind. Eng. Chem. Fundam.* **22**, 193–201.
Gidaspow, D., Syamlal, M., & Seo, Y. C. 1986. Hydrodynamics of fluidization: supercomputer generated vs. experimental bubbles. *J. Powder & Bulk Solids Tech.* **10**(3), 19–23.
Hoomans, B. P. B., Kuipers, J. A. M., Briels, W. J., & van Swaaij, W. P. M. 1996. Discrete particle simulation of a two-dimensional gas-fluidized bed: a hard sphere approach. *Chem. Eng. Sci.* **51**, 99–118.

Jackson, R. 1993. Gas–solid flow in pipes. In *Particulate Two-Phase Flow*, ed. M. C. Roco, pp. 701–742. Butterworth-Heinemann.

Kunii, D. & Levenspiel, O. 1969. *Fluidization Engineering*. John Wiley & Sons.

Leva, M. 1959. *Fluidization*. McGraw-Hill.

Matsumoto, S. 1986. Stability analysis of choking in pneumatic conveying. In *Encyclopedia of Fluid Mechanics, Vol. 4, Solids and Gas–Solids Flows*, ed. N. P. Cheremisinoff, pp. 485–496. Gulf Publishing Company.

Othmer, D. F. 1956. *Fluidization*. Reinhold Publishing Corp.

Pritchett, J. W., Blake, T. R., & Garg, S. K. 1978. A numerical model of gas fluidized beds. *AIChE Symp. Ser.* **176** (74), 134–148.

Richardson, J. F. & Zaki, W. N. 1954. Sedimentation and fluidization. Part I. *Trans. Inst. Chem. Eng.* **32**, 35–53.

Tsuji, Y., Kawaguchi, T., & Tanaka, T. 1993. Discrete particle simulation of two-dimensional fluidized bed. *Powder Technol.* **77**, 79–87.

Zenz, F. A. & Othmer, D. F. 1960. *Fluidization and Fluid–Particle Systems*. Reinhold Publishing Corp.

2

Equations of motion

2.1 Averaged Equations

A complete description of the motion of an assembly of solid particles immersed in a Newtonian fluid is provided, in principle, by solving the Navier–Stokes equation for the interstitial fluid motion and the equations of linear and angular momentum for each particle, with the former being coupled to the latter by the no-slip boundary condition imposed on the surface of each particle. Although this can now be accomplished computationally when only a few particles are present it remains impractical for the enormous number of particles contained in most systems of commercial interest, such as fluidized beds, pneumatic transport lines, and sedimentation vessels. Even if a complete solution could be generated for these systems it would provide far more detailed information than could be used directly, making some method of filtering or averaging necessary to generate useful results. For example, the complete solution would give the instantaneous traction exerted by the system on every point of its boundary while, for practical purposes, only the total force experienced by relatively large areas might be of importance.

For reasons of this sort there have been many attempts to derive equations relating some suitably averaged form of the physical variables, which would provide the information needed while filtering out irrelevant detail. There is, of course, no guarantee that such equations can be found, but an extensive literature has grown up around the process of averaging the equations of motion. (See, for example, Slattery (1967), Anderson & Jackson (1967), Whitaker (1969), Wallis (1969), Delhaye (1969), Drew (1971), Drew & Segel (1971), Hinch (1977), Nigmatulin (1979), Drew (1983), Arnold et al. (1989), Joseph & Lundgren (1990), Wallis (1991), Drew & Lahey (1993), Zhang & Prosperetti (1994, 1997), and Jackson (1997).) Some authors have used local spatial averages taken over regions small in extent compared to macroscopic length scales of interest; others have averaged at each point of space over an ensemble of "macroscopically equivalent" systems. The process of averaging itself is purely formal and poses no difficulty, but it generates averaged quantities more numerous than the available equations and consequently there remains a closure problem; the averaged quantities must all be expressed in terms of a subset equal in number to the number of equations. This proves to be formidably difficult and has been achieved only for simple limiting cases. However, it is possible to manipulate the averaged equations so that the closure problem is confined to a small number of well-defined terms that can be related explicitly to variables describing the detailed motion of fluid and particles, thus making clear the information at the "microscopic" level needed to effect closure.

Here the procedure will be developed in terms of local space averages. These are somewhat less easy to manipulate mathematically than ensemble averages, but they have the advantage of being closer to the physical processes of measurement. Also, for the type of systems we address, ensemble averaging has a potentially serious problem of principle. To illustrate this consider, as an example of an averaged variable, the bulk density. From the point of view of spatial averaging the bulk density at a point can be defined as the ratio of mass to volume for a sphere centred at the point in question, large enough to contain many particles, but small compared to macroscopically interesting length scales. From the point of view of ensemble averaging, however, the bulk density is the arithmetic mean of the actual density of the material occupying the point in question, for each of a large number of "macroscopically equivalent" systems. Two systems are understood to be macroscopically equivalent if the entire histories of all macroscopically adjustable variables (such as pressure drops and the positions of solid boundaries) are identical for both systems. Thus, the process of ensemble averaging smooths out any phenomena that cannot be controlled reproducibly by manipulating the macroscopic variables. For example, in a steadily bubbling fluidized bed where bubbles are generated spontaneously, rather than injected in a controlled way, the ensemble average bulk density at a point will correspond at all times to a particle concentration intermediate between that in the dense phase and that in the interior of bubbles, depending on the fraction of the members for which the point in question lies within a bubble. The local space average, in contrast, will be representative of the dense phase concentration at times when the point in question lies in the dense phase, and it is representative of the concentration within a bubble when the point lies within a bubble. The two averages are clearly not the same, and the ensemble average, as defined, could not be used to describe the instantaneous distribution of particle concentration within, around, and between bubbles. If the values of ensemble average variables at a point are the same as local averages for a single system, chosen at random from the ensemble, the system is said to be ergodic. The above example suggests that many two-phase systems will not have this property, and for such cases ensemble average quantities cannot be related to actual measurements on a test system.

Local space averages are defined in terms of a weighting function, $g(r)$, which is a monotone decreasing function of r, the radial distance from the point in question, and whose integral over the whole of space exists and is normalized as follows:

$$4\pi \int_0^\infty g(r) r^2 \, dr = 1. \tag{2.1}$$

For example, we could take

$$g(r) = \begin{cases} 3/4\pi l_0^3 & \text{for } r \le l_0, \\ 0 & \text{for } r > l_0, \end{cases}$$

where l_0 is a constant with dimensions of length. This sort of weighting function is said to define "hard" spatial averages. For our purpose it is more convenient to assume that g is differentiable as often as needed for our manipulations and to define its *radius, l,* by

$$\int_0^l g(r) r^2 \, dr = \int_l^\infty g(r) r^2 \, dr.$$

Such a weighting function is said to define "soft" averages and l can be referred to as the averaging radius. Clearly the hard weighting function can be approximated, as closely as desired, by a sequence of soft weighting functions, and its radius is given by $l = l_0/2^{1/3}$. The overall average value, at position $\mathbf{x}$ and time t, of any point function f of position and time is then defined by

$$\langle f \rangle (\mathbf{x}, t) = \int_V f(\mathbf{y}, t) g(|\mathbf{x} - \mathbf{y}|) \, dV_y, \tag{2.2}$$

where the domain of integration is the whole system. From this it appears that $\langle f \rangle$ depends on the particular form chosen for the weighting function. However, this dependence becomes progressively weaker the larger the ratio of the smallest macroscopic length scale to the particle size, provided the radius of the weighting function is properly chosen. Specifically, if the particle radius is denoted by a and the macroscopic length scale by L, we would expect the value of the average to be insensitive to the radius of g and its particular algebraic form, provided $L \gg l \gg a$. Such a choice of l is possible only if $L \gg a$, of course, and there is then said to be *separation of scales* between the macroscopic problem and the detailed motion on the scale of a single particle. Without this separation the results of local averaging become sensitive to the particular form of the weighting function.

In addition to the overall average, introduced above, we can define averages associated with each of the phases separately, as follows.

(i) Fluid phase averages   The *void fraction*, or fraction of space occupied by fluid, in the neighbourhood of point $\mathbf{x}$ is defined by

$$\varepsilon(\mathbf{x}) = \int_{V_f} g(|\mathbf{x} - \mathbf{y}|) \, dV_y, \tag{2.3}$$

where V_f indicates that part of the whole system occupied by fluid. (From now on the time dependence will not be exhibited explicitly, unless this is necessary to avoid confusion.) The *fluid phase average*, $\langle f \rangle^f$, of point property f is then defined by

$$\varepsilon(\mathbf{x})\langle f \rangle^f(\mathbf{x}) = \int_{V_f} f(\mathbf{y}) g(|\mathbf{x} - \mathbf{y}|) \, dV_y. \tag{2.4}$$

The averaging process, so defined, does not commute with space or time differentiation but expressions for the averages of the derivatives are easily obtained (Anderson & Jackson, 1967), namely

$$\varepsilon(\mathbf{x})\langle \partial f/\partial x_k \rangle^f = \frac{\partial}{\partial x_k}[\varepsilon(\mathbf{x})\langle f \rangle^f(\mathbf{x})]$$
$$- \sum_p \int_{S_p} f(\mathbf{y}) n_k(\mathbf{y}) g(|\mathbf{x} - \mathbf{y}|) \, ds_y \tag{2.5}$$

and

$$\varepsilon(\mathbf{x})\langle \partial f/\partial t \rangle^f = \frac{\partial}{\partial t}[\varepsilon(\mathbf{x})\langle f \rangle^f(\mathbf{x})]$$
$$+ \sum_p \int_{S_p} f(\mathbf{y}) n_k(\mathbf{y}) u_k(\mathbf{y}) g(|\mathbf{x} - \mathbf{y}|) \, ds_y. \tag{2.6}$$

Here the integrals are taken over the surface of particle p and the summation is over all the particles; n_k denotes the kth component of the unit outward normal to the particle surface and u_k is the kth component of the point velocity. (Both here and in what follows the summation convention is adopted for repeated suffixes.) Each of the relations is valid provided the distance of $\mathbf{x}$ from the nearest point on the system boundary is large compared to the radius of the weighting function.

(ii) Solid phase averages   These are analogous to fluid phase averages but the domain of integration is now the disjoint region comprising the interiors of all the particles. Thus the solids volume fraction is defined by

$$\phi(\mathbf{x}) = \sum_p \int_{v_p} g(|\mathbf{x} - \mathbf{y}|) \, dv_y, \qquad (2.7)$$

where v_p is the interior of particle p, and the *solid phase average*, $\langle f \rangle^s$, of point property f is given by

$$\phi(\mathbf{x})\langle f \rangle^s(\mathbf{x}) = \sum_p \int_{v_p} f(\mathbf{y}) g(|\mathbf{x} - \mathbf{y}|) \, dv_y. \qquad (2.8)$$

For the solid phase averages of space and time derivatives there are obvious analogues of Equations (2.5) and (2.6).

Note that the overall average, defined in Equation (2.2), is related to the phase averages by

$$\langle f \rangle = \varepsilon \langle f \rangle^f + \phi \langle f \rangle^s \qquad (2.9)$$

and it is also useful to define a mass average, $\langle f \rangle^m$, by

$$\bar{\rho}\langle f \rangle^m = \rho_f \varepsilon \langle f \rangle^f + \rho_s \phi \langle f \rangle^s, \qquad (2.10)$$

where $\bar{\rho} = \varepsilon \rho_f + \phi \rho_s$ and ρ_f and ρ_s are the densities of the fluid and solid materials.

(iii) Particle phase averages   For simplicity in what follows we will confine our attention to identical, spherical, rigid particles of radius a. Then to specify completely the motion of the solid material it is necessary only to give the velocity of the centre of each particle, together with its angular velocity. As a consequence of this the dynamical equations reduce to a momentum balance for the centre of mass motion and an angular momentum equation, and only the resultant force and resultant moment of the tractions exerted at the particle surface, rather than the complete stress distribution within the particle, feature in these equations. Because of this, as pointed out by Anderson & Jackson (1967), it is convenient when dealing with the mechanics of the particulate phase to use averages formed from these overall properties of the particles, rather than the solid phase averages defined by (2.8). To this end we first define the number $n(\mathbf{x})$ of particles per unit volume (i.e., the number density) at position $\mathbf{x}$ by

$$n(\mathbf{x}) = \sum_p g(|\mathbf{x} - \mathbf{x}^p|), \qquad (2.11)$$

where $\mathbf{x}^p$ is the position of the centre of particle p at the time in question. Then if f^p is any property of particle p as a whole, for example, the centre of mass velocity, the *particle phase average* of f is defined by

$$n(\mathbf{x})\langle f \rangle^p(\mathbf{x}) = \sum_p f^p g(|\mathbf{x} - \mathbf{x}^p|). \qquad (2.12)$$

While there is no analogue of Equation (2.5) for this type of average the analogue of (2.6) is

$$n(\mathbf{x})\langle \partial f / \partial t \rangle^P(\mathbf{x}) = \frac{\partial}{\partial t}[n(\mathbf{x})\langle f \rangle^P(\mathbf{x})] + \frac{\partial}{\partial x_k} \sum_p f^p u_k^p g(|\mathbf{x} - \mathbf{x}^p|),$$

(2.13)

where u_k^p is the kth component of the velocity of the centre of particle p.

Note finally that, for each of the averaging processes defined above, a second averaging does not leave the averaged variable unchanged. However, the difference between the results of successive averagings is of order a^2/L^2 compared with each of them, and it can therefore be neglected when there is separation of scales.

The averaging procedures described above can now be applied to the equations of motion, starting with the equation of mass conservation. For the fluid, assumed to be incompressible, this takes the form of the continuity equation $\partial u_k / \partial x_k = 0$. Setting $f \equiv u_k$ in (2.5) and $f \equiv 1$ in (2.6), adding, and using the continuity equation we find that

$$\frac{\partial \varepsilon}{\partial t} + \frac{\partial}{\partial x_k}(\varepsilon \langle u_k \rangle^f) = 0,$$

(2.14)

which is the local averaged form of the continuity equation for the fluid. Similarly, for the solid phase

$$\frac{\partial \phi}{\partial t} + \frac{\partial}{\partial x_k}(\phi \langle u_k \rangle^s) = 0.$$

(2.15)

In terms of particle phase averages the corresponding result describes the conservation of total number of particles, and it is found by setting $f \equiv 1$ in (2.13), giving

$$\frac{\partial n}{\partial t} + \frac{\partial}{\partial x_k}(n \langle u_k \rangle^P) = 0.$$

(2.16)

The point momentum equation for the fluid is

$$\rho_f \left[\frac{\partial u_i}{\partial t} + \frac{\partial}{\partial y_k}(u_i u_k) \right] = \frac{\partial \sigma_{ik}}{\partial y_k} + \rho_f g_i,$$

(2.17)

where σ_{ik} and g_i are the components of the stress tensor and the specific gravity force vector, respectively. Then, multiplying both sides by $g(|\mathbf{x} - \mathbf{y}|)$, integrating over V_f, using (2.6) with $f \equiv u_i$ and (2.5) with $f \equiv u_i u_k$ on the left-hand side, and using (2.5) with $f \equiv \sigma_{ik}$ on the right-hand side, we obtain

$$
\rho_f \left[\frac{\partial}{\partial t} (\varepsilon \langle u_i \rangle^f) + \frac{\partial}{\partial x_k} (\varepsilon \langle u_i u_k \rangle^f) \right]
$$

$$
= \frac{\partial}{\partial x_k} (\varepsilon \langle \sigma_{ik} \rangle^f) - \sum_p \int_{S_p} \sigma_{ik}(\mathbf{y}) n_k(\mathbf{y}) g(|\mathbf{x} - \mathbf{y}|) \, ds_y + \rho_f \varepsilon g_i, \qquad (2.18)
$$

which is the averaged form of the fluid phase momentum balance. Note that this equation is not closed, since it is not expressed entirely in terms of averaged quantities of the types defined earlier but contains a sum of integrals of the traction forces exerted by the fluid on the surface of each particle, with the force on each surface element weighted by the value of the function g at that element.

In the case of a particle we must consider both the motion of the centre of mass and rotation about the centre of mass. These respond to traction forces exerted on the surface of the particle by the fluid, and by other particles with which it is in contact. Thus the momentum equation for particle p is

$$
\rho_s v u_i^p = \int_{S_p} \sigma_{ik}(\mathbf{y}) n_k(\mathbf{y}) \, ds_y + \sum_{q \neq p} f_i^{pq} + \rho_s v g_i,
$$

where v is the particle volume and $\mathbf{u}^p$ denotes the centre of mass velocity. The areas of contact between the particle in question and its touching neighbours are assumed to be small fractions of the total surface area and are idealized to points, so f_i^{pq} denotes the ith component of the force exerted by particle q on particle p at their point of contact. (Though the summation extends over all other particles q, the contact force vanishes, of course, for all but those few particles in contact with p.) Correspondingly the integral of the traction exerted by the fluid can be extended over the whole surface of p, since the solid–solid contacts are confined to a few isolated points. Because this integral represents the resultant force exerted on particle p by the fluid, it can be written as f_i^{pf} in a manner consistent with the notation for the solid–solid contact forces. Multiplying the above equation through by $g(|\mathbf{x} - \mathbf{x}^p|)$, summing over p, then invoking (2.13) we obtain the averaged momentum equation for the particle phase:

$$
\rho_s v \left[\frac{\partial}{\partial t} (n \langle u_i \rangle^p) + \frac{\partial}{\partial x_k} (n \langle u_i u_k \rangle^p) \right] = n \langle f_i^f \rangle^p + n \langle f_i^s \rangle^p + \rho_f v n g_i,
$$

$$
\qquad\qquad (2.19)
$$

where

$$
n \langle f_i^f \rangle^p = \sum_p g(|\mathbf{x} - \mathbf{x}_p|) \int_{S_p} \sigma_{ik}(\mathbf{y}) n_k(\mathbf{y}) \, ds_y \qquad (2.20)
$$

and

$$n\langle f_i^s \rangle^P = \sum_p g(|\mathbf{x} - \mathbf{x}_p|) \sum_{q \neq p} f_i^{pq}. \tag{2.21}$$

Equations (2.20) and (2.21) represent the particle phase averages of the resultant forces exerted on the particles by the fluid and by other particles, respectively.

The angular momentum equation for particle p is

$$I\dot{\omega}_i^p = \varepsilon_{ilm} \left[\int_{S_p} (y_l - x_l^p) \sigma_{mk}(\mathbf{y}) n_k(\mathbf{y}) \, ds_y + \sum_{q \neq p} (y_l^{pq} - x_l^p) f_m^{pq} \right],$$

where I is the moment of inertia of the particle about its centre (namely $2a^2 \nu \rho_s / 5$), $\mathbf{y}^{pq}$ is the position of the point of contact between particles p and q, and ε_{ilm} is the Kronecker permutation symbol. The two terms in the square brackets represent the moments of the fluid traction forces and the solid–solid contact forces exerted on particle p. The particle phase average of this equation, obtained in the same way as (2.19), is

$$I \left[\frac{\partial}{\partial t} (n \langle \omega_i \rangle^P) + \frac{\partial}{\partial x_k} (n \langle \omega_i u_k \rangle^P) \right]$$

$$= a\varepsilon_{ilm} \sum_p g(|\mathbf{x} - \mathbf{x}_p|) \left[\int_{S_p} \sigma_{mk}(\mathbf{y}) n_l(\mathbf{y}) n_k(\mathbf{y}) \, ds_y + \sum_{q \neq p} f_m^{pq} n_l^{pq} \right], \tag{2.22}$$

where $\mathbf{n}^{pq}$ is the unit outward normal to the surface of particle p at its point of contact with particle q.

The term (2.20), appearing on the right-hand side of the particle phase momentum equation (2.19), differs from the sum appearing on the right-hand side of the fluid phase momentum equation (2.18) notwithstanding Newton's third law, which might lead one to expect the two to be the same, apart from sign. This paradox can be resolved by expanding the weighting function g, on the surface of particle p, in a Taylor series about the centre of the particle. By this device it can be shown (Jackson, 1997) that (2.18) and (2.19) can be reduced to the following forms:

$$\rho_f \varepsilon \frac{D_f \langle u_i \rangle^f}{Dt} = \frac{\partial}{\partial x_k} \left\{ \varepsilon \langle \sigma_{ik} \rangle^f + n \langle s_{ik}^f \rangle^P - \frac{1}{2} \frac{\partial}{\partial x_l} (n \langle s_{ikl}^f \rangle^P) - \rho_f \varepsilon \langle u_i' u_k' \rangle^f \right\}$$

$$- n \langle f_i^f \rangle^P + \rho_f \varepsilon g_i \tag{2.23}$$

and

$$\rho_s \phi \frac{D_p \langle u_i \rangle^P}{Dt} = \frac{\partial}{\partial x_k} \left\{ n \langle s_{ik}^s \rangle^P - \frac{1}{2} \frac{\partial}{\partial x_l} (n \langle s_{ikl}^s \rangle^P) - \rho_s \phi \langle u_i' u_k' \rangle^P \right\}$$

$$+ n \langle f_i^f \rangle^P + \rho_s \phi g_i \tag{2.24}$$

after omitting terms of order a^2/L^2 or smaller relative to one or another of the terms retained. In these equations

$$n\langle s_{ik}^f \rangle^P(\mathbf{x}) = a \sum_p g(|\mathbf{x} - \mathbf{x}_p|) \int_{S_p} t_i n_k \, ds,$$

$$n\langle s_{ikl}^f \rangle^P(\mathbf{x}) = a^2 \sum_p g(|\mathbf{x} - \mathbf{x}_p|) \int_{S_p} t_i n_k n_l \, ds,$$

$$n\langle s_{ik}^s \rangle^P(\mathbf{x}) = a \sum_p g(|\mathbf{x} - \mathbf{x}_p|) \sum_{q \neq p} f_i^{pq} n_k^{pq},$$

$$n\langle s_{ikl}^s \rangle^P(\mathbf{x}) = a^2 \sum_p g(|\mathbf{x} - \mathbf{x}_p|) \sum_{q \neq p} f_i^{pq} n_k^{pq} n_l^{pq},$$

where $t_i = \sigma_{ia} n_a$ is the traction exerted by the fluid on the surface of a particle. Also

$$\frac{D_f}{Dt} \equiv \frac{\partial}{\partial t} + \langle u_k \rangle^f \frac{\partial}{\partial x_k}, \qquad \frac{D_p}{Dt} \equiv \frac{\partial}{\partial t} + \langle u_k \rangle^p \frac{\partial}{\partial x_k}$$

and

$$\langle u_i' u_k' \rangle^f \equiv \langle (u_i - \langle u_i \rangle^f)(u_k - \langle u_k \rangle^f) \rangle^f;$$

$$\langle u_i' u_k' \rangle^p \equiv \langle (u_i^p - \langle u_i \rangle^p)(u_k^p - \langle u_k \rangle^p) \rangle^p.$$

Using this notation the particle phase angular momentum balance (2.22) can be written as

$$nI \frac{D_p \langle \omega_i \rangle^p}{Dt} = \varepsilon_{ilm} \left[n\langle s_{ml}^f \rangle^p + n\langle s_{ml}^s \rangle^p \right] - \frac{\partial}{\partial x_k} \{ nI \langle \omega_i' u_k' \rangle^p \}, \qquad (2.25)$$

where

$$\langle \omega_i' u_k' \rangle^p \equiv \langle (\omega_i^p - \langle \omega_i \rangle^p)(u_k^p - \langle u_k \rangle^p) \rangle^p.$$

This is as far as the formal process of local averaging can take us, though the momentum equations can be presented in many different ways, since any three independent linear combinations of (2.23), (2.24), and (2.25) will serve equally well, and the momentum equation for the particle phase can be rewritten in terms of $\langle \mathbf{u} \rangle^s$, rather than $\langle \mathbf{u} \rangle^p$. One important variant is obtained by substituting the "mixture momentum equation" for either (2.23) or (2.24). This is the equation generated by averaging the point momentum balance, which must be satisfied everywhere both inside and outside the particles, over the whole space occupied by the system. The result (Jackson, 1997) is

$$\bar{\rho}\frac{D_m \langle u_i \rangle^m}{Dt} = \frac{\partial}{\partial x_k}\left\{ \varepsilon \langle \sigma_{ik} \rangle^f + \frac{n}{2}\left(\langle s_{ik}^f \rangle^p + \langle s_{ki}^f \rangle^p + \langle s_{ik}^s \rangle^p + \langle s_{ki}^s \rangle^p\right)\right.$$

$$- \frac{\partial}{\partial x_l}\left[\frac{n}{2}\left(\langle s_{ikl}^f \rangle^p + \langle s_{ikl}^s \rangle^p\right)\right] - \bar{\rho}\langle u_i' u_k' \rangle^m$$

$$\left.+ \frac{\rho_s a^2}{5}n\nu(\delta_{ik}\langle \omega_l \omega_l \rangle^p - \langle \omega_i \omega_k \rangle^p)\right\} + \bar{\rho}g_i, \qquad (2.26)$$

where

$$\frac{D_m}{Dt} \equiv \frac{\partial}{\partial t} + \langle u_k \rangle^m \frac{\partial}{\partial x_k}$$

and

$$\langle u_i' u_k' \rangle^m \equiv \langle (u_i - \langle u_i \rangle^m)(u_k - \langle u_k \rangle^m) \rangle^m$$

and, once again, terms are neglected that are of order a^2/L^2 or smaller relative to those retained. Note the appearance of the mass average velocity on the left-hand side of this equation and the fact that the resultant force exerted between fluid and particles cancels, as it should.

The momentum equations (2.23) and (2.24) are of the following form:

$$\rho_f \varepsilon \frac{D_f \langle u_i \rangle^f}{Dt} = \frac{\partial S_{ik}^f}{\partial x_k} - n\langle f_i^f \rangle^p + \rho_f \varepsilon g_i \qquad (2.27)$$

and

$$\rho_s \phi \frac{D_p \langle u_i \rangle^p}{Dt} = \frac{\partial S_{ik}^p}{\partial x_k} + n\langle f_i^f \rangle^p + \rho_s \phi g_i, \qquad (2.28)$$

where S_{ik}^f and S_{ik}^p may be regarded as effective stress tensors associated with the fluid and particle phases, respectively. The process of averaging has related these tensors to details of the interactions between fluid and particles and between particles and particles but, in doing so, it reveals clearly the formidable difficulties to be faced in closing the equations in terms of averaged variables.

2.2 Buoyancy

Before going further it is convenient to simplify the notation, introducing the following symbols:

$$\mathbf{u} = \langle \mathbf{u} \rangle^f; \quad \mathbf{v} = \langle \mathbf{u} \rangle^p; \quad \mathbf{f} = \langle \mathbf{f}^f \rangle^p.$$

Then Equations (2.27) and (2.28) take the form

$$\rho_f \varepsilon \frac{D_f \mathbf{u}}{Dt} = \nabla \cdot \mathbf{S}^f - n\mathbf{f} + \rho_f \varepsilon \mathbf{g} \qquad (2.27')$$

and

$$\rho_s \phi \frac{D_p \mathbf{v}}{Dt} = \nabla \cdot \mathbf{S}^p + n\mathbf{f} + \rho_s \phi \mathbf{g}, \qquad (2.28')$$

where $\mathbf{f}$ represents the average value of the resultant force exerted by the fluid on a particle.

The form of one contribution to this local average force exerted by the fluid on a particle, namely the buoyancy, is usually regarded as fairly obvious, so $\mathbf{f}$ is next decomposed without further ado into a sum of the buoyancy force and all remaining contributions. Unfortunately, however, this decomposition is not completely unambiguous. In this connection the description of buoyancy by a sixteenth century shipwright is interesting:

The syde [of a ship] being rounde and full, it is the more boyenter a great deale. (W. Bourne. Treas. for Trav., 1578)

This picturesque description, while imprecise, is no more confused than some of the discussion of this topic in the current literature.

The difficulty is, perhaps, best illustrated by a sequence of simple examples. First, consider the classical hydrostatic problem of the force exerted by a fluid at rest on an immersed body that is also at rest. This can be written in two ways:

$$\mathbf{f} = -v\nabla p = -\rho_f v\mathbf{g}, \qquad (i)$$

where p denotes the fluid pressure and v the volume of the body. Their equivalence follows from the hydrostatic expression for the pressure gradient, namely $\nabla p = \rho_f \mathbf{g}$. Because of Galilean invariance these expressions remain correct if the fluid moves with a constant and spatially uniform velocity $\mathbf{u}_0$ and the immersed body moves with the same velocity.

Second, let us generalise this to a situation in which the fluid again has the constant velocity $\mathbf{u}_0$ at all points sufficiently far from the particle, but the particle may be in arbitrary motion relative to the fluid. Then, in addition to the Archimedean buoyancy force above, there will be a contribution to $\mathbf{f}$ resulting from the motion of the body through the fluid and we can write

$$\mathbf{f} = \mathbf{f}_a - v\nabla p_0 = \mathbf{f}_b - \rho_f v\mathbf{g}, \qquad (ii)$$

where p_0 is the pressure field at large distances from the immersed body. In each of the equations (ii) the second term serves to define the buoyancy force and it is easy to see that the two definitions are equivalent, since the momentum equation applied to the fluid far from the body gives $\nabla p_0 = \rho_f \mathbf{g}$, and consequently $\mathbf{f}_a = \mathbf{f}_b$.

Third, let us generalise further to a situation in which the velocity $\mathbf{u}_0$ of the distant fluid is no longer constant but is subject to a spatially uniform acceleration $\mathbf{a}_0 = d\mathbf{u}_0/dt$. Then the momentum equation applied to this fluid gives $\nabla p_0 = \rho_f(\mathbf{g} - \mathbf{a}_0)$. Therefore, if $\mathbf{f}$ were decomposed as in (ii) above, $\mathbf{f}_a$ and $\mathbf{f}_b$ would no longer be equal; in other words each equation (ii) would define a different buoyancy force. However, if we adopt a modified decomposition given by

$$\mathbf{f} = \mathbf{f}_1 - \nu\nabla p_0 = \mathbf{f}_2 - \rho_f \nu(\mathbf{g} - \mathbf{a}_0) \tag{iii}$$

then $\mathbf{f}_1$ and $\mathbf{f}_2$ are equal. Thus, $-\nu\nabla p_0$ and $-\rho_f \nu(\mathbf{g} - \mathbf{a}_0)$ provide equivalent expressions for the buoyancy force. This is not surprising, since the specific gravity force seen by an observer moving with the fluid is no longer $\mathbf{g}$, but $\mathbf{g} - \mathbf{a}_0$.

Fourth, consider a flow of fluid through an array of particles with uniform void fraction ε. The local average fluid velocity $\mathbf{u}_0$, as defined in this chapter, is spatially uniform , but not constant, and the local average acceleration of the fluid is $\mathbf{a}_0 = d\mathbf{u}_0/dt$. Then the local averaged momentum equation (2.27') for the fluid gives

$$\rho_f \varepsilon \mathbf{a}_0 = -\nabla p_0 - n\mathbf{f} + \rho_f \varepsilon \mathbf{g}.$$

Now, in terms of the decomposition defined by the first of equations (iii) above, this becomes

$$\rho_f \mathbf{a}_0 = -\nabla p_0 - \frac{n\mathbf{f}_1}{\varepsilon} + \rho_f \mathbf{g},$$

while in terms of the second of (iii) it becomes

$$\rho_f \mathbf{a}_0 = -\nabla p_0 - n\mathbf{f}_2 + \rho_f \mathbf{g}.$$

Comparing these two equations we see that $\mathbf{f}_1$ and $\mathbf{f}_2$ are no longer the same; indeed $\mathbf{f}_2 = \mathbf{f}_1/\varepsilon$. Thus, in this case, $-\nu\nabla p_0$ and $-\rho_f \nu(\mathbf{g} - \mathbf{a}_0)$ provide different interpretations of what we mean by the buoyancy force. The physical origin of the difference is not difficult to trace. When a fluid flows through an assembly of particles there is a contribution to its local average pressure gradient from the relative motion. But this contribution, in turn, exerts a force on the immersed particles that may be attributed either to the buoyancy term or to the other term in the decomposition. In the case of the first form of the decomposition (iii) it is attributed to the buoyancy term, since this is proportional to the gradient of the local average pressure. For the second form, however, it is attributed to the term $\mathbf{f}_2$, and this is the reason why $\mathbf{f}_2$ is larger than $\mathbf{f}_1$ by the factor $1/\varepsilon$.

Thus, in our fourth example there is no unique, unambiguous way of defining the buoyancy force. Either of the alternative decompositions (iii) is acceptable but, of course, one must take care to formulate $\mathbf{f}_1$ and $\mathbf{f}_2$ differently, in such a way as to match the different forms for the buoyancy term, so that the value of $\mathbf{f}$ is the same in both cases. It is this sort of ambiguity that has given rise to much confusion in the literature.

These considerations carry over into the case of more general motions described by the momentum equations (2.27′) and (2.28′). The appropriate generalization of the first of equations (iii) is

$$nf = \phi \nabla \cdot \mathbf{S}^f + n\mathbf{f}_1. \tag{2.29}$$

Using this to eliminate nf permits (2.27′) and (2.28′) to be written as

$$\rho_f \varepsilon \frac{D_f \mathbf{u}}{Dt} = \varepsilon \nabla \cdot \mathbf{S}^f - n\mathbf{f}_1 + \rho_f \varepsilon \mathbf{g} \tag{2.30}$$

and

$$\rho_s \phi \frac{D_p \mathbf{v}}{Dt} = \nabla \cdot \mathbf{S}^p + \phi \nabla \cdot \mathbf{S}^f + n\mathbf{f}_1 + \rho_s \phi \mathbf{g}. \tag{2.31}$$

The corresponding generalization of the second of (iii) is

$$nf = -\rho_f \phi \left(\mathbf{g} - \frac{D_f \mathbf{u}}{Dt} \right) + n\mathbf{f}_2, \tag{2.32}$$

where $\mathbf{f}_2 = \mathbf{f}_1 / \varepsilon$. Using (2.32) we may then reformulate the momentum balances (2.27′) and (2.28′) as

$$\rho_f \frac{D_f \mathbf{u}}{Dt} = \nabla \cdot \mathbf{S}^f - n\mathbf{f}_2 + \rho_f \mathbf{g} \tag{2.33}$$

and

$$\rho_s \phi \frac{D_p \mathbf{v}}{Dt} = \nabla \cdot \mathbf{S}^p + n\mathbf{f}_2 + (\rho_s - \rho_f)\phi \mathbf{g} + \rho_f \phi \frac{D_f \mathbf{u}}{Dt}. \tag{2.34}$$

Although these differ in appearance from (2.30) and (2.31) it is easy to see that they are equivalent. Recalling that $\mathbf{f}_2 = \mathbf{f}_1 / \varepsilon$, (2.33) is obtained simply by dividing (2.30) by ε, while (2.34) is found by forming a linear combination of (2.30) and (2.31) to eliminate $\nabla \cdot \mathbf{S}^f$.

If we were to select an alternative decomposition in terms of the hydrostatic form for the buoyancy force, namely

$$nf = -\rho_f \phi \mathbf{g} + n\mathbf{f}_3, \tag{2.35}$$

the term $\rho_f \phi D_f \mathbf{u}/Dt$ in (2.32) would be lumped in with the contribution $n\mathbf{f}_3$. With the partition (2.35) the momentum equations take the form

$$\rho_f \varepsilon \frac{D_f \mathbf{u}}{Dt} = \nabla \cdot \mathbf{S}^f - n\mathbf{f}_3 + \rho_f \mathbf{g} \tag{2.36}$$

and

$$\rho_s \phi \frac{D_p \mathbf{v}}{Dt} = \nabla \cdot \mathbf{S}^p + n\mathbf{f}_3 + (\rho_s - \rho_f)\phi\mathbf{g}. \tag{2.37}$$

These are equivalent, of course, to (2.33) and (2.34) and differ from them in appearance only because the term $\rho_f \phi D_f \mathbf{u}/Dt$ is buried within $n\mathbf{f}_3$, rather than exhibited explicitly as part of the buoyancy force.

Now consider the case of a fluid propelled by a pressure gradient through an assembly of particles held at rest by constraining forces, for example, a packed bed. The volume fraction ϕ is independent of position, as is the fluid averaged velocity $\mathbf{u}$, and we assume an absence of gravitational force. Then (2.32) reduces to $n\mathbf{f} = n\mathbf{f}_2$, indicating that the total force exerted on the particles by the fluid is a direct measure of $n\mathbf{f}_2$. Furthermore $\nabla \cdot \mathbf{S}^f = -\nabla p^f$ so the fluid phase momentum balance (2.33) reduces to $n\mathbf{f}_2 = -\nabla p^f$, showing that a measurement of the pressure gradient also gives $n\mathbf{f}_2$. But fluid can also be impelled through the bed of particles by a body force, such as gravity, even in the absence of any pressure gradient , and in this case (2.29) gives $n\mathbf{f} = n\mathbf{f}_1$. Thus, a measurement of the force exerted on the particles by the fluid will now give $n\mathbf{f}_1$, rather than $n\mathbf{f}_2$. It is often overlooked that the interaction force between particles and fluid is not determined uniquely by their relative motion but, as we have just seen, it also depends on the nature of the forces responsible for that motion.

The forms of the momentum equations corresponding to the definitions of buoyancy embodied in (2.29), (2.32), and (2.35), respectively, are all equivalent if care is taken to include the correct terms in the associated force contributions $n\mathbf{f}_1$, $n\mathbf{f}_2$, and $n\mathbf{f}_3$. They merely mirror different ways of partitioning the fluid–particle interaction force between a buoyancy contribution and the remainder. However, these are by no means the only ways of dividing up $n\mathbf{f}$ that can be found in the literature. One popular alternative is to replace the buoyancy term $\phi\nabla \cdot \mathbf{S}^f$ in (2.29) by $-\phi\nabla p^f$, which has the curious feature of singling out just the isotropic part of the tensor $\mathbf{S}^f$ for inclusion in the buoyancy force. However, this also could give momentum equations equivalent to (2.30) and (2.31) provided the action on the particles of the gradient of the deviatoric part of $\mathbf{S}^f$ is picked up in the other term of the decomposition.

A number of workers, most notably Foscolo & Gibilaro (1984, 1987), have argued strongly for yet another partition, which replaces the fluid density in (2.35) by the mean density of the suspension, giving

$$n\mathbf{f} = -\bar{\rho}\phi\mathbf{g} + n\mathbf{f}_4, \tag{2.38}$$

where $\bar{\rho} = \varepsilon\rho_f + \phi\rho_s$. The momentum equations then become

$$\rho_f \varepsilon \frac{D_f \mathbf{u}}{Dt} = \nabla \cdot \mathbf{S}^f - n\mathbf{f}_4 + [\rho_f \varepsilon(1 + \phi) + \rho_s \phi^2]\mathbf{g} \tag{2.39}$$

and

$$\rho_s \phi \frac{D_p \mathbf{v}}{Dt} = \nabla \cdot \mathbf{S}^p + n\mathbf{f}_4 + (\rho_s - \rho_f)\varepsilon\phi\mathbf{g}. \tag{2.40}$$

The partition (2.38) can also be written

$$n\mathbf{f} = -\rho_f \phi\mathbf{g} - (\rho_s - \rho_f)\phi^2\mathbf{g} + n\mathbf{f}_4.$$

Comparing this with (2.35) we obtain

$$n\mathbf{f}_4 = n\mathbf{f}_3 + \phi^2(\rho_s - \rho_f)\mathbf{g}.$$

Although this decomposition can also, in principle, give correct momentum balances if an appropriate form is chosen for $n\mathbf{f}_4$, its proponents have generally failed to do this, and consequently their proposed momentum equations are untenable, as we shall see.

The shortcomings of certain momentum equations that have been based on the partitions (2.35) and (2.38) are revealed rather simply by the following analogue of Galileo's famous "Pisa" experiment. A uniform dispersion of particles is initially held at rest in a body of fluid that is also at rest within a container, with the whole residing in a uniform gravitational field. At time zero the container is allowed to fall freely and, simultaneously, the constraints holding the particles in place are removed. Then clearly the container and both the phases fall with the common acceleration due to the gravitational field . Any momentum equations for which this simple motion is not a solution are unarguably defective.

As a first example of this test consider the momentum equations proposed by Murray (1965). In the present notation these can be written as

$$\rho_f \varepsilon \frac{D_f \mathbf{u}}{Dt} = -\nabla p^f + \varepsilon\nabla \cdot \sigma^f$$
$$- \left[\beta(\phi)(\mathbf{u} - \mathbf{v}) + C\rho_f\phi\frac{d(\mathbf{u} - \mathbf{v})}{dt} - \rho_f\phi\mathbf{g} \right] + \rho_f\varepsilon\mathbf{g}$$

and

$$\rho_s \phi \frac{D_p \mathbf{v}}{Dt} = \phi\nabla \cdot \sigma^s + \left[\beta(\phi)(\mathbf{u} - \mathbf{v}) + C\rho_f\phi\frac{d(\mathbf{u} - \mathbf{v})}{dt} - \rho_f\phi\mathbf{g} \right] + \rho_s\phi\mathbf{g},$$

where σ^f and σ^s are viscous stress tensors of the standard Newtonian form in terms of the velocity fields $\mathbf{u}$ and $\mathbf{v}$, respectively, and C is a virtual mass coefficient. The terms bracketed represent the fluid–particle interaction force and Murray specifically identifies $-\rho_f\phi\mathbf{g}$ as the buoyancy force. It is easily checked that neither of these equations is satisfied by the test motion. The reason for this is clear. The buoyancy force chosen corresponds to the partition (2.35), but with this partition we have seen that a term of the form $\rho_f\phi D_f\mathbf{u}/Dt$ must be included as part of the rest of the interaction force. This is missing from Murray's equations. Had it been included the equations would have passed the test.

As an example of fallacious equations based on the partition (2.38) we consider the following pair, due to Foscolo & Gibilaro (1987):

$$\rho_f \varepsilon \left(\frac{\partial u}{\partial t} + u \frac{\partial u}{\partial z} \right) = -\frac{\partial p^f}{\partial z} - (\rho_s - \rho_f) g \phi$$
$$\times \left[\frac{\varepsilon(u - v)}{v_t} \right]^{4.8/n} \varepsilon^{-3.8} - \bar{\rho} \phi g - \rho_f \varepsilon g$$

and

$$\rho_s \phi \left(\frac{\partial v}{\partial t} + v \frac{\partial v}{\partial z} \right) = -3.2 g d \phi (\rho_s - \rho_f) \frac{\partial \phi}{\partial z} + (\rho_s - \rho_f) g \phi$$
$$\times \left[\frac{\varepsilon(u - v)}{v_t} \right]^{4.8/n} \varepsilon^{-3.8} + \bar{\rho} \phi g - \rho_s \phi g.$$

These are specific to one-dimensional motion along the vertical z axis, but clearly neither one is satisfied by the free-fall test motion. (More recent versions of these authors' equations are not subject to this criticism.)

The physical origin of the difficulty associated with the partition (2.38) is easy to trace. The motivation for this partition is to recognize that, when the fluid supports the weight of the suspended particles, there is an extra contribution to its pressure gradient, which now becomes $\bar{\rho} \mathbf{g}$, rather than $\rho_f \mathbf{g}$. However, this is true only if the particles are, indeed, supported by the fluid in a state of rest or uniform motion. For more general states of motion, where the particles are accelerating, their weight is not balanced by the pressure gradient in the fluid. Thus the connection between $\bar{\rho} \mathbf{g}$ and the fluid pressure gradient fails in general. A particularly simple example of this is provided by the free-fall test motion.

2.3 Explicit Closures of the Equations

The averaged equations of motion can be expressed by the two conservation of mass equations, (2.14) and (2.16), two momentum equations, (2.23) and (2.24), and an angular momentum equation for the particle phase, (2.25). To close these in general it is necessary to express all the following quantities in terms of the averaged variables:

$$\langle \sigma_{ik} \rangle^f, \quad \langle f_i^f \rangle^p, \quad \langle s_{ik}^f \rangle^p, \quad \langle s_{ik}^s \rangle^p, \quad \langle s_{ikl}^f \rangle^p, \quad \langle s_{ikl}^s \rangle^p,$$
$$\langle u_i' u_k' \rangle^f, \quad \langle u_i' u_k' \rangle^p, \quad \langle \omega_i' u_k' \rangle^p. \tag{2.41}$$

For rigid particles in an incompressible Newtonian fluid the first quantity is easy to evaluate (see Joseph & Lundgren, 1990) by noting that

$$\int_V \left(\frac{\partial u_i}{\partial y_k} + \frac{\partial u_k}{\partial y_i} \right) g(|\mathbf{x} - \mathbf{y}|) \, dV$$

$$= \int_{V_f} \left(\frac{\partial u_i}{\partial y_k} + \frac{\partial u_k}{\partial y_i} \right) g(|\mathbf{x} - \mathbf{y}|) \, dV \qquad (2.42)$$

since the rate of deformation vanishes everywhere inside the particles. On the left-hand side of this the order of overall spatial averaging and spatial differentiation can be reversed, since these operations commute, while on the right-hand side we can write

$$\frac{\partial u_i}{\partial y_k} + \frac{\partial u_k}{\partial y_i} = \frac{\sigma_{ik} + p\delta_{ik}}{\mu},$$

which is valid everywhere in the fluid. Thus

$$\frac{\partial \langle u_i \rangle}{\partial x_k} + \frac{\partial \langle u_k \rangle}{\partial x_i} = \frac{\varepsilon}{\mu} (\langle \sigma_{ik} \rangle^f + \langle p \rangle^f \delta_{ik})$$

or

$$\langle \sigma_{ik} \rangle^f = -\langle p \rangle^f \delta_{ik} + \frac{\mu}{\varepsilon} \left(\frac{\partial \langle u_i \rangle}{\partial x_k} + \frac{\partial \langle u_k \rangle}{\partial x_i} \right). \qquad (2.43)$$

However, the remaining terms (2.41) can be evaluated only in certain, very limited circumstances.

Considerable attention has been devoted to closing the equations of motion for the situation in which the fluid is inviscid. This is relevant when the dispersed phase consists of gas bubbles, so that a no-slip boundary condition need not be imposed at the interface. Zhang & Prosperetti (1994) have derived the complete closure for a dilute dispersion, correct to first order in the volume fraction ϕ, and have commented in some detail on the relation of their results to other work on this case. In addition to treatments based on averaging Geurst (1986) has proposed an interesting and quite different approach through a variational principle. However, because we focus here on solid particles, this whole body of work will not be discussed further. The interested reader should refer to the above article by Zhang & Prosperetti and to other references found in it.

The relative importance of those terms in (2.41) representing interactions between the fluid and the particles, and those representing contact forces between particles, depends on the average value of a Stokes number $St = \rho_s a v_1 / \mu$, where v_1 is the relative velocity of approach of a pair of particles. If this is large compared to $\ln(a/l_f)$, where l_f is the mean free path for molecules of the fluid, the lubrication film of fluid separating the particles breaks down and contact forces dominate the interaction. At the opposite extreme of small Stokes number the lubrication films are robust and fluid–particle forces dominate. Clearly the large Stokes number case is most likely to be encountered for dense particles in a gas, whereas the Stokes number may be small for particles of moderate density in a viscous liquid. In addition to the Stokes number, the value of the Reynolds number for relative motion of the fluid and the particles is important. This is defined by $Re = \rho_f a v_2 / \mu$, where v_2 is the average relative velocity of the particles and the fluid in their neighbourhood, and if Re is small compared to unity inertia associated with the motion of fluid around and between the particles can be neglected. Complete closures for the momentum equations have been derived when both the Stokes number and the Reynolds number are small, in the sense just defined and, in addition, the solids concentration ϕ is very small (Jackson, 1997; Zhang & Prosperetti, 1997). Closures have also been proposed for large Stokes number (Koch, 1990; Buyevich, 1994; Buyevich & Kapbasov, 1994; Koch & Sangani, 1999). The latter are of most direct interest for the gas–particle mixtures discussed in much of the remainder of this book. However, since the small Stokes number closures may be relevant for aerosols of small, light particles, both types of closure will be described, starting with the small Stokes number case.

2.3.1 Closure for Small Stokes Number

The results of Zhang & Prosperetti (1997) and those of Jackson (1997) are both limited to dispersions so dilute that hydrodynamic interactions between particles can be neglected. (This is the limiting case for which Einstein (1906, 1911) derived the effective viscosity of a suspension.) Zhang & Prosperetti approached the problem from the point of view of ensemble averaging, while Jackson built on the structure of the equations derived above by volume averaging. Encouragingly, both approaches led to essentially the same results. We shall, therefore, start from the momentum equations in the form of (2.27) and (2.28), reverting to suffix notation for the moment to match these equations. To close Equations (2.27) and (2.28) expressions are needed for the stress tensors S_{ik}^p and S_{ik}^f, together with the interaction force $n\langle f_i^f \rangle^p$. For small Stokes number there are no solid–solid contacts and in the absence of hydrodynamic interactions $\langle u_i' u_k' \rangle^p = 0$, so $S_{ik}^p = 0$. For the remaining quantities we simply quote the results, namely

$$
S_{ik}^f = -\delta_{ik}\langle p \rangle^f + \mu\left(\frac{\partial \langle u_i \rangle}{\partial x_k} + \frac{\partial \langle u_k \rangle}{\partial x_i}\right)
$$
$$
+ \frac{5\mu\phi}{2}\left(\frac{\partial \langle u_i \rangle^f}{\partial x_k} + \frac{\partial \langle u_k \rangle^f}{\partial x_i}\right) - \frac{\partial}{\partial x_k}\left[\frac{3\mu\phi}{4}(\langle u_i \rangle^f - \langle u_i \rangle^p)\right]
$$
$$
+ 3\mu\phi\left[\frac{1}{2}\left(\frac{\partial \langle u_i \rangle^f}{\partial x_k} - \frac{\partial \langle u_k \rangle^f}{\partial x_i}\right) - \varepsilon_{iak}\langle \omega_a \rangle^p\right] \tag{2.44}
$$

and

$$n\langle f_i^f \rangle^p = -\rho_f \phi \left[g_i - \frac{D_f \langle u_i \rangle^f}{Dt} \right] + \frac{9\mu\phi}{2a^2}$$
$$\times \left[(\langle u_i \rangle^f - \langle u_i \rangle^p) + \frac{a^2}{6} \frac{\partial^2 \langle u_i \rangle^f}{\partial x_k^2} \right]. \quad (2.45)$$

Equations (2.44) and (2.45) are correct to the first order in ϕ and they neglect terms that are small and of order a^2/L^2 compared to those exhibited.

Certain terms in these expressions are familiar. The first term in (2.45) represents the buoyancy force, as given by (2.32). This vindicates our choice of (2.32) as a proper way to define buoyancy and it is consistent with the result of Maxey & Riley (1983). The second and third terms are the Stokes drag force and the Faxen force, respectively. Together they represent $n\mathbf{f}_2$ in (2.32). The first term in (2.44) is simply the averaged fluid pressure. The second and third are viscous terms, with the latter representing the Einstein (1906, 1911) correction for a dilute suspension. The fourth term involves the relative, or "slip", velocity between the two phases, while the last term involves the "rotational slip", in other words, the difference between the angular velocity associated with the average fluid velocity field and the average angular velocity of rotation of the particles. Using these results Equations (2.27) and (2.28) become

$$\rho_f \frac{D_f \langle u_i \rangle^f}{Dt} = \frac{\partial S_{ik}^f}{\partial x_k} - \frac{9\mu\phi}{2a^2} \left[(\langle u_i \rangle^f - \langle u_i \rangle^p) + \frac{a^2}{6} \frac{\partial^2 \langle u_i \rangle^f}{\partial x_k^2} \right] + \rho_f g_i,$$
$$(2.46)$$

with S_{ik}^f given by (2.44), and

$$\rho_s \frac{D_p \langle u_i \rangle^p}{Dt} = \frac{9\mu}{2a^2} \left[(\langle u_i \rangle^f - \langle u_i \rangle^p) + \frac{a^2}{6} \frac{\partial^2 \langle u_i \rangle^f}{\partial x_k^2} \right]$$
$$+ (\rho_s - \rho_f) g_i + \rho_f \frac{D_f \langle u_i \rangle^f}{Dt}. \quad (2.47)$$

The equation of motion of the mixture as a whole, that is, the closure of Equation (2.26) for this case, is quite complicated in general. However, in the common situation where the rotational slip, $(\langle \boldsymbol{\omega} \rangle^p - \frac{1}{2}\nabla \times \langle \mathbf{u} \rangle^f)$, is small it reduces approximately to

$$\bar{\rho} \frac{D_m \langle u_i \rangle^m}{Dt} = -\frac{\partial \langle p^f \rangle}{\partial x_i} + \frac{\partial}{\partial x_k} \left\{ \mu \left(\frac{\partial \langle u_i \rangle}{\partial x_k} + \frac{\partial \langle u_k \rangle}{\partial x_i} \right) \right.$$
$$+ \frac{5\mu\phi}{2} \left(\frac{\partial \langle u_i \rangle^f}{\partial x_k} + \frac{\partial \langle u_k \rangle^f}{\partial x_i} \right) - \frac{\partial}{\partial x_k} \left[\frac{3\mu\phi}{4} (\langle u_i \rangle^f - \langle u_i \rangle^p) \right]$$
$$\left. - \frac{\rho_s \phi \rho_f \varepsilon}{\bar{\rho}} (\langle u_i \rangle^f - \langle u_i \rangle^p)(\langle u_k \rangle^f - \langle u_k \rangle^p) \right\} + \bar{\rho} g_i. \quad (2.48)$$

The last term in the braces arises from the fact that differences among $\langle \mathbf{u} \rangle^m$, $\langle \mathbf{u} \rangle^f$, and $\langle \mathbf{u} \rangle^p$ contribute to $\langle u_i' u_k' \rangle^m$, even though $\langle u_i' u_k' \rangle^f$ and $\langle u_i' u_k' \rangle^p$ are both negligible. It is an inertial contribution to the effective stress tensor and might therefore be expected to be negligible in the present case, where the Stokes approximation has been invoked for the relative motion of fluid and particles. The term $\partial/\partial x_k [3\mu\phi(\langle u_i \rangle^f - \langle u_i \rangle^p)/4]$, however, is not inertial in origin. Nevertheless it has not appeared in published discussions of the rheology of dilute suspensions since these have focussed on cases where there is no difference between the average velocities of the fluid and the particles or, if such a difference exists, it corresponds simply to a uniform relative motion, as in sedimentation.

It is interesting to note that Drew et al. (1990) have claimed that volume averaging leads, incorrectly, to inclusion of the buoyancy force $\rho_f \phi g_i$ twice in the momentum equation for the particles. Equation (2.47) shows that this is not so; the buoyancy force appears just once, where it should. The last term in (2.47) enters because the partition of $n\mathbf{f}$ represented by (2.32) replaces g_i in the buoyancy term by the apparent gravitational acceleration, $g_i - D_f \langle u_i \rangle^f / Dt$, which we have seen to be appropriate when the fluid is accelerating. It is true that various incorrect conclusions about the buoyancy force have been drawn from arguments based on volume averaging or, indeed, ensemble averaging, when they have not been carried through to complete closure with sufficient care. However, since ensemble and volume averaging are both formal mathematical procedures it is clear that neither can, in itself, lead to erroneous results if applied correctly.

Though the fluid phase momentum balance (2.46) includes the terms needed to make it correct to first order in ϕ, the corresponding momentum balance (2.47) for the particle phase does not. Indeed, on multiplying it through by the particle volume we recover simply the equation of motion of an isolated particle at low Reynolds number. This is not evidence of inconsistency in the treatment of the suspension as a whole. The mixture momentum equation can be obtained by linear combination of (2.46) and (2.47), multiplying the former by ε and the latter by ϕ (though this is not as straightforward as it might seem at first sight) and the contribution from the particle momentum equation is then seen to be small, of order ϕ. If (2.47) were supplemented by corrections of order ϕ these would contribute to the mixture momentum equation only at order ϕ^2.

2.3.2 Closure for Large Stokes Number

When the Stokes number is large the particles can interact by direct collision. Then the particle phase stress tensor $\mathbf{S}^p$ is expected to play a major role in the dynamics and, as we see from Equation (2.24), this tensor represents both stress transmission by contact forces and momentum transfer due to fluctuations in particle velocities about their local average. These are the same as the mechanisms of stress transmission in a granular material (particles with no interstitial fluid), for which equations of motion are known in certain circumstances. Except at particle concentrations approaching close packing a granular material resembles a molecular gas in all ways but one, namely the inelastic nature of collisions between macroscopic particles. Thus, one might anticipate that the equations of motion of gases would also serve for granular materials, with some modification to account for the inelasticity of collisions. This analogy has been exploited widely (see, for example, Haff (1983), Jenkins & Savage (1983), Lun et al. (1984), Jenkins (1987), Lun (1991), and Goldshtein & Shapiro (1995)). In the case of a granular material the momentum balance for the particle phase contains both an effective pressure and an effective viscosity, entirely analogous to the pressure and viscosity of a molecular gas. For a gas these depend on the temperature, which must be determined by a separate equation of energy balance. In the same way for a granular material a "particle temperature" may be defined by $T = \frac{1}{3} \langle u'_k u'_k \rangle^p$, and this influences the values of both the effective pressure and the effective viscosity appearing in $\mathbf{S}^p$. It must therefore be determined from a supplementary equation of balance for the "pseudothermal" energy of the particle velocity fluctuations. This differs from the energy balance equation for a molecular gas only by the inclusion of a term to represent energy losses in inelastic collisions. However, in the situation of interest to us where an interstitial fluid fills the space between the particles, other terms must be added to the equations of momentum and pseudothermal energy balance to represent the fluid–particle interaction force, the generation of particle velocity fluctuations by fluctuating forces exerted on the particles by the fluid, and their damping by viscous resistance to motion of the particles. Thus the momentum balances retain the form defined by (2.27′) and (2.28′), but these must be supplemented by a pseudothermal energy balance equation:

$$\frac{3}{2} \rho_s \phi \frac{D_p T}{Dt} = \mathbf{S}^p : \nabla \mathbf{v} - \nabla \cdot \mathbf{q} + Q_+ - Q_- - Q_c. \tag{2.49}$$

The first term on the right-hand side of this represents the generation of pseudothermal energy by working of the particle phase stresses and the second represents its accumulation by conduction, with $\mathbf{q}$ denoting the flux vector. The terms Q_+, Q_-, and Q_c account for generation of pseudothermal energy as a result of the relative motion of the phases (represented by $\mathbf{w} = \mathbf{u} - \mathbf{v}$), its dissipation by viscous forces, and its dissipation by collisions, respectively.

There is also a question of whether the velocity fluctuations can be described adequately by a single scalar measure, indicated above by T. A preferred direction is defined by $\mathbf{u} - \mathbf{v}$ so, if Q_+ contributes significantly to the generation of velocity fluctuations, one might expect there to be circumstances in which the mean square fluctuations in velocity components parallel and perpendicular to $\mathbf{u} - \mathbf{v}$ differ from each other. Then there would be not one, but two "particle temperatures" and, correspondingly, two balance equations would be needed to determine them.

There have been three significant attempts to provide fundamentally based, high Stokes number closures, one valid only under the restriction that ϕ should be small enough that terms of $O(\phi^2)$ can be neglected, and the other two claimed to apply up to higher concentrations, until networks of sustained contacts between particles begin to be established.

To close (2.27′), (2.28′), and (2.49) expressions for $n\mathbf{f}$, $\mathbf{S}^p$, $\mathbf{S}^f$, $\mathbf{q}$, Q_+, Q_-, and Q_c are needed. Under the restrictions that terms of $O(\phi^2)$ can be neglected compared to terms of $O(\phi)$, that $St \gg \phi^{-3/2}$, and that collisions between particles should be perfectly elastic Koch (1990) derived the following closure, which takes account of the generation of particle velocity fluctuations by fluctuating hydrodynamic interactions:

$$\mathbf{S}^f = -p^f\mathbf{I} + \mu\nabla^2\mathbf{u}, \quad \mathbf{S}^p = -p^p\mathbf{I} = -\rho_s\phi T\mathbf{I}, \quad \mathbf{q} = 0,$$

$$n\mathbf{f} = \left[1 + 3(\phi/2)^{1/2}\right](9\mu\phi\mathbf{w}/2a^2), \quad Q_c = 0, \tag{2.50}$$

$$Q_+ = \frac{81\mu^2\phi|\mathbf{w}|^2}{8a^3\rho_s(\pi T)^{1/2}}, \quad Q_- = \left[1 + 3(\phi/2)^{1/2}\right](27\mu\phi T/2a^2).$$

In contrast to the above quite simple result, a much more elaborate closure due to Buyevich and coworkers is claimed to be applicable at higher concentrations, though it is expected to fail if the volume fraction ϕ is too small. This closure is described, in its most complete form, in two papers (Buyevich, 1994; Buyevich & Kapbasov, 1994) but the principles involved date back to earlier work by the same author (Buyevich, 1971a,b; 1972a,b,c) and the mechanism for generating fluctuations in particle velocity is elaborated in subsequent works (Buyevich, 1997; Buyevich & Kapbasov, 1999). The main idea is that the generation of fluctuations in particle velocities by the relative motion of the two phases is primarily a result of random fluctuations in the void fraction, on the scale of the volume of space associated with a single particle (that is, a volume $1/n$, where n is the mean number of particles per unit volume). The fluctuations in void fraction induce fluctuations in the drag force exerted by the fluid on a particle, and these in turn generate fluctuations in the velocity component directed parallel to the drag force. The resulting fluctuations are, of course, anisotropic, but frequent collisions between particles are assumed to eliminate the anisotropy, leaving essentially isotropic fluctuations that can be characterized by the granular temperature defined above. The closure refers only to the case of large Stokes number and, as noted above, for a given value of the Stokes number it is not expected to be valid if the volume fraction ϕ is too small. (There is a similar limitation on Koch's closure, namely $St \gg \phi^{-3/2}$.) The closure is also inappropriate at the largest values of ϕ where the particles can form extended structures capable of transmitting stress via sustained solid–solid contacts. The derivation is long and involves a number of assumptions and approximations. Here we simply quote the final results. As in the case of Koch's closure, these refer to the equations of motion in the form of (2.27'), (2.28'), and (2.49):

$$n\mathbf{f} = \rho_s\phi[F_1(\phi) + F_2(\phi)|\mathbf{w}|]\mathbf{w} - \bar{\rho}\phi\mathbf{g},$$

$$\mathbf{S}^p = -p^p\mathbf{I} + \mu^p[\nabla\mathbf{v} + (\nabla\mathbf{v})^T - 2/3(\nabla\cdot\mathbf{v})\mathbf{I}], \qquad (2.51)$$

$$\mathbf{q} = -\lambda^p\nabla T$$

and also

$$Q_c = \frac{12}{\pi^{1/2}}(1 - e)\frac{\rho_s\phi^2\chi(\phi)}{a}T^{3/2} = \alpha_c T^{3/2},$$

$$Q_- = \rho_s\phi[3F_1(\phi) + 4F_2(\phi)|\mathbf{w}|]T = \alpha_- T, \qquad (2.52)$$

$$Q_+ = [\alpha_-\sqrt{T^*} + \alpha_c T^*]\sqrt{T}.$$

In these equations e denotes an effective coefficient of restitution for collisions between particles and the remaining undefined scalar quantities are expressed as follows:

$$F_1(\phi) = \frac{9}{2}\frac{\rho_f}{\rho_s}\frac{\nu}{a^2}\frac{1}{(1 - \phi)^{5/2}}, \quad F_2(\phi) = \frac{3}{16}\frac{\rho_f}{\rho_s}\frac{1}{a}\left(\frac{1 - \phi}{1 - 1.17\phi^{2/3}}\right)^2,$$

$$p^p = [1 + 4\phi\chi(\phi)]\rho_s\phi T,$$

$$\mu^p = 4\phi\left\{\frac{1}{4\phi\chi(\phi)} + 0.8 + 0.76(4\phi\chi(\phi))\right\}\frac{5}{48}a\rho_s(\pi T)^{1/2}, \qquad (2.53)$$

$$\lambda^p = 4\phi\left\{\frac{1}{4\phi\chi(\phi)} + 1.2 + 0.75(4\phi\chi(\phi))\right\}\frac{25}{64}a\rho_s(\pi T)^{1/2}.$$

Here ν denotes the kinematic viscosity of the fluid ($\nu = \mu/\rho_f$) and the function $\chi(\phi)$ is the radial distribution function at contact. Two different expressions are proposed for this, namely

$$\chi(\phi) = \frac{2 - \phi}{2(1 - \phi)^3} \qquad (2.54)$$

and

$$\chi(\phi) = \frac{(\phi/\phi_m)^{1/3}}{4\phi[1 - (\phi/\phi_m)^{1/3}]}. \tag{2.55}$$

The former, due to Carnahan & Starling (1969), is expected to be valid up to moderately high particle concentrations. The latter should give a more realistic representation of $\chi(\phi)$ for the highest concentrations near $\phi = \phi_m$, where ϕ_m denotes the volume fraction at random close packing.

The most complicated part of the closure is the evaluation of the parameter T^*, which, like T, has the dimensions of energy per unit mass. The paper of Buyevich & Kapbasov (1994) is devoted entirely to this question, and their final result is

$$T^* = 6.145 \times 10^{-2} \left(\frac{1 - 0.56f}{3 + 4f} M\right)^2 \langle\phi'^2\rangle w^2 \quad \text{with } f = F_2(\phi)w/F_1(\phi), \tag{2.56}$$

where $w = |\mathbf{w}|$. In the above

$$M = \frac{1}{1 - \phi} + \frac{d(\ln F_1)/d\phi + fd(\ln F_2)/d\phi + (1 - \rho_f/\rho_s)(g/F_1 w)(\mathbf{i} \cdot \mathbf{w}/w)}{1 + 2f}, \tag{2.57}$$

where $\mathbf{i}$ denotes the unit vector in the upward vertical direction and g is the acceleration due to gravity. The factor $\langle\phi'^2\rangle$ represents the mean square fluctuation in the volume fraction, on the scale of the volume associated with a single particle. Two expressions are given for this corresponding to the two expressions for $\chi(\phi)$in (2.54) and (2.55):

$$\langle\phi'^2\rangle = \frac{\phi^2}{1 + \frac{2\phi(4-\phi)}{(1-\phi)^4}} \tag{2.58}$$

and

$$\langle\phi'^2\rangle = \frac{\phi^2\left[1 - (\phi/\phi_m)^{1/3}\right]}{\left[1 + \frac{1}{3}\frac{(\phi/\phi_m)^{1/3}}{1-(\phi/\phi_m)^{1/3}}\right]}, \tag{2.59}$$

with (2.58) applicable at moderate particle concentrations and (2.59) at concentrations approaching random close packing. This completes Buyevich's closure.

The third, and most recent attempt at an explicit closure is due to Koch & Sangani (1999), who consider particles whose Reynolds number is small at their terminal velocity of fall, v_t, and whose Stokes number ($St = v_t^2/ga$) is fairly large, in the sense that $St \gg \phi^{-3/4}$. It is therefore intended to be valid over a wider range of values for the Stokes number, extending to values low enough that the velocity distribution is no longer determined entirely by direct collisions between pairs of particles. For simplicity we shall present the equations only in their one-dimensional form, applicable to average motion confined to the z direction. Though the full set consists of two continuity equations and two momentum equations it is easy, in this geometry, to use the mixture continuity equation to eliminate the fluid velocity. The precise form of the particle phase momentum balance then depends on the particular frame of reference from which the system is viewed; the two common choices are the fluid bed reference frame, in which the average particle velocity vanishes, and the sedimentation reference frame, in which the average mixture velocity $\phi v + (1 - \phi)u$ vanishes. Although neither choice has any significant advantage for a suspension of unbounded extent, in interpreting the equations it is necessary to be aware which has been made. Koch & Sangani choose the sedimentation frame; they also render their variables dimensionless by scaling velocities with v_t, space coordinates with a, time with a/v_t, and stresses with $\rho_s v_t^2$. Then, in terms of these dimensionless variables, the particle phase momentum equation takes the form

$$\phi \left[\frac{\partial v}{\partial t} + v \frac{\partial v}{\partial z} \right] = \frac{\phi}{St} - \frac{\phi}{St} R_{\text{drag}} v - \frac{\partial P_m}{\partial z} + \frac{\partial}{\partial z} \left[\mu_c \frac{\partial v}{\partial z} \right]. \qquad (2.60)$$

Here the first term on the right-hand side represents gravity, the second is the drag force, the third is the gradient of a stress P_m and contains both a pressure gradient and viscous contributions, while the last is an additional viscous stress generated by collisions between particles. Also

$$P_m = (\phi + 8B/5)T_\parallel + (12/5)BT, \qquad \mu_c = \frac{48}{5\sqrt{\pi}} BT^{1/2},$$

$$B = \phi^2 \eta \chi, \qquad \eta = (1 + e)/2, \qquad (2.61)$$

with χ representing the hard sphere radial distribution function at contact, for which the following expression (due to Ma & Ahmadi, 1988) is adopted:

$$\chi = \frac{1 + 2.5\phi + 4.5094\phi^2 + 4.515439\phi^3}{[1 - (\phi/0.64356)^3]^{0.67802}}. \qquad (2.62)$$

$T_\parallel$ and $T_\perp$ represent the particle temperatures that measure fluctuations in vertical and horizontal components of velocity, respectively, while $T = \frac{1}{3}T_\parallel + \frac{2}{3}T_\perp$. Finally, the drag coefficient is related to volume fraction by

$$R_{\text{drag}} = \begin{cases} \dfrac{1 + 3(\phi/2)^{1/2} + (135/64)\phi \ln \phi + 17.14\phi}{1 + 0.681\phi - 8.48\phi^2 + 8.16\phi^3} & \text{for } \phi < 0.4, \\[2ex] \dfrac{10\phi}{1 - \phi^3} + 0.7 & \text{for } \phi \geq 0.4. \end{cases} \qquad (2.63)$$

For small ϕ the first expression agrees to $O(\phi)$ with theoretical equations for the drag in dilute fixed beds, whereas for larger values of ϕ it is fitted to results of direct simulations of the motion of an assembly of particles in a gas. The spatial domain simulated has to be limited in size; otherwise the results are obscured by the well-known instability of a uniform suspension, which will be discussed in detail in Chapter 4. The second expression in (2.63), applicable when $\phi \geq 0.4$, is the well-known equation of Carman for drag in a closely packed bed, supplemented by the addition of 0.7 to ensure that it takes over from the first expression at $\phi = 0.4$ without a discontinuity.

To complete the closure we need balance equations for the energies of velocity fluctuations in the vertical and horizontal directions to determine $T_{\parallel}$ and $T_{\perp}$. These equations take the form

$$\frac{1}{2}\phi\left[\frac{\partial T_{\parallel}}{\partial t} + v\frac{\partial T_{\parallel}}{\partial z}\right] = \frac{1}{3}\frac{\partial}{\partial z}\left(\kappa\frac{\partial T_{\parallel}}{\partial z}\right) - P_m\frac{\partial v}{\partial z} + \mu_c\left(\frac{\partial v}{\partial z}\right)^2 + S_{\parallel} - \Gamma_{\parallel}$$

(2.64)

and

$$\frac{3}{2}\phi\left[\frac{\partial T}{\partial t} + v\frac{\partial T}{\partial z}\right] = \frac{\partial}{\partial z}\left(\kappa\frac{\partial T}{\partial z}\right) - P_T\frac{\partial v}{\partial z} + \mu_c\left(\frac{\partial v}{\partial z}\right)^2 + S - \Gamma,$$

(2.65)

where $S_{\parallel}$ and $\Gamma_{\parallel}$ are source and sink terms for the energy of fluctuations in the vertical component of velocity, while S and Γ are corresponding source and sink terms for the total energy of velocity fluctuations. Also

$$P_T = 4BT(3\eta - 2)$$

(2.66)

and κ is the thermal conductivity of a gas with temperature $T_{\parallel}$:

$$\kappa = \frac{8}{\sqrt{\pi}}\phi^2 T_{\parallel}^{1/2}\left(1 + \frac{25\pi}{512\phi^2\chi^2}\right).$$

(2.67)

The source and sink terms are given by

$$S_{\parallel} - \Gamma_{\parallel} = \frac{S_{\parallel}^* v^2 \phi}{St^2 T^{1/2}} - \frac{\phi R_{\text{diss}} T_{\parallel}}{St} - \frac{24}{5\sqrt{\pi}}BT^{1/2}$$
$$\times\left[(2 - \eta)T_{\parallel} - \frac{1}{3}(1 + 2\eta)T\right]$$

(2.68)

and

$$S - \Gamma = \frac{S^* v^2 \phi}{St^2 T^{1/2}} - \frac{3\phi R_{\text{diss}} T}{St} - \frac{24}{\sqrt{\pi}}BT^{3/2}(1 - \eta).$$

(2.69)

Here the first and second terms represent generation and dissipation, respectively, due to hydrodynamic stresses, the third term in (2.69) represents dissipation due to inelastic collisions, and the third term in (2.68) includes both dissipation and redistribution of energy between vertical and horizontal velocity fluctuations as a result of collisions.

It remains to specify R_{diss}, $S_\parallel^*$, and S^*. For R_{diss} an expression due to Sangani et al. (1996) is invoked for the viscous energy dissipation in a suspension of particles with Maxwell velocity distribution, namely

$$R_{\text{diss}} = 1 + 3(\phi/2)^{1/2} + (135/64)\phi \ln\phi + 11.26\phi$$
$$\times (1 - 5.1\phi + 16.57\phi^2 - 21.77\phi^3) - \phi\chi\ln\varepsilon_m. \qquad (2.70)$$

The last term in this calls for some comment. Denoting the gap between the surfaces of two closely approaching particles of radius a by $2a\varepsilon$, we find the rate of viscous energy dissipation in the gap to increase in proportion to $1/\varepsilon$ as $\varepsilon \to 0$. Consequently, to retain a finite value for the overall rate of dissipation in the suspension, it is necessary to impose a lower bound ε_m on the value of ε, and this appears in the last term of (2.70). For gaps narrower than this it is assumed that the continuum lubrication flow breaks down. In a gas this occurs when the gap width becomes comparable with the molecular mean free path length. This situation has been analysed in detail by Sundararajakumar & Koch (1996), who showed that (2.70) gives a correct value for the viscous dissipation rate with the choice $\varepsilon_m = 4.88\lambda/a$, where λ denotes the mean free path.

Finally, $S_\parallel^*$ and $S^*(= S_\parallel^* + 2S_\perp^*)$ are found using expressions for $S_\parallel^*$ and $S_\perp^*$ with the same temperature dependence as was derived for dilute suspensions by Koch (1990) and a dependence on concentration contained in a multiplicative factor $R_s R_{\text{drag}}^2$:

$$S_\parallel^* = \frac{3}{16\sqrt{\pi}}\left(\frac{T}{T_\parallel}\right)^{1/2} R_s R_{\text{drag}}^2 \left[(\beta^5 + \beta)\ln\left(\frac{\beta+1}{\beta-1}\right) - \frac{2}{3}\beta^2 - 2\beta^4 \right],$$
$$(2.71)$$

$$S_\perp^* = \frac{3}{16\sqrt{\pi}}\left(\frac{T}{T_\parallel}\right)^{1/2} R_s R_{\text{drag}}^2 \left[\frac{1}{2}(\beta^3 - \beta^5)\ln\left(\frac{\beta+1}{\beta-1}\right) - \frac{2}{3}\beta^2 + \beta^4 \right],$$
$$(2.72)$$

with

$$\beta^2 = \frac{T_\parallel}{T_\parallel - T_\perp} = \frac{2}{3}\frac{T_\parallel}{T_\parallel - T} \qquad (2.73)$$

and

$$R_s = \frac{1}{\chi(1 + 3.5\phi^{1/2} + 5.9\phi)}. \tag{2.74}$$

The expression (2.74), like (2.63), is found by fitting to the results of numerical simulations of the motion of a suspension confined to a small spatial domain. Note that $S_\parallel^*$ and $S_\perp^*$ both remain bounded in the limit $\beta \to \infty$, corresponding to isotropy of the particle velocity fluctuations. Indeed, $S_\parallel^* \to 12/5$ and $S_\perp^* \to 2/15$ and the equations reduce, as they should, to limiting forms for high Stokes number, which were also derived separately by Koch & Sangani.

Though explicit, the closures of Buyevich and of Koch & Sangani are both very complicated. The source of much of their complexity is the mechanism by which pseudothermal motion of the particles is generated as a result of the difference between the average velocities of the two phases. Gidaspow (1994, Table 10.1) has formulated a set of equations similar to those of Buyevich, but decidedly simpler in form, by simply omitting the term Q_+ from the pseudothermal energy balance, thereby avoiding the problem of providing for its closure. However, as a consequence of this omission, there is no mechanism in Gidaspow's equations for thermalizing the particles in a uniform fluidized suspension.

Both the equations of Buyevich and those found by Koch raise certain questions. First note that Buyevich's equations fail the free-fall test formulated above in Section 2.2. This is because the buoyancy force has been expressed as $-\bar{\rho}\phi\mathbf{g}$; replacing this expression by $-\rho_f\phi(\mathbf{g} - D_f\mathbf{u}/Dt)$ would immediately resolve this difficulty, as we have shown. However, there is another anomaly, associated with the pseudothermal energy balance, that requires a little more explanation.

This second anomaly is one that Buyevich's energy equation inherits from the granular material energy balance, of which it is a generalization. Consider an assembly of particles in plane shear ($\mathbf{v} = (\Gamma y, 0, 0)$) in the absence of gravity. The volume concentration ϕ is independent of position and the interstitial space is occupied by a gas. Then $\mathbf{u} = \mathbf{v}$ everywhere and the energy balance (2.49) reduces to

$$0 = \mu^p \Gamma^2 - Q_- - Q_c$$

or

$$[H(\phi)a\rho_s T^{1/2}]\Gamma^2 = 3\rho_s\phi F_1(\phi)T + \frac{12}{\pi^{1/2}}(1-e)\frac{\rho_s\phi^2\chi(\phi)}{a}T^{3/2},$$

where, referring to (2.53), we have written $\mu_p = H(\phi)a\rho_s T^{1/2}$. The above has the form

$$A\Gamma^2 T^{1/2} = B\phi T + C\phi^2 T^{3/2},$$

which has two real, nonnegative roots in T, namely $T = 0$ and

$$T^{1/2} = \frac{-B + \sqrt{B^2 + 4AC\Gamma^2}}{2C\phi}. \tag{2.75}$$

The solution $T = 0$ is easily seen to be an unstable steady state, in the sense that dT/dt becomes positive as soon as T is increased incrementally. Indeed, it is the initial condition for a solution of the energy equation with $T \propto t^2$. Thus the root given by (2.75) is the physically significant one. Consider its behaviour when ϕ becomes small. As $\phi \to 0$

$$A \to \frac{5\pi^{1/2}a\rho_s}{48}, \quad B \to \frac{27}{2}\frac{\rho_f}{\rho_s}\frac{\nu}{a^2}, \quad C \to \frac{12}{\pi^{1/2}}\frac{(1-e)\rho_s}{a};$$

so T tends to infinity like $1/\phi^2$. Consequently the shear stress also tends to infinity as the concentration of particles tends to zero. This anomaly is familiar in the literature of fast flow of granular materials (see, for example, Lun et al. (1983)). It arises from the behaviour of the effective particle phase viscosity as the particle concentration tends to zero. At small particle concentration the viscosity can be written in terms of a mean free path length, l_p, for the particle assembly in the standard way, namely

$$\mu^p = \frac{5}{16}mnl_p(\pi T)^{1/2},$$

where m is the mass of a particle and n the number per unit volume. This agrees with the limiting form of (2.53), as $\phi \to 0$, if we take the usual expression $l_p = \frac{1}{4}\pi a^2 n$ for the mean free path. With this choice μ_p has a nonvanishing low-concentration limit, so the rate of generation of pseudothermal energy by shear work remains positive in this limit. The terms Q_- and Q_c, however, tend to zero with the concentration, leaving nothing to balance the rate of generation, so the temperature increases without bound.

This difficulty can be resolved by noting that the above expression for l_p increases without bound as $n \to 0$. This is appropriate for particles *in vacuo*, but in the presence of an interstitial gas there is a bounded value for the stopping length of a moving particle as a result of viscous drag, and this gives an upper bound for meaningful values of l_p. It follows that l_p remains bounded when $n \to 0$ in the presence of a gas, so μ_p becomes proportional to n or, equivalently, to ϕ. The argument above then shows that T remains bounded when $\phi \to 0$. A modification of the expression for μ^p to account for viscous stopping, as suggested by Boelle et al. (1995), would therefore remove this difficulty from the theory.

Turning now to Koch's (1990) closure, we note that Q_c vanishes because collisions between particles are regarded as perfectly elastic. To the first order in ϕ, which is the highest order of significance in Koch's closure, his expression for Q_- agrees with that of Buyevich. However, his expression for Q_+ is quite different. This is obvious since, according to Buyevich, $Q_+ \propto T^{1/2}$, whereas Koch's closure (2.50) gives $Q_+ \propto T^{-1/2}$. Thus with Koch's closure Q_+ diverges as $T \to 0$ and consequently $D_p T / Dt \to \infty$; the fluid–particle interaction rapidly thermalizes a suspension that starts from $T = 0$.

Perhaps the difficulty with Buyevich's equations when $\phi \to 0$ and their failure to agree with Koch's closure at this limit should not be regarded as serious defects, since Buyevich explicitly disclaims their applicability at very small values of ϕ. Nevertheless, one is left to wonder how the transition from $Q_+ \propto T^{1/2}$ (Buyevich) to $Q_+ \propto T^{-1/2}$ (Koch) occurs as ϕ decreases towards zero.

One other issue must be raised in relation to Koch's (1990) closure: the absence of any deviatoric terms from the effective stress tensor $\mathbf{S}^p$. If the bound on the mean free path length is neglected we have seen that

$$\mu^p \to \frac{5}{48} a \rho_s (\pi T)^{1/2} \quad \text{as } \phi \to 0.$$

Thus the deviatoric terms are of order zero in ϕ at this limit. Since Koch's closure aims to be correct to the first order in ϕ it is therefore unclear why these terms do not appear. Even if the viscous stopping length is taken into account, so that μ^p becomes proportional to ϕ as $\phi \to 0$, the viscous part of $\mathbf{S}^P$ is of order ϕ and should presumably appear in the closure.

2.3.3 *Other Comments on Explicit Closures*

One other attempt at an explicit closure must be mentioned, though this is for-mulated for motion confined to one dimension, with the single space coordinate z measured vertically upward. Then Foscolo & Gibilaro (1987) have proposed momentum equations of the following form:

$$\rho_f \varepsilon \left(\frac{\partial u}{\partial t} + u \frac{\partial u}{\partial z} \right) = -\frac{\partial p^f}{\partial z} - (\rho_s - \rho_f) g \phi$$

$$\times \left[\frac{\varepsilon(u - v)}{v_t} \right]^{4.8/n} \varepsilon^{-3.8} - \bar{\rho} \phi g - \rho_f \varepsilon g \qquad (2.76)$$

and

$$\rho_s \phi \left(\frac{\partial v}{\partial t} + v \frac{\partial v}{\partial z} \right) = -3.2 g d \phi (\rho_s - \rho_f) \frac{\partial \phi}{\partial z} + (\rho_s - \rho_f) g \phi$$

$$\times \left[\frac{\varepsilon(u - v)}{v_t} \right]^{4.8/n} \varepsilon^{-3.8} + \bar{\rho} \phi g - \rho_s \phi g. \qquad (2.77)$$

Table 2.1. *Typical values for Reynolds and
Stokes numbers*

d	ε	
	0.5	1.0
50 μm	$Re = 0.045$	$Re = 0.50$
	$St = 74.3$	$St = 832$
	$\ln(a/l_f) = 5.52$	$\ln(a/l_f) = 5.52$
200 μm	$Re = 2.75$	$Re = 22.1$
	$St = 4587$	$St = 36,796$
	$\ln(a/l_f) = 6.9$	$\ln(a/l_f) = 6.9$
1 mm	$Re = 130$	$Re = 417$
	$St = 217,211$	$St = 694,555$
	$\ln(a/l_f) = 8.5$	$\ln(a/l_f) = 8.5$

The above authors put these equations forward as universal forms for the momentum balances, valid for both solid–liquid and solid–gas mixtures, and without restriction on the value of the particle concentration; indeed, they are used to distinguish between the behaviour of liquid and gas fluidized beds through a linear stability analysis. However, though relatively simple, they have some unsatisfactory features. First, they contain no terms representing viscous stresses and hence, as we shall see in Chapter 4, they predict that instabilities of vanishingly short wavelength will grow most rapidly. More seriously, as pointed out in Section 2.2, they fail the free-fall test and cannot, therefore, represent valid equations of motion.

Finally, the validity of each of the closures discussed above is confined to certain ranges of the Reynolds and Stokes numbers. To gain some feeling for the values of these groups to be expected in gas–particle suspensions Table 2.1 shows values of Re and St for three sizes of particles with $\rho_s = 2.0$ (typical of mineral material), fluidized by air at ambient conditions. For each particle size Re and St are presented for a quite dense fluidized bed with $\varepsilon = 0.5$, and also for the limiting case of a highly expanded bed with $\varepsilon \rightarrow 1$. The values of both the Reynolds and the Stokes numbers are based on the interstitial velocity of the fluidizing air, and Buyevich's expression is used for the force exerted on the particles by the gas.

The table also shows values of $\ln(a/l_f)$ for comparison with the Stokes numbers. We see that the latter is much larger than the former in all cases, bearing out our assumption that collisions between particles will have a significant influence on $\mathbf{S}^P$. The Reynolds numbers are small enough to justify the neglect of fluid inertia in the motion of the gas relative to the particles only for the smallest of the three particle sizes.

2.4 Empirical Closures

We now return to the general form of the momentum equations, expressed as (2.30) and (2.31), or equivalently as (2.33) and (2.34). The general problem of closure is to provide expressions for the tensors $\mathbf{S}^f$ and $\mathbf{S}^P$ and the vector $n\mathbf{f}_1$ (or $n\mathbf{f}_2$) that are applicable over the whole range of Reynolds and Stokes numbers and up to high values of the particle volume fraction. We have already glimpsed the complexity of this, and it seems unlikely that the problem will be solved in the foreseeable future. For practical purposes, therefore, we must be content with something less. What we need are not necessarily the exact equations of motion, but equations that are good enough to describe phenomena of interest to the accuracy required. Thus, the problem can be reformulated as that of finding closures that are as simple as possible, while incorporating just enough physics to accomplish this, for a given range of phenomena and a specified precision. This is the opposite of the "Principle of Equipresence", which essentially requires that the equations should contain every sort of term that does not violate certain general principles of continuum mechanics. The difficulty with such "equipresent" equations is that they contain far more parameters than can be discriminated by any reasonable set of experimental measurements. Even with ruthless pruning of the terms included this remains a problem for two-phase systems.

Let us, then, return to Equations (2.30) and (2.31), or equivalently (2.33) and (2.34), and examine in turn the terms on their right-hand sides for which closures are needed. It is easiest to start with the fluid–particle interaction force $n\mathbf{f}_1$ (or $n\mathbf{f}_2$), since there is experimental evidence bearing directly on this. It is generally agreed that the main contributors are a term depending on the particle concentration and the relative velocity $(\mathbf{u} - \mathbf{v})$, a second term depending on the concentration and the relative acceleration, and a third term representing a force normal to the direction of $(\mathbf{u} - \mathbf{v})$. These are referred to as the drag force, the virtual mass force, and the lift force, respectively. Other contributions, usually neglected, include the Faxen force (see (2.45)) and a history-dependent term analogous to the Basset (1888) force for the motion of an isolated particle. (A comprehensive listing of possible contributors is given by Drew & Lahey (1993).) Here we shall write $n\mathbf{f}_1 = \mathbf{F}_D + \mathbf{F}_V + \mathbf{F}_L$, representing the drag, virtual mass, and lift contributions, and discuss these separately.

The drag force acts in the direction of the relative velocity $\mathbf{u} - \mathbf{v}$ and therefore has the general form $\mathbf{F}_D = (\mathbf{u} - \mathbf{v})F(\phi, |\mathbf{u} - \mathbf{v}|)$, where the form of the scalar function F is to be determined and will depend on the particle size and shape and the properties of the fluid, in addition to the arguments indicated. The experimental information with most direct bearing on the form of $\mathbf{F}_D$ is the observed relation between the sedimentation velocity v_s of a dispersion of the particles in the fluid of interest and their concentration, measured by ε. This was investigated extensively by Richardson & Zaki (1954), who found the following empirical relation for a suspension of uniform, spherical particles, infinite in extent:

$$v_s = v_t \varepsilon^n, \tag{2.78}$$

with

$$n = \begin{cases} 4.65 & \text{for } Re_t < 0.2, \\ 4.35\,Re_t^{-0.03} & \text{for } 0.2 < Re_t < 1, \\ 4.45\,Re_t^{-0.1} & \text{for } 1 < Re_t < 500, \\ 2.39 & \text{for } 500 < Re_t, \end{cases} \tag{2.79}$$

where $Re_t = 2a\rho_f v_t/\mu$, the Reynolds number for an isolated particle falling at its terminal velocity v_t in the fluid. (Richardson & Zaki also investigated the influence of the diameter of the container holding the suspension, and their correlations for n include this.)

The relation (2.78) is sufficient to determine the functional form for $\mathbf{F}_D$ completely, at both small and large values of $|\mathbf{u} - \mathbf{v}|$. For small values $\mathbf{F}_D$ is proportional to the relative velocity so we can write $\mathbf{F}_D = \beta_1(\mathbf{u} - \mathbf{v})$. Applying this to the case of steady sedimentation, where $\mathbf{S}^p = 0$ and $\mathbf{S}^f = -p_f\mathbf{I}$, reduces Equations (2.30) and (2.31) to

$$0 = -\varepsilon\frac{dp_f}{dz} - \beta_1(u - v) - \rho_f\varepsilon g,$$

$$0 = -\phi\frac{dp_f}{dz} + \beta_1(u - v) - \rho_s\phi g,$$

where z is a coordinate in the upward vertical direction and u and v are the z components of $\mathbf{u}$ and $\mathbf{v}$, respectively. Since the total volume flux vanishes, $\varepsilon u + \phi v = 0$. Then, eliminating u and dp_f/dz between this and the above pair of equations, and noting that the sedimentation velocity v_s is simply $-v$, we find that

$$\frac{\beta_1 v_s}{\varepsilon^2} = (\rho_s - \rho_f)\phi g.$$

Combining this with (2.78) then gives

$$\beta_1 = \frac{(\rho_s - \rho_f)\phi g}{v_t\varepsilon^{n-2}}, \quad \mathbf{F}_D = \frac{(\rho_s - \rho_f)\phi g}{v_t\varepsilon^{n-2}}(\mathbf{u} - \mathbf{v}), \tag{2.80}$$

with n determined from (2.79). At the opposite extreme where the value of $|\mathbf{u} - \mathbf{v}|$ is sufficiently large, the regime of fluid flow relative to the particles is Newtonian, so we expect that $\mathbf{F}_D = \alpha_1(\mathbf{u} - \mathbf{v})|\mathbf{u} - \mathbf{v}|$. Then an argument analogous to that above gives

$$\alpha_1 = \frac{(\rho_s - \rho_f)\phi g}{v_t^2 \varepsilon^{2n-3}}, \quad \mathbf{F}_D = \frac{(\rho_s - \rho_f)\phi g}{v_t^2 \varepsilon^{2n-3}}|\mathbf{u} - \mathbf{v}|(\mathbf{u} - \mathbf{v}). \quad (2.81)$$

More generally, if $\mathbf{F}_D$ can be factorized in the form $\mathbf{F}_D = \gamma_1(\phi)(\mathbf{u} - \mathbf{v}) h(|\mathbf{u} - \mathbf{v}|)$, where h may be any function of $|\mathbf{u} - \mathbf{v}|$, the same reasoning leads from the Richardson–Zaki relation to the form of the function γ_1. The result is

$$\gamma_1 = \frac{(\rho_s - \rho_f)\phi g}{v_t \varepsilon^{n-2} h(v_t \varepsilon^{n-1})},$$

$$\mathbf{F}_D = \frac{(\rho_s - \rho_f)\phi g}{v_t \varepsilon^{n-2} h(v_t \varepsilon^{n-1})} h(|\mathbf{u} - \mathbf{v}|)(\mathbf{u} - \mathbf{v}). \quad (2.82)$$

Unfortunately, however, for values of the relative velocity intermediate between the viscous and Newtonian regimes a factorization of this sort is not possible, and consequently the complete algebraic form of the drag force law cannot be deduced unambiguously from the Richardson–Zaki equation.

Perhaps the best known attempt to provide an empirical expression for $\mathbf{F}_D$, valid over the whole range of values of Reynolds numbers (based on the relative velocity) and particle concentrations, is by Ishii & Zuber (1979). This takes, as its starting point, the following well-known expression for the drag force on a single particle:

$$\mathbf{f}_D = C_D(Re)\frac{\pi a^2}{2}\rho_f|\mathbf{w}|\mathbf{w}, \quad (2.83)$$

where $\mathbf{w}$ denotes the velocity of the fluid relative to the particle, and the drag coefficient, C_D, is a function of the Reynolds number $Re = 2a\rho_f|\mathbf{w}|/\mu$. Various empirical expressions for C_D can be found in the literature; Ishii & Zuber chose the following:

$$C_D = \begin{cases} \dfrac{24}{Re}(1 + 0.1Re^{0.75}) & \text{for } Re < 1,000, \\ 0.45 & \text{for } Re > 1,000. \end{cases} \quad (2.84)$$

To generalise these relations from a single particle to an assembly of many particles Ishii & Zuber suggested that Re should merely be replaced by a modified Reynolds number, based on an effective viscosity for the mixture, rather than the fluid viscosity. Thus they replace Re by $Re_m = 2a\rho_f|\mathbf{w}|/\mu_m$, with

$$\mu_m = \mu\left(1 - \frac{\phi}{\phi_m}\right)^{-2.5\phi_m},$$

where ϕ_m denotes the maximum possible value of ϕ for the particles in question. The drag force per unit total volume is then given by $\mathbf{F}_D = n\mathbf{f}_D$.

It is clear that the above procedure cannot give a correct expression for the drag force up to the highest possible concentration of the particles. From the above expression, when $\phi \to \phi_m$ then $\mu_m \to \infty$, and $Re_m \to 0$ for all values of the fluid velocity. Consequently, $|\mathbf{F}_D| \to \infty$ for all nonvanishing values of the velocity. But this is absurd; we know that a close packed assembly of particles retains a finite permeability to fluid flow. As a further consequence of this behaviour the Ishii–Zuber correlation predicts that the sedimentation velocity tends to zero on approaching close packing – again a clearly incorrect result. One way of avoiding this difficulty would be to replace ϕ_m, in the expression above for μ_m, by some suitably chosen quantity ϕ_m', larger than ϕ_m. There is also a problem when $|\mathbf{w}|$ is increased, with ϕ held fixed at some value smaller than ϕ_m. Once $|\mathbf{w}|$ is large enough that $Re_m > 1,000$, C_D assumes the constant value 0.45, independent of the actual value of ϕ. This does not seem to be realistic.

A different approach to $\mathbf{F}_D$ is typified by Buyevich's expression in (2.51) above. The gap between the low and high Reynolds number forms for $\mathbf{F}_D$ is spanned simply by forming a linear combination of the two, writing

$$\mathbf{F}_D = \frac{9\mu\phi}{2a^2} \frac{1}{\varepsilon^{5/2}} \mathbf{w} + \frac{3\rho_f\phi}{16a} \left(\frac{\varepsilon}{1 - 1.17\phi^{2/3}} \right)^2 |\mathbf{w}|\mathbf{w}. \qquad (2.85)$$

Then the first term dominates when $|\mathbf{w}|$ is small and the second when it is large. The first term agrees with the form (2.80) deduced from the Richardson–Zaki relation, provided n is set equal to 4.5 in (2.80). However, the second term, which comes from a semitheoretical argument due to Goldstik (1972), is not the same as the large Reynolds number result (2.81) from the Richardson–Zaki relation.

Another popular expression for $\mathbf{F}_D$, which, like (2.85), expresses it as the sum of low and high Reynolds number contributions, is derived from the Ergun (1952) equation for the pressure drop in flow through a packed bed, namely

$$\nabla p = -\frac{75\mu}{2a^2} \frac{\phi^2}{\varepsilon^2} \mathbf{w} - \frac{1.75\rho_f}{2a} \frac{\phi}{\varepsilon} |\mathbf{w}|\mathbf{w}.$$

But for this situation (2.30) gives $0 = -\varepsilon \nabla p - n\mathbf{f}_1 = -\varepsilon \nabla p - \mathbf{F}_D$. Combining this with the above gives

$$\mathbf{F}_D = \frac{75\mu}{2a^2} \frac{\phi^2}{\varepsilon} \mathbf{w} + \frac{1.75\rho_f}{2a} \phi |\mathbf{w}|\mathbf{w}. \qquad (2.86)$$

This expression for $\mathbf{F}_D$ is clearly wrong for small values of ϕ. For example, when $|\mathbf{w}|$ is small enough that the first term dominates and $\phi \ll 1$ it gives $\mathbf{F}_D = 75\mu\phi^2\mathbf{w}/2a^2$, rather than the correct expression $\mathbf{F}_D = 9\mu\phi\mathbf{w}/2a^2$ that follows from Stokes's law. Some authors use (2.86) for large values of ϕ and replace it with something that behaves properly for smaller ϕ. However, it should be borne in mind that the Ergun equation was intended only to estimate the pressure drop for packed beds. Its explicit dependence on ϕ is such as to accommodate the variations in packing density resulting, for example, from different particle shapes or methods of compaction, but it was never intended to represent correctly the effect of the large variations in ϕ found in suspensions.

We now turn to the virtual mass and lift forces. A suitable form for $\mathbf{F}_V$ is suggested by the exact expression derived by Zhang & Prosperetti (1994) for an inviscid fluid at the limit of low particle concentration, namely

$$\mathbf{F}_V = \frac{1}{2}\rho_f\phi\left(\frac{D_f\mathbf{u}}{Dt} - \frac{D_p\mathbf{v}}{Dt}\right).$$

At higher values of ϕ this should presumably be replaced by

$$\mathbf{F}_V = C_V(\phi)\rho_f\phi\left(\frac{D_f\mathbf{u}}{Dt} - \frac{D_p\mathbf{v}}{Dt}\right), \tag{2.87}$$

where C_V is a concentration-dependent virtual mass coefficient. Zuber (1964) has suggested that

$$C_V(\phi) = \frac{1}{2} + \frac{3}{2}\phi \tag{2.88}$$

at moderate values of ϕ. Zhang & Prosperetti also gave an expression for the lift force in the form

$$\mathbf{F}_L = C_L(\phi)\rho_f\phi(\nabla \times \mathbf{u}) \times (\mathbf{v} - \mathbf{u}) \tag{2.89}$$

with $C_L \to \frac{1}{2}$ when $\phi \to 0$.

The sum of $\mathbf{F}_V$ and $\mathbf{F}_L$ (though not these terms individually) is frame indifferent if $C_V = C_L$; indeed, this was the basis for choosing $C_L = \frac{1}{2}$ in the limit of small ϕ.

Two cautionary remarks should be made in relation to Equations (2.87) and (2.89) and the demand that $C_V(\phi) = C_L(\phi)$. First, they are consistent with a rigorous calculation by Auton, Hunt & Prud'homme (1988) of the force on an isolated *spherical* particle in an inviscid fluid. However, these authors also calculated the force on a cylindrical particle, and in that case the virtual mass and lift coefficients are found to be different ($C_V = 1$ and $C_L = 2$), making the combination of the virtual mass and lift forces no longer frame indifferent. This is not surprising; frame indifference is a reasonable requirement for *constitutive* relations, but the terms in question are of dynamical origin and are not related to the properties of a material. Second, the expression (2.89) for the lift force is specific to the inviscid case. At the opposite limit of small (but nonvanishing) Reynolds number there is also a lift force on an isolated particle in a nonuniform velocity field, as calculated by Saffman (1965), but this is quite different in form, being proportional in magnitude to the product of the square root of the shear rate with the relative velocity between the fluid and the particle.

Finally, the analysis of Maxey & Riley (1983), for the limiting case of zero Reynolds number, gives the following expression for the virtual mass force on an isolated spherical particle:

$$\frac{1}{2}\rho_f \nu \frac{D_P}{Dt}(\mathbf{u} - \mathbf{v}),$$

where ν denotes the particle volume. This appears to be inconsistent with (2.87) but, as Maxey & Riley point out, it differs from

$$\frac{1}{2}\rho_f \nu \left(\frac{D_f \mathbf{u}}{Dt} - \frac{D_p \mathbf{v}}{Dt} \right)$$

only by an amount that can be neglected under the assumption of low Reynolds number.

To summarize our discussion of the fluid–particle interaction force, $n\mathbf{f}_1$, we note that it is commonly represented as the sum of three terms. For the first term, or drag force, there is a substantial body of experimental evidence that forms the basis for useful empirical expressions. The algebraic form of the second, or virtual mass, term is well established, though other forms continue to appear in the literature. The last term represents a lift force normal to the relative velocity. Its algebraic form is uncertain, since it depends on the particle shape and is known to take quite different forms in the inviscid and low Reynolds number cases.

In contrast with the situation for the fluid–particle interaction force there is very little experimental evidence that bears directly on the tensors $\mathbf{S}^p$ and $\mathbf{S}^f$, despite a number of investigations of the rheological properties of both liquid and gaseous suspensions of solid particles. In the case of a gas these consist of rheometric measurements of stress in sheared fluidized beds. A good, though old, account of these is given by Schugerl (1971). It is important to emphasize that not all techniques available to study the rheology of fluids are suitable for use with fluidized suspensions. For example, a falling sphere technique would generate regions of defluidized material adjacent to the sphere which would be expected to have a marked influence on the measurements. Indeed, any method that introduces into the bed solid surfaces oriented other than vertically is clearly disqualified. Thus, Couette or oscillating pendulum viscometers are the only devices that can be used.

There is difficulty in interpreting the results partly because the fluidized bed is unlikely to be a uniform suspension. As we shall see, instability in gas fluidized beds often appears immediately as the bed becomes fluidized, and it generates bubbles of essentially particle-free gas rising through the bed. Measured values of the effective viscosity are very large – usually several poises – but there could be a number of reasons for this. Momentum transport by particles displaced laterally during the passage of a bubble will contribute to the measured shear stress, as will frictional forces from any parts of the bed that are not fully fluidized. A second difficulty is deciding how the measured stress should be apportioned between $\mathbf{S}^p$ and $\mathbf{S}^f$, though in gas fluidized beds of large, heavy particles it is natural to attribute the measured stresses to $\mathbf{S}^p$, together with the other possible contributing mechanisms just mentioned. Despite the fact that the mechanisms of stress transmission are presumably quite different in gas and liquid suspensions, and that bubbles are usually absent from the latter, it is striking that measured shear stresses point to effective viscosities of about the same size for dense suspensions of both types. Many of these ambiguities could be eliminated if measurements could be made on particle assemblies in the absence of gravity, rather than fluidized beds, and this is a promising direction for future experimentation.

In view of these uncertainties many of the existing applications of the theory have been based on very simple assumptions about the stress tensors. Most often these have been taken to be of Newtonian form, with $\mathbf{S}^f$ related to the fluid average velocity field $\mathbf{u}$ and $\mathbf{S}^p$ to the particle average velocity field $\mathbf{v}$. Thus

$$\mathbf{S}^f = -p^f \mathbf{I} + \mu^f \left[\nabla \mathbf{u} + (\nabla \mathbf{u})^T - \frac{2}{3}(\nabla \cdot \mathbf{u})\mathbf{I} \right] \tag{2.90}$$

and

$$\mathbf{S}^p = -p^p \mathbf{I} + \mu^p \left[\nabla \mathbf{v} + (\nabla \mathbf{v})^T - \frac{2}{3}(\nabla \cdot \mathbf{v})\mathbf{I} \right], \tag{2.91}$$

where $\mathbf{I}$ denotes the unit tensor.

Closure for these tensors then requires only expressions for the scalar quantities p^f, p^p, μ^f, and μ^p and, furthermore, if the compressibility of the fluid is neglected p^f becomes an independent variable that need not be specified constitutively.

However, there is clear indication, even in the limiting cases for which complete closures were given above, that the stress tensors do not take the simple forms (2.90) and (2.91). For example, Equation (2.44) shows that gradients of both $\mathbf{u}$ and $\mathbf{v}$ should appear in the expression for $\mathbf{S}^f$, even in the limiting case of low concentration, small Reynolds number, and small Stokes number. At higher concentrations of particles it is apparent that (2.90) is not an adequate form for the fluid phase stress tensor since, with a given nonuniform shear field for $\mathbf{u}$, the force experienced by a fluid element will clearly differ depending on whether the neighbouring particles are held in fixed positions or are allowed to move as impelled by the fluid. In the former case the fixed assembly of particles constitutes a porous material, which will screen the influence of shear in the fluid velocity field quite effectively. In the latter case there will be no such screening. The proper way of representing the mechanical influence on $\mathbf{S}^f$ of gradients in both the particle and fluid velocity fields remains undetermined at all but the lowest concentration of particles. Nevertheless, as we shall see, equations using closures of the form of (2.90) and (2.91) have had marked success in treating a number of problems.

The simplest way of completing these closures is to assume that p^p, μ^f, and μ^p depend only on the concentration of the particles, represented by ϕ. p^p and μ^p are expected to be monotone increasing functions of ϕ that diverge as $\phi \to \phi_m$, the volume fraction at random close packing. When $\phi \to 0$, p^p is expected to become proportional to ϕ, while μ^f and μ^p should approach constant values. Various expressions, with these general features, have been used by different workers and, as noted above, the available experimental evidence is inadequate for distinguishing among them.

A more elaborate approach recognizes the contributions to $\mathbf{S}^f$ and $\mathbf{S}^p$ from momentum transfer by velocity fluctuations, and hence the explicit dependence of p^p, μ^f, and μ^p on the granular temperature, T, as well as the particle concentration. An additional balance equation is then needed to determine T. The closures of Koch and Buyevich, discussed earlier, are of this type. Gidaspow (1994, Chapter 10) and Balzer et al. (1995, 1996) have used equations based on this sort of closure as a basis for computations of flow in conventional and circulating fluidized beds. However, the difficult question of pseudothermal energy generation by interaction between the particles and the fluid is not resolved in these works.

At high values of the particle concentration, approaching random close packing, it is certain that the Newtonian form (2.91) gives an inadequate representation of $\mathbf{S}^p$. Most particles are then forced into contact with more than one neighbour simultaneously and interaction is by normal and tangential forces between particles at these points of sustained contact. Then the particle assembly forms a structure with yield stress and the closure for the stress tensor might be expected to be of the type studied in soil mechanics. For certain types of soil, for example clays, it is well known that such structures can persist over extended intervals of solids volume fraction, whereas for others, typically sands, they are confined to a narrow interval below random close packing. Stresses of this type may play a significant role in the mechanics of dense fluidized beds, especially for particles such as fluidized cracking catalyst (FCC). Because they are particularly important in relation to the transition from packed to fluidized beds an account of their mechanical consequences will be deferred until Chapter 3.

2.5 Approximations

In some circumstances there is rapid relaxation between the fluid and particle velocity fields; in other words, the inertia associated with their relative motion can be neglected. To explore this rewrite the momentum balances (2.30) and (2.31) as

$$\frac{D_f \mathbf{u}}{Dt} = \frac{1}{\rho_f} \nabla \cdot \mathbf{S}^f - \frac{n\mathbf{f}_1}{\rho_f \varepsilon} + \mathbf{g},$$

$$\frac{D_p \mathbf{v}}{Dt} = \frac{1}{\rho_s \phi} \nabla \cdot \mathbf{S}^p + \frac{1}{\rho_s} \nabla \cdot \mathbf{S}^f + \frac{n\mathbf{f}_1}{\rho_s \phi} + \mathbf{g}.$$

Subtracting these then gives

$$\frac{D_f \mathbf{u}}{Dt} - \frac{D_p \mathbf{v}}{Dt} + \left(\frac{1}{\rho_s \phi} + \frac{1}{\rho_f \varepsilon} \right) n\mathbf{f}_1 = -\frac{1}{\rho_s \phi} \nabla \cdot \mathbf{S}^p + \left(\frac{1}{\rho_f} - \frac{1}{\rho_s} \right) \nabla \cdot \mathbf{S}^f.$$

Now write $n\mathbf{f}_1 = \mathbf{F}_D + \mathbf{F}_V + \mathbf{F}_L$, with

$$\mathbf{F}_D = F(\phi, |\mathbf{u} - \mathbf{v}|)(\mathbf{u} - \mathbf{v}) \quad \text{and} \quad \mathbf{F}_V = \rho_f \phi C_V(\phi) \left(\frac{D_f \mathbf{u}}{Dt} - \frac{D_p \mathbf{v}}{Dt} \right),$$

giving

$$\left[1 + \rho_f \phi C \left(\frac{1}{\rho_s \phi} + \frac{1}{\rho_f \varepsilon} \right) \right] \left(\frac{D_f \mathbf{u}}{Dt} - \frac{D_p \mathbf{v}}{Dt} \right) + \left(\frac{1}{\rho_s \phi} + \frac{1}{\rho_f \varepsilon} \right)$$

$$\times [F(\mathbf{u} - \mathbf{v}) + \mathbf{F}_L] = -\frac{1}{\rho_s \phi} \nabla \cdot \mathbf{S}^p + \left(\frac{1}{\rho_f} - \frac{1}{\rho_s} \right) \nabla \cdot \mathbf{S}^f,$$

which can also be written in the form

$$\tau \left(\frac{D_f \mathbf{u}}{Dt} - \frac{D_p \mathbf{v}}{Dt} \right) + (\mathbf{u} - \mathbf{v}) + \frac{\mathbf{F}_L}{F} = \frac{-\frac{1}{\rho_s \phi} \nabla \cdot \mathbf{S}^p + \left(\frac{1}{\rho_f} - \frac{1}{\rho_s} \right) \nabla \cdot \mathbf{S}^f}{F \left(\frac{1}{\rho_s \phi} + \frac{1}{\rho_f \varepsilon} \right)},$$

(2.92)

where

$$\tau = \frac{\rho_s \phi}{F} \left\{ \frac{1}{\left(1 + \frac{\rho_s \phi}{\rho_f \varepsilon} \right)} + \frac{\rho_f}{\rho_s} C(\phi) \right\}.$$

(2.93)

This quantity has the dimensions of time and represents a relaxation time for the relative velocity of the fluid and the particles. A straightforward scaling argument then indicates that the first of the three terms on the left-hand side of (2.92) may be omitted with small error provided $\tau \ll t_m$, where t_m is the shortest time scale of significance for the macroscopic motion. Equation (2.92) then reduces to

$$F(\mathbf{u} - \mathbf{v}) + \mathbf{F}_L = \frac{1}{\left(1 + \frac{\rho_s \phi}{\rho_f \varepsilon}\right)} \left[-\nabla \cdot \mathbf{S}^p + \left(\frac{\rho_s}{\rho_f} - 1\right)\phi \nabla \cdot \mathbf{S}^f \right], \quad (2.94)$$

which will be referred to as the *short relaxation time* approximation.

Note that τ can be split into two contributions, $\tau = \tau_1 + \tau_2$, where

$$\tau_1 = \frac{\rho_s \phi}{F} \frac{1}{\left(1 + \frac{\rho_s \phi}{\rho_f \varepsilon}\right)} \quad \text{and} \quad \tau_2 = \frac{\rho_s \phi}{F} \frac{\rho_f}{\rho_s} C(\phi). \quad (2.95)$$

Clearly each of these is much smaller that t_m when $\tau \ll t_m$, and the condition $\tau_2 \ll t_m$ indicates that the virtual mass force $\mathbf{F}_V$ may be neglected compared to the drag force $\mathbf{F}_D$ in the equations of motion. Thus, when $\tau \ll t_m$, the particle phase momentum balance (2.32) may be truncated as

$$\rho_s \phi \frac{D_p \mathbf{v}}{Dt} = \nabla \cdot \mathbf{S}^p + \phi \nabla \cdot \mathbf{S}^f + F(\mathbf{u} - \mathbf{v}) + \mathbf{F}_L + \rho_s \phi \mathbf{g}.$$

Using (2.94) we can reduce this to

$$\bar{\rho} \frac{D_p \mathbf{v}}{Dt} = \nabla \cdot (\mathbf{S}^p + \mathbf{S}^f) + \bar{\rho} \mathbf{g}. \quad (2.96)$$

Thus, when the short relaxation time approximation is valid, the particle phase velocity satisfies the same momentum equation as a single-phase fluid with the mean density $\bar{\rho}$ of the mixture, subject to a stress that is the sum of the fluid and particle phase stress tensors. Equation (2.96) can further be used to eliminate $\nabla \cdot \mathbf{S}^p$ from (2.94), with the result

$$F(\mathbf{u} - \mathbf{v}) + \mathbf{F}_L = \varepsilon \nabla \cdot \mathbf{S}^f + \rho_f \varepsilon \left(\mathbf{g} - \frac{D_p \mathbf{v}}{Dt} \right). \quad (2.97)$$

This merely indicates that the motion of the fluid relative to the particles is impelled by the sum of the stress gradient in the fluid phase, acting over the void area, together with a modified gravity force, as seen by an observer moving with the particles. Equations (2.96) and (2.97) provide a physically appealing form for the momentum equations with the short relaxation time approximation. However, other pairs, such as (2.31) and (2.94), can serve equally well.

We must now examine the conditions under which the short relaxation time condition is satisfied, to see whether it is likely to be of practical value. To this end consider the drag force

$$\mathbf{F}_D = n\mathbf{f}_D = F\mathbf{w} \quad \text{where} \quad \mathbf{w} = \mathbf{u} - \mathbf{v}.$$

For a given value of $|\mathbf{w}|$ the drag force on a single member of an assembly of particles is larger that that on an identical isolated particle, $|\mathbf{f}_D| \geq |\mathbf{f}_D^0|$.

Table 2.2. *Scrutiny of (2.98) in eight systems*

System	τ_{St}	lhs (2.98) @ $\phi = 0.05$	lhs (2.98) @ $\phi = 0.5$
1 mm/air	6.75 s	77 ms	6.0 ms
100 μm/air	67.5 ms	0.77 ms	0.06 ms
1 mm/H$_2$O	120 ms	135 ms	65 ms
100 μm/H$_2$O	1.2 ms	1.4 ms	0.65 ms

Furthermore, at a given value of $|\mathbf{w}|$ the magnitude of $\mathbf{f}_D^0$ is greater than, or equal to, the Stokes drag, namely $6\pi a\mu|\mathbf{w}|$. Thus

$$|\mathbf{F}_D| \geq 6\pi a\mu|\mathbf{w}|n = \frac{6\pi a\mu\phi}{(4\pi a^3/3)}|\mathbf{w}| = \frac{9\mu\phi}{2a^2}|\mathbf{w}|;$$

consequently $F \geq 9\mu\phi/2a^2$. Then from (2.93) it follows that

$$\tau \leq \tau_{St}\left\{\frac{1}{\left(1 + \frac{\rho_s\phi}{\rho_f\varepsilon}\right)} + \frac{\rho_f}{\rho_s}C(\phi)\right\},$$

where $\tau_{St} = 2a^2\rho_s/9\mu$ is the Stokes relaxation time for an isolated particle in an infinite body of fluid. From this it follows that the short relaxation time condition is certainly satisfied if

$$\tau_{St}\left\{\frac{1}{\left(1 + \frac{\rho_s\phi}{\rho_f\varepsilon}\right)} + \frac{\rho_f}{\rho_s}C(\phi)\right\} \ll t_m. \tag{2.98}$$

To see how this criterion fares in practice consider two types of particle, namely glass beads of diameters 1 mm and 100 μm, moving either in air at ambient conditions or in water. The density of the glass is taken to be 2,200 kg/m^3; the density and viscosity of air at these conditions are 1.3 kg/m^3 and 1.81×10^{-5} kg/(m·s) respectively, while the density and viscosity of water are 1,000 kg/m^3 and 10^{-3} kg/(m·s). Table 2.2 shows both the Stokes relaxation time and the value of the left-hand side of (2.98) for each of these cases, evaluated at $\phi = 0.05$ and at $\phi = 0.5$.

It is seen that (2.98) is satisfied for the 100 μm beads, either in air or in water, for any suspension with $\phi > .05$, provided the shortest macroscopic time scale exceeds 10 ms. For the 1 mm beads, however, (2.98) is satisfied only if the shortest macroscopic time scale exceeds a significant fraction of a second.

Note that the left-hand side of inequality (2.98) approaches $\tau_{St}(1+(\rho_f/2\rho_s))$ when $\phi \to 0$. This corresponds to the relaxation of the velocity of an isolated particle to that of an infinite body of surrounding fluid. When $\phi \to \phi_m$, however, the left-hand side of (2.98) approaches a smaller value that depends on the ratio of ρ_f to ρ_s. These observations reflect the fact that the velocity of each phase relaxes towards the centre of mass velocity of the mixture provided there are no constraining forces. Thus, when the mass fraction of solids in the mixture is large, it is the gas velocity that relaxes towards the solids velocity, rather than the other way round. It is, therefore, wrong to regard the relaxation time for an isolated particle as the proper time scale for relaxation between the averaged velocities of fluid and particles in a concentrated mixture.

If viscous terms such as those contained in the closures (2.90) and (2.91) for S^f and S^p, respectively, are omitted from the equations of motion, then these equations become ill-posed. As we shall see disturbances of short wavelength then grow rapidly, and their rate of growth may even increase without bound as the wavelength approaches zero. This appears to have caused considerable concern in the early 1980s and it was addressed by a number of papers. (See, for example, Lahey et al. (1980) and the review article by Drew (1983).) However, it is not unexpected physically. Viscous dissipation, which increases as the wavelength decreases owing to increasing rates of deformation, is the physical mechanism that controls the growth rate at short wavelengths and sets a dominant length scale for growing disturbances. As we shall see in Chapter 4 the inclusion of viscous stresses of any magnitude renders the equations well posed, and when the magnitude of these stresses is physically realistic there is no insuperable difficulty facing numerical integration of the full equations.

It would seem to be appropriate to include, in a section describing approximations, a discussion of scaling for the equations of motion. However, this is much less straightforward than the corresponding question for a single-phase fluid because of the existence of intrinsic scales of time and length associated with instabilities of the dispersion. Therefore the issue of scaling in relation to each class of problems will be addressed separately in the following chapters.

2.6 Discussion of Averaging Procedures

As indicated at the beginning of this chapter several different averaging methods have been applied to the equations of motion, and there has been some discussion of the relative merits of these. Certain authors (see, for example, Arnold et al. (1989)) have gone so far as to suggest that one method generates "wrong" results, and they have provided specific examples to support this. However, a moment's consideration shows that such a view cannot be correct. Each averaging method is merely a formal procedure and, as such, it can generate "wrong" results only if some error is made in its application. Nevertheless, there are certainly circumstances in which equations derived, in part, by an averaging procedure yield results that do not match reasonable physical expectations. This can happen because the procedure is applied to a system where the type of averages it generates are not the physically relevant quantities, although they may be misinterpreted as such. It can also happen because the closures that are postulated to reduce the formally averaged equations to explicit form turn out to be inappropriate. Despite claims to the contrary no averaging procedure is immune from these problems; indeed, there is little to choose between their susceptibilities to misapplication or misinterpretation, though the resulting difficulties may become apparent in different physical situations.

To illustrate this consider an example quoted by Arnold et al. (1989) to bring out potential difficulties associated with local volume averaging. These authors focus attention on a two-dimensional stagnation flow with velocity components given by

$$u_x = Ax, \quad u_y = -Ay, \quad u_z = 0.$$

Then $\langle u_x \rangle(\mathbf{x}) \neq u_x(\mathbf{x})$ and $\langle u_y \rangle(\mathbf{x}) \neq u_y(\mathbf{x})$ for a general point $\mathbf{x}$, where the angular brackets denote local volume averages centred on point $\mathbf{x}$. These differences between the local averages and the point values of the velocity components decrease with decreasing radius of the averaging region and tend to zero when this radius approaches zero. As a consequence of the above it follows that $\langle u'_x u'_x \rangle$ and $\langle u'_y u'_y \rangle$ take nonvanishing values, where $u'_x = u_x - \langle u_x \rangle$ and $u'_y = u_y - \langle u_y \rangle$, though, once again, these quantities tend to zero as the averaging radius approaches zero. Arnold et al. then claim that "the turbulent stress calculated from the volume average is non-zero, even though the flow is laminar" and conclude that volume averaging is incorrect. However, this is merely a physical misidentification of the mathematical quantities in question. In general quantities such as $\langle u'_x u'_x \rangle$, as evaluated by volume averaging, may or may not have a physical significance related to turbulence, or any other source of real statistical fluctuations, depending on the length scales associated with the flow in question. As was emphasized at the beginning of this chapter, though volume averaging can always be carried out formally, it has an unambiguous physical significance only when there is a clear separation of length scales in the system of interest, so that the averaging radius can be chosen much smaller than the large scale and much bigger than the small scale. Then, and only then, calculated averages will be essentially independent of the averaging radius and should represent something physically significant. In the example of Arnold et al. this condition is not satisfied; there is no separation of scales in the postulated flow so the computed averages, such as $\langle u'_x u'_x \rangle$, have values that are determined entirely by the averaging radius. The application of volume averaging, though formally possible, makes no sense in this example.

Although this resolves the perceived difficulty with volume averaging, some explanation was needed, whereas ensemble averaging yields the obviously correct result directly. In this flow every member of the ensemble of realizations is identical. Thus the velocity components and their ensemble averages coincide and, correspondingly, quantities such as $\langle u'_x u'_x \rangle$ vanish. However, though the advantage in this case lies with ensemble averaging, this is by no means always so. As an illustration consider the definition of void fraction, that is, the fraction of space not occupied by particles. From its name alone one might expect that this quantity would have to be defined with reference to some finite neighbourhood of a point of interest and this is so in volume averaging, as seen from Equation (2.3), where the neighbourhood in question is defined by the averaging radius. Nevertheless an ensemble average void fraction can be defined at a point of space, without reference to any neighbourhood of that point, in the following way. In each of the individual realizations of the system that constitute the ensemble a given point of space lies either in a particle or in the void space. In the former case the void fraction in this realization is assigned the value zero, while in the latter case it is assigned the value unity. The ensemble average void fraction is then simply defined as the fraction of the realizations in which it has the value unity, in other words, the fraction of the realizations for which the point in question lies in the void space. This is a perfectly reasonable definition. However, note that the process described also allows one to calculate the mean square fluctuation of the void fraction about its ensemble average value, namely $\langle (\varepsilon_n - \langle \varepsilon \rangle)^2 \rangle$, where ε_n is the void fraction at the point in question for the nth realization of the system and the angular brackets now denote ensemble averaging. Indeed, it follows immediately that $\langle (\varepsilon_n - \langle \varepsilon \rangle)^2 \rangle = \langle \varepsilon \rangle (1 - \langle \varepsilon \rangle)$. But this does not correspond to anything that we would recognize as physically meaningful fluctuations in void fraction. Consider, for example, a regular array of spherical particles in crystalline close packing, for which the void fraction is 0.26. The position of this assembly is not defined, so in our ensemble of realizations it can be imagined to be thrown down at random. Then ensemble averaging gives the correct value for the void fraction but also finds that $\langle \varepsilon'^2 \rangle \equiv \langle (\varepsilon_n - \langle \varepsilon \rangle)^2 \rangle = 0.19$. Though this is formally correct it clearly has no relation to what we might think of physically as fluctuations in void fraction. Indeed, the close packed assembly of spheres is the archetype of systems we regard as having spatially uniform void fraction. This parallels the example of Arnold et al., but it illustrates a case where ensemble averaging, rather than volume averaging, throws up an averaged fluctuation that, despite being formally correct, is not physically significant. In systems with truly nonuniform void fraction the average mean square fluctuation in this quantity will depend on the size of the averaging region, and this is reflected correctly by volume averaging.

The conclusion to be drawn from these examples is that the question of whether ensemble or volume averaging is the "correct" procedure is without meaning. Each must be viewed circumspectly, questioning whether its results are likely to correspond to the physical quantities of interest in the particular problem under consideration.

Finally, though a significant part of this chapter has been devoted to the problem of closure for averaged equations, it must be emphasized that there is no guarantee that such a closure is possible, even in principle. Large assemblies of particles are described by statistical distributions of position and velocity that change with time. This change determines, among other things, the change in the average velocity in the neighbourhood of each point of space, but there is no guarantee that such temporal variations can be related to local values of the average velocities themselves (that is, the first moments of the velocity distribution) and their spatial gradients. Indeed, there are well-known situations where this is not the case, such as Knudsen streaming of gases through channels at very low pressure, and these certainly arise also with assemblies of macroscopic particles.

References

Anderson, T. B. & Jackson, R. 1967. A fluid mechanical description of fluidized beds. Equations of motion. *Ind. Eng. Chem. Fundam.* **6**, 527–539.

Arnold, G. S., Drew, D. A., & Lahey, R. T. 1989. Derivation of constitutive equations for interfacial force and Reynolds stress for a suspension of spheres using ensemble averaging. *Chem. Eng. Comm.* **86**, 43–54.

Auton, T. R., Hunt, J. C. R., & Prud'homme, M. 1988. The force exerted on a body in inviscid, unsteady, non-uniform, rotational flow. *J. Fluid Mech.* **197**, 241–257.

Balzer, G., Boelle, A., & Simonin, O. 1995. Eulerian gas–solid flow modelling of dense fluidized bed. *Fluidization VIII*, May 1995.

Balzer, G., Simonin, O., Boelle, A., & Lavieville, J. 1996. A unifying modelling approach for the numerical prediction of dilute and dense gas–solid two phase flow. *Circulating Fluidized Bed V*, Beijing, May 28–June 1, 1996.

Basset, A. B. 1888. *Treatise on Hydrodynamics*. Deighton Bell.

Boelle, A., Balzer, G., & Simonin, O. 1995. Second-order prediction of the particle-phase stress tensor of inelastic spheres in simple shear dense suspensions. *Proc. 6th Int. Symp. on gas-solid Flows*, ASME FED **228**, 9–18.

Buyevich, Yu. A. 1971a. On the fluctuations of concentration in disperse systems. The random number of particles in a fixed volume. *Chem. Eng. Sci.* **26**, 1195–1201.

Buyevich, Yu. A. 1971b. Statistical hydrodynamics of disperse systems. Part 1. Physical background and general equations. *J. Fluid Mech.* **49**, 489–507.

Buyevich, Yu. A. 1972a. Statistical hydrodynamics of disperse systems. Part 2. Solutions of the kinetic equations for suspended particles. *J. Fluid Mech.* **52**, 345–355.

Buyevich, Yu. A. 1972b. Statistical hydrodynamics of disperse systems. Part 3. Pseudo-turbulent structure of homogeneous suspensions. *J. Fluid Mech.* **56**, 313–336.

Buyevich, Yu. A. 1972c. On the fluctuations in concentration in disperse systems II. *Chem. Eng. Sci.* **27**, 1699–1708.

Buyevich, Yu. A. 1994. Fluid dynamics of coarse dispersions. *Chem. Eng. Sci.* **49**, 1217–1228.

Buyevich, Yu. A. 1997. Particulate pressure in monodisperse fluidized beds. *Chem. Eng. Sci.* **52**, 123–140.

Buyevich, Yu. A. & Kapbasov, Sh. K. 1994. Random fluctuations in a fluidized bed. *Chem. Eng. Sci.* **49**, 1229–1243.

Buyevich, Yu. A. & Kapbasov, Sh. K. 1999. Particulate pressure in disperse flow. *Int. J. Fluid Mech. Res.* **26**, 72–97.

Carnahan, N. F. & Starling, K. 1969. Equation of state for non-attracting rigid spheres. *J. Chem. Phys.* **51**, 635.

Delhaye, J. M. 1969. General equations of two-phase systems and their applications to air–water bubble flow and to steam–water flashing flow. ASME–AIChE Heat Transfer Conference, Minneapolis.

Drew, D. A. 1971. Averaged field equations for two-phase media. *Stud. Appl. Math.* **50**, 133–166.

Drew, D. A. 1983. Mathematical modelling of two-phase flow. *Annu. Rev. Fluid Mech.* **15**, 261–291.

Drew, D. A. & Segel, L. A. 1971. Averaged equations for two-phase flows. *Stud. Appl. Math.* **50**, 205–231.

Drew, D. A., Arnold, G. S., & Lahey, R. T. 1990. Relation of microstructure to constitutive equations. In *Two Phase Flows and Waves*, ed. D. D. Joseph & D. G. Schaeffer, pp. 45–55. Springer-Verlag.

Drew, D. A. & Lahey, R. T. 1993. Analytical modelling of multiphase flow. In *Particulate Two-Phase Flow*, ed. M. C. Roco, pp. 510–566. Butterworth-Heinemann.

Einstein, A. 1906, 1911. Eine neue Bestimmung der Molekuldimensionen. *Ann. Phys.* **19**, 289; **34**, 591.

Ergun, S. 1952. Fluid flow through packed columns. *Chem. Eng. Progr.* **48**, 89–94.

Foscolo, P. U. & Gibilaro, L. G. 1984. A fully predictive criterion for the transition between particulate and aggregative fluidization. *Chem. Eng. Sci.* **39**, 1667–1675.

Foscolo, P. U. & Gibilaro, L. G. 1987. Fluid dynamic stability of fluidized suspensions. The particle bed model. *Chem. Eng. Sci.* **42**, 1489–1500.

Geurst, J. A. 1986. Variational principles and two-fluid hydrodynamics of bubbly liquid/gas mixtures. *Physica* **135A**, 455–486.

Gidaspow, D. 1994. *Multiphase Flow and Fluidization*. Academic Press Inc.

Goldshtein, A. & Shapiro, M. 1995. Mechanics of collisional motion of granular materials. Part I. General hydrodynamic equations. *J. Fluid Mech.* **282**, 75–114.

Goldstik, M. A. 1972. Elementary theory of a fluidized bed. *Zh. Prikl. Mekh. Tekhn. Fiz.* **6**, 106–112 (in Russian).

Haff, P. K. 1983. Grain flow as a fluid mechanical phenomenon. *J. Fluid Mech.* **134**, 401–430.

Hinch, E. J. 1977. An averaged equation approach to particle interactions in a fluid suspension. *J. Fluid Mech.* **83**, 695–720.

Ishii, M. & Zuber, N. 1979. Drag coefficient and relative velocity in bubbly, droplet or particulate flows. *AIChE J.* **25**, 843–855.

Jackson, R. 1997. Locally averaged equations of motion for a mixture of identical spherical particles and a Newtonian fluid. *Chem. Eng. Sci.* **52**, 2457–2469.

Jenkins, J. T. 1987. Rapid flows of granular materials. In *Non-Classical Continuum Mechanics*, pp. 213–225. Cambridge University Press.

Jenkins, J. T. & Savage, S. B. 1983. A theory for the rapid flow of identical, smooth, nearly elastic, spherical particles. *J. Fluid Mech.* **130**, 187–202.

Joseph, D. D. & Lundgren, T. S. 1990. Ensemble averaged and mixture theory equations for incompressible fluid–particle suspensions. *Int. J. Multiphase Flow* **16**, 35–42.

Koch, D. L. 1990. Kinetic theory for a monodisperse gas–solid suspension. *Phys. Fluids A* **2**, 1711–1723.

Koch, D. L. & Sangani, A. S. 1999. Particle pressure and marginal stability limits for a homogeneous monodisperse gas fluidized bed: kinetic theory and numerical simulation. *J. Fluid Mech.* **400**, 229–263.

Lahey, R. T., Cheng, L. Y., Drew, D. A., & Flaherty, J. E. 1980. The effect of virtual mass on the numerical stability of accelerating two-phase flows. *Int. J. Multiphase Flow* **6**, 281–294.

Lun, C. K. K. 1991. Kinetic theory for granular flow of dense, slightly inelastic, slightly rough spheres. *J. Fluid Mech.* **233**, 539–559.

Lun, C. K. K., Savage, S. B., Jeffrey, D. J., & Chepurniy, N. 1984. Kinetic theories for granular flow: Inelastic particles in Couette flow and slightly inelastic particles in a general flow field. *J. Fluid Mech.* **140**, 223–256.

Ma, D. & Ahmadi, G. 1988. A kinetic model for rapid granular flows of nearly elastic particles, including interstitial fluid effects. *Powder Technol.* **56**, 191–207.

Maxey, M. R. & Riley, J. J. 1983. Equation of motion for a small rigid sphere in a nonuniform flow. *Phys. Fluids* **26**, 883–889.

Murray, J. D. 1965. On the mathematics of fluidization. Part I. Fundamental equations and wave propagation. *J. Fluid Mech.* **21**, 465–493.

Nigmatulin, R. I. 1979. Spatial averaging in the mechanics of heterogeneous and dispersed systems. *Int. J. Multiphase Flow.* **5**, 353–385.

Richardson, J. F. & Zaki, W. N. 1954. Sedimentation and fluidization. Part I. *Trans. Inst. Chem. Eng.* **32**, 35–53.

Saffman, P. G. 1965. The lift on a small sphere in a slow shear flow. *J. Fluid Mech.* **22**, 385–400.

Sangani, A. S., Mo, G., Tsao, H-K., & Koch, D. L. 1996. Simple shear flows of dense gas–solid suspensions at finite Stokes numbers. *J. Fluid Mech.* **313**, 309–341.

Schugerl, K. 1971. Rheological behavior of fluidized systems. Chapter 6 of *Fluidization*, ed. J. F. Davidson & D. Harrison. Academic Press.

Slattery, J. C. 1967. Flow of viscoelastic fluids through porous media. *AIChE J.* **13**, 1066–1071.

Sundararajakumar, R. R. & Koch, D. L. 1996. Non-continuum lubrication flows between particles colliding in a gas. *J. Fluid Mech.* **313**, 283–308.

Wallis, G. B. 1969. *One-Dimensional Two-Phase Flow*. McGraw-Hill.

Wallis, G. B. 1991. The averaged Bernoulli equation and macroscopic equations of motion for the potential flow of a two-phase dispersion. *Int. J. Multiphase Flow* **17**, 683–695.

Whitaker, S. 1969. Advances in the theory of fluid motion in porous media. *Ind. Eng. Chem.* **61**, 14–28.

Zhang, D. Z. & Prosperetti, A. 1994. Averaged equations for inviscid disperse two-phase flow. *J. Fluid Mech.* **267**, 185–219.

Zhang, D. Z. & Prosperetti, A. 1997. Momentum and energy equations for disperse two-phase flows and their closure for dilute suspensions. *Int. J. Multiphase Flow.* **23**, 425–453.

Zuber, N. 1964. On dispersed two-phase flow in the laminar flow regime. *Chem. Eng. Sci.* **19**, 897–903.

3

Fluidization and defluidization

3.1 Introduction

In this chapter we address one of the most basic problems in the mechanics of fluid–particle systems, namely the transition from a packed to a fluidized bed when an assembly of particles supported below by a porous plate is subjected to an increasing upward flow of fluid. The reverse transition, when the fluid flow is decreased, will also be addressed. We shall refer to these as the processes of fluidization and defluidization, respectively; they are of particular interest because they carry the bed through the interval of high particle concentration adjacent to maximum random packing, where stresses transmitted at points of sustained contact between particles are important. These were mentioned in Chapter 2 but detailed discussion of their modelling was deferred. If we are to give an adequate account of fluidization and defluidization it can no longer be avoided, so the first part of this chapter is devoted to a rather extended discussion of this contribution to stress in the particle phase. Once this foundation has been laid the processes of fluidization and defluidization are treated in Section 3.4.

The discussion of the particle phase stress tensor $\mathbf{S}^p$ in Chapter 2 was limited to stresses generated by collisions and momentum transfer associated with random fluctuations in the velocities of individual particles. However, at high concentrations, each particle may be in contact with more than one neighbour at the same time, so that connected structures are formed, involving many particles. Within these, stress is transmitted by normal forces at points of mutual contact and the associated tangential forces due to friction. The existence of these structures does not necessarily preclude deformation of the particle assembly since contacts between particles may break and reform, particles that touch may roll about their point of contact, or the particle surfaces may slip relative to each other at this point. Typically, in this situation, the particle assembly will

65

exhibit a yield phenomenon; when applied stresses are small the whole assembly behaves as an elastic solid, while under larger stress it deforms, with the particles in motion relative to each other. At lower concentrations, where the particles interact through binary collisions, kinetic theory provides an expression for $\mathbf{S}^p$, but at present we have no satisfactory way of constructing the stress tensor from microscopic considerations in this much more complicated situation.

Though there have been several attempts to predict the conditions for initial yield from microscopic considerations (Rowe, 1962; Horne, 1965, 1969; Thornton, 1979; Kanatani, 1981; Christoffersen et al., 1981; Goddard & Didwania, 1998) most results in this area have come from empirical continuum models of stress in the particulate material. There is an extensive literature devoted to models of this type and their applications because of their importance in geophysics and soil mechanics. Here we shall make no attempt to review this work exhaustively but will sketch the salient observations and describe just one of the stress models devised to account for them. It will then be seen why the incorporation of this type of stress contribution leads to great difficulties in finding solutions of the equations of motion of gas–particle systems for most situations of practical interest. However, by making reasonable simplifying assumptions, it proves possible to deduce the influence of these stresses on the transition between packed and fluidized beds as the flow of fluidizing gas is increased or decreased.

3.2 The Observed Behaviour of Dense Particle Assemblies under Deformation

Consider a layer of a concentrated assembly of particles (henceforth referred to as a "granular material" for brevity) sheared between rough parallel plates in the absence of gravity. This can be regarded as an idealization of practical tests in a shear cell, where the material is confined within an annular trough whose plane surfaces rotate relative to each other about an axis of symmetry normal to the annulus. A constant normal force N is applied to the plates, which are otherwise free to move normal to their own plane, and T denotes the tangential force per unit area. The initial bulk density of the granular material is ρ_0 and s denotes the relative displacement of the plates in the plane of shear. These definitions are illustrated by Figure 3.1. As T is increased progressively from zero the granular material at first deforms reversibly, behaving like an elastic solid with a shear modulus similar to that of the material from which the particles are composed. The plate displacement s is proportional to T and the line relating them is retraced exactly if T is decreased. This interval of elastic behaviour is bounded by a value T_Y of the shear stress that depends on both ρ_0 and N and is called

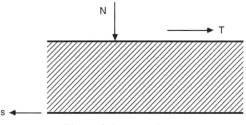

Figure 3.1. Plane shear.

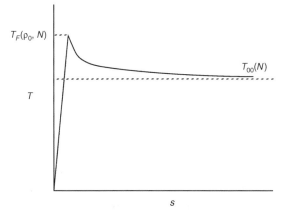

Figure 3.2. Failure in plane shear.

the yield stress. When T is increased beyond T_Y what happens depends on the relative values of N and ρ_0. If the granular material is densely packed initially and the normal stress N is small the relation between T and the displacement s is as sketched in Figure 3.2. This type of behaviour is called failure and the the corresponding yield stress is denoted by $T_F(\rho_0, N)$. With further shearing beyond the yield point the shear stress needed to maintain displacement of the bounding plate decreases and approaches an asymptotic value for large s that is independent of the initial density of packing but does depend on the normal stress, increasing with the value of N. We denote it by $T_\infty(N)$. The kinematics of failure is sketched in Figure 3.3(i), which shows the displacement of the material as a function of position. Shearing is seen to be confined to a thin layer parallel to the plane of shear. Figure 3.3(ii) indicates how T_F depends on N, for several fixed values of the initial packing density ρ_0. Each curve is referred to as a failure locus.

The decrease in the value of T needed to maintain shearing after failure clearly indicates that the material weakens during shear, and this is why the

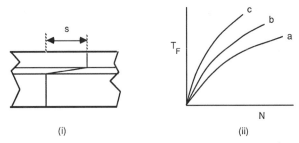

(i)   (ii)

Figure 3.3. (i) Kinematics of failure. (ii) Failure loci; $\rho_0^a < \rho_0^b < \rho_0^c$.

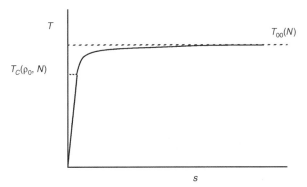

Figure 3.4. Consolidation in plane shear.

shearing is confined to a thin layer. Once shearing is initiated in this layer the material there becomes less resistant to shear than the material outside, so failure cannot spread beyond the initial layer. If measurements of the local bulk density could be made they would show a significantly decreased density within the shear layer but, because this constitutes only a small fraction of the material between the plates, the separation of the plates is increased only slightly.

In the opposite situation, where the granular material is initially packed relatively loosely and the applied normal stress is large, the behaviour beyond the yield point is quite different from that just described. The variation of T with s, which is sketched in Figure 3.4, is referred to as consolidation and the corresponding yield stress is denoted by $T_C(\rho_0, N)$. Beyond the yield point the shear stress needed to maintain deformation now increases, approaching an asymptotic value $T_\infty(N)$ when $s \rightarrow \infty$. This is the same function of N as for the case of failure described above. The kinematics of yield for this case is sketched in Figure 3.5(i), which should be compared with Figure 3.3(i). In contrast to the former case, during consolidation the shear spreads

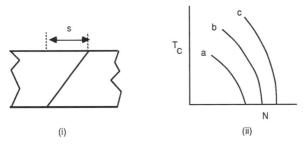

Figure 3.5. (i) Kinematics of consolidation. (ii) Consolidation loci; $\rho_0^a < \rho_0^b < \rho_0^c$.

uniformly throughout the layer of granular material. Figure 3.5(ii) shows consolidation loci that relate T_C to N for fixed initial packing density. The increase in T required to maintain shearing beyond the yield point shows that the strength of the material becomes greater as it is deformed. This explains why the material shears uniformly. If any layer were to shear less than its neighbours it would become weaker than them, so shear would immediately be localized there until its strength caught up with that of the neighbouring material. If the bulk density were measured it would be found to increase as shear continues, and this is the source of the observed increase in strength. Since the increase in density occurs everywhere in the material it is reflected in a corresponding percentage decrease in the separation between the bounding plates.

The failure loci of Figure 3.3(ii) should be compared with the consolidation loci of Figure 3.5(ii). The former start from the origin and have positive slope. This is a consequence of the dilation that accompanies shear in these circumstances. Because the dilation is opposed by the normal load N on the plates, as N increases it becomes more difficult to induce the material to shear, and there is a corresponding increase in T_F. When N is zero there is nothing to oppose dilation, so shear can occur in response to a vanishingly small shear stress. In the case of consolidation, on the other hand, the volume contracts during shear, and this is encouraged by an increase in N. Thus T and N work together to cause deformation and an increase in the normal load allows deformation to begin at a smaller value of T_C. Each consolidation locus intersects the axis $T_C = 0$ at a finite value of N, which represents the value of the normal stress that is capable of inducing consolidation even in the absence of a shear stress. The relation between this value of N and ρ_0 will be important in understanding the mechanics of defluidization.

For most particulate materials the consolidation behaviour described above is difficult to observe since the consolidation induced by the earth's gravity is itself almost all that can be achieved without applying very large normal loads that could fracture the particles. An exception to this statement is provided by certain materials, for example clay, that are capable of sustaining connected networks of particle–particle contacts over a relatively wide range of bulk densities. Over this density range, they exhibit yield behaviour of the sort just described.

Finally, there is evidence that T_∞, the asymptotic value of the shear stress after large deformation, is the same function of N for both failure and consolidation experiments and that it is well represented by a straight line through the origin, $T_\infty = N \tan \phi$, where the angle ϕ is characteristic of the particular granular material. This form was originally proposed by Coulomb (1776) so it is said to characterize a Coulomb material.

This completes our brief account of the observed behaviour of granular materials in shear. We turn now to constitutive stress models consistent with this behaviour.

3.3 Constitutive Relations for Dense, Slowly Deforming Granular Materials

Constitutive relations for the tensor $\mathbf{S}^p$ have been presented in Section 2.3 for particle assemblies whose concentration is sufficiently low that interactions occur by binary collisions. The relations to be described now apply at the opposite limit, where the concentration is sufficiently high that the particles are held in a network of semipermanent contacts with their neighbours, and stress is transmitted through normal and tangential forces at these points of mutual contact. Unfortunately, we have no satisfactory constitutive models that span the intermediate situation between these extremes, though Savage (1982) suggested the simple expedient of combining additively the stresses predicted by the separate models for the two limiting cases, and this idea was exploited with some success in studies of plane shear by Johnson & Jackson (1987) and of flow down an inclined chute by Johnson et al. (1990).

Continuum models for slowly deforming, dense materials have been based on ideas from plasticity theory and take various forms. Here we shall confine attention to just one of these, with origins in the work of Drucker & Prager (1952), subsequently developed by Shield (1955), Jenike & Shield (1959), and extensively by the Cambridge school of soil mechanics, Roscoe et al. (1958), Schofield & Wroth (1968), Roscoe (1970). This springs from a geometric representation of the stress tensor commonly used in plasticity and soil mechanics but less familiar to the fluid mechanics community.

The conceptual basis for the model is most easily presented for motions confined to two dimensions though the ideas, once formulated, are easily extended to a third dimension. Then if x and y are Cartesian coordinates the (symmetric)

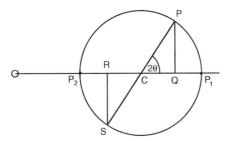

Figure 3.6. Mohr circle construction.

stress tensor is completely defined by its three independent elements p_{xx}, p_{yy}, and p_{xy}. (Noncohesive granular materials cannot support tensile stress so it is convenient to re-define the stress tensor in the compressive sense, writing $\mathbf{P}^p = -\mathbf{S}^p$. The elements of $\mathbf{P}^p$ are then denoted by p_{xx}, p_{yy}, and p_{xy}.) Alternatively $\mathbf{P}^p$ can be determined by specifying the values of the major and minor principal stresses, p_1 and p_2, together with the orientation of the principal stress axes, quantified by an angle γ measured clockwise from the positive x-axis to the direction of the major principal stress axis. Given the values of p_1, p_2, and γ, it is possible to deduce the traction on any plane surface of specified orientation. An elegant way of doing this geometrically is provided by the Mohr circle construction illustrated in Figure 3.6. A circle is drawn with centre C on a horizontal line. Its points of intersection with this line are labeled P_1 and P_2 and a third point of the line is labeled O. The diameter of the circle and the position of O are chosen such that the distances OP_1 and OP_2 are equal to the magnitudes, p_1 and p_2, of the major and minor principal stresses, respectively. Then the tangential and normal components of the traction (measured in the compressive sense) on a plane whose normal makes an angle θ with the direction of the major principal stress, are given by the distances PQ and OQ, respectively. The distance OR gives the normal stress on a plane oriented at right angles to the first plane. This construction immediately provides the link between the alternative descriptions of the stress tensor in terms of (p_1, p_2, γ) or (p_{xx}, p_{yy}, p_{xy}), as shown in Figure 3.7. From the geometry of the diagram the relations sought are

$$p_{xx} = p + q \cos 2\gamma, \quad p_{yy} = p - q \cos 2\gamma, \quad p_{xy} = -q \sin 2\gamma, \quad (3.1)$$

where $p = (p_1 + p_2)/2$ and $q = (p_1 - p_2)/2$.

In terms of the description of stress using p_1, p_2, and γ or equivalently p, q, and γ, a constitutive theory of yield must do two things: 1) it must define the relation between p_1, p_2, and ρ (or p, q, and ρ) for the material to yield, and 2)

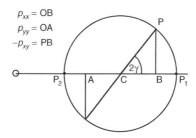

$p_{xx} = OB$
$p_{yy} = OA$
$-p_{xy} = PB$

Figure 3.7. Mohr construction for elements of the stress tensor.

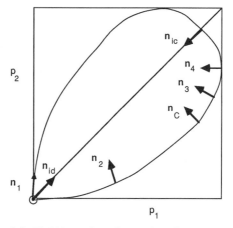

Figure 3.8. Yield locus for a fixed value of the bulk density.

it must relate the orientation and magnitude of the principal stresses to the kinematics of deformation of the material.

The observed behaviour summarized in Figures 3.2 to 3.5 can be modelled by adopting a viewpoint put forward by the Cambridge school of soil mechanics and discussed at length by Schofield & Wroth (1968). For a given value of the bulk density ρ both the failure and consolidation loci in the (T, N)-plane appear as images of different arcs of a single *yield locus* in the (p_1, p_2)-plane, with a shape of the sort sketched in Figure 3.8. This locus in necessarily symmetric about the principal diagonal, because the numbering of the principal stresses can be interchanged. However, if p_1 is taken to be the major principal stress the arc lying below the principal diagonal represents the combinations of stress for which material of this bulk density will yield. When the point representing the principal stresses lies between this arc and the diagonal, the material behaves as an elastic solid.

Figure 3.8 also shows the directions of inward normals to the yield locus at several points along its length, and these may be related to the kinematics of the deformation that occurs at yield in a way proposed by Drucker & Prager (1952). This relation, called the *plastic potential flow rule*, can be stated as follows: (a) the minor principal axis of the rate of deformation tensor is aligned with the major principal axis of stress (defined in the compressive sense, as above), and (b) the ratio of the principal rates of deformation is the same as the ratio of the components of the inward normal to the yield locus in the (p_1, p_2)-plane.

Consider statement (b) in relation to the normal $\mathbf{n}_2$ in Figure 3.8. This vector has one positive component and one negative component with the former larger in magnitude than the latter. This is reflected in the fact that the projection of $\mathbf{n}_2$ along the direction of the principal diagonal (i.e., the $(1, 1)$-direction) is positive. Thus, the sum of the components of $\mathbf{n}_2$ is positive and, according to (b), it follows that the sum of the principal rates of deformation is positive. In other words, $\nabla \cdot \mathbf{v} > 0$, so the material dilates as it deforms. By the same argument yield at the point corresponding to the normal $\mathbf{n}_3$ is accompanied by consolidation, with $\nabla \cdot \mathbf{v} < 0$. There is a unique point on the arc where its normal $\mathbf{n}_C$ is orthogonal to the principal diagonal, and here yield occurs with no change in volume. Schofield and Wroth (1968) refer to this as a *critical state*; it divides the yield locus into two parts – an arc between the critical point and the origin where deformation is accompanied by dilation, and the remainder of the locus where there is consolidation.

Several other points on the yield locus correspond to particular types of deformation. The normal $\mathbf{n}_{ic}$ at the upper intersection of the yield locus with the principal diagonal points in the $(-1, -1)$-direction, so the principal rates of deformation are equal and negative and the deformation is an isotropic consolidation. At the point where the normal $\mathbf{n}_4$ is horizontal, on the other hand, the minor principal rate of deformation is negative while the major rate of deformation vanishes, so the corresponding deformation is a one-dimensional consolidation. At the origin of the (p_1, p_2)-plane the yield locus has a discontinuity in slope, so here the normal can be drawn in any direction pointing into the first quadrant. In particular $\mathbf{n}_1$, directed along the p_2-axis, corresponds to one-dimensional dilation, while $\mathbf{n}_{id}$ directed along the principal diagonal represents isotropic dilation.

The above yield locus has an image in the (p, q)-plane obtained simply by the substitutions $p_1 = p + q$, $p_2 = p - q$. Adopting the plastic potential flow rule it is also possible to generate the yield locus in the shear stress – normal stress plane by drawing a set of Mohr circles corresponding to all points of the (p_1, p_2)-locus. Their envelope is then the (T, N) yield locus, which is the union

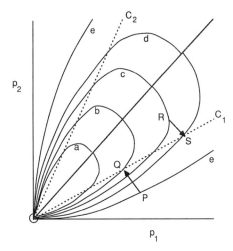

Figure 3.9. Yield loci $\rho_e > \rho_d > \rho_c > \rho_b > \rho_a$.

of the failure locus and the consolidation locus for material of this particular bulk density.

All of these considerations apply to material with a given bulk density, and Figure 3.8, therefore, exhibits just one of an infinite set of yield loci, one for each value of the density. Several of these are sketched in Figure 3.9 for a sequence of values of the bulk density. They might be expected to be similar in shape and nested within each other, as shown. The critical states for all the loci are connected by broken lines. To conform with the observation, mentioned earlier, that the ratio of shear stress to normal stress becomes constant after extensive shear, these must be a pair of straight lines through the origin, OC_1 and OC_2. A vector OP joining the origin to any point P of a yield locus has components that are the principal stresses, defined in the compressive sense. The same vector with reversed direction therefore gives the principal stresses defined in the tensile sense. Then, since the components of the inward normal to the locus at P are proportional to the principal rates of deformation, the scalar product of PO with this normal is proportional to the rate of dissipation, per unit volume, in the deforming material. This cannot be negative, so it follows that the line OP must pass outward through the yield locus at point P. Because this must be true for every point on each of the loci, their shape must be convex, as drawn.

Now consider the initiation of yield in a material of initial bulk density ρ_e. The corresponding yield locus is labeled "e" in Figure 3.9 and the mean stress $p (= (p_1 + p_2)/2)$ is such that yield occurs at the point P on the dilational part of the locus. Then, if the mean stress is held constant the material weakens as yield continues and the point representing the principal stresses moves along the line PQ in the $(-1, 1)$-direction, indicating a reduction in the shear stress $(p_1 - p_2)/2$. This continues until the point Q is reached on the critical line OC_1. The bulk density has then fallen to ρ_b but Q is the critical state for the yield locus b so deformation then continues without any further change of volume, and Q continues to represent the principal stresses. This is just the sort of dilational behaviour observed in practice and described earlier. Consolidation, on the other hand, is represented by a path such as RS. This starts at point R on the consolidational arc of the yield locus for material of bulk density ρ_c. As deformation proceeds the bulk density increases, the material strengthens and, if the mean stress is held constant, the representative point moves along the line RS to the point S where it intersects OC_1. The bulk density has then increased to ρ_d and S represents the critical state for material of this density, so deformation can continue with no further change in density. Thus, the picture presented in Figure 3.9, together with the plastic potential flow rule, gives a good account of the general features of plastic deformation, including the observation that the stresses approach asymptotic values with increasing deformation.

Despite its wholly geometric formulation the model provides a true constitutive relation, in the sense that it permits the stress tensor to be found given the bulk density and the rate of deformation tensor. To see this note first that the orientation of the principal rate of deformation axes determines the orientation of the principal stress axes because of the assumption that these are aligned: the so-called coaxiality condition. The ratio of the principal rates of deformation determines the direction of a vector in the (p_1, p_2)-plane whose components are in the same ratio, and the plastic potential flow rule then requires one to seek that point on the yield locus, for the bulk density in question, where the inward normal is parallel to this vector. The coordinates of this point then give the values of the principal stresses. Knowing these, and the orientation of the principal axes of stress, the stress tensor is completely determined.

This description immediately reveals one important feature of the constitutive relation. If the principal rates of deformation are both multiplied by any common positive factor, the above process clearly leads to exactly the same stress tensor. In other words, the constitutive relation is of *degree zero* in the rates of deformation. Because of this, the predicted behaviour of a dense granular material is often decidedly counter-intuitive to someone with a background in Newtonian fluid mechanics.

Though the constitutive relation has been framed in geometric terms, it is a straightforward matter to translate the geometry into equations. The yield loci relate p_1, p_2, and ρ or, equivalently, p, q, and ρ, and it is convenient to use the latter variables, writing the relation as

$$q = f(p, \rho). \tag{3.2}$$

Then, the condition that the principal axes of stress and rate of deformation should be aligned (i.e., the 'coaxiality condition') can be written

$$\cos 2\gamma \left(\frac{\partial v_x}{\partial y} + \frac{\partial v_y}{\partial x} \right) = \sin 2\gamma \left(\frac{\partial v_y}{\partial y} - \frac{\partial v_x}{\partial x} \right) \tag{3.3}$$

and the plastic potential flow rule takes the form

$$\cos 2\gamma \left(\frac{\partial v_y}{\partial y} + \frac{\partial v_x}{\partial x} \right) = \frac{\partial f}{\partial p} \left(\frac{\partial v_y}{\partial y} - \frac{\partial v_x}{\partial x} \right). \tag{3.4}$$

(3.2), (3.3), and (3.4) are three equations to determine p, q, and γ in terms of the bulk density and the spatial derivatives of the velocity components. Then (3.1) relates the components of the stress tensor explicitly to these variables, completing the constitutive relation.

Though the above ideas have been formulated entirely in two dimensions, their extension to three-dimensional motions is straightforward in principle. The yield loci of Figure 3.9 generalise to a set of yield surfaces in the space of the three principal stresses p_1, p_2, and p_3. A ray through the origin in the $(1, 1, 1)$-direction then plays the same role as the principal diagonal in Figure 3.9 but now there arises a new question, namely, the shape of sections of the yield surfaces by planes normal to $(1, 1, 1)$. Because the numbering of the coordinate axes is arbitrary, these shapes must be invariant under interchange of the axes. The simplest shapes consistent with this requirement are circles and hexagons. (The hexagons need not be regular but they must be invariant under rotations through $2\pi/3$.) Yield surfaces with circular or hexagonal sections are said to be of von Mises or Tresca forms, respectively. The coaxiality condition and the plastic potential flow rule remain the same in geometric terms but, of course, their translation into algebraic form becomes more complicated than (3.3) and (3.4).

Generalisation of the model to the case of cohesive granular materials that are capable of sustaining a finite *tensile* stress without yielding is straightforward. Each yield locus (or surface) is displaced in the $(-1, -1)$-direction, with those for highest bulk density being displaced most. Then a part of every yield locus lies in each quadrant of the (p_1, p_2)-plane and, for points of the locus outside the first quadrant, at least one of the principal stresses represents a tension.

Finally, we must point out that the constitutive model just described is by no means the only one available for dense granular flow. In particular, there is reason to suppose that the coaxiality condition is not satisfied in general, and there are "nonassociated" flow rules that permit misalignment between the principal axes of stress and those of deformation rate.

Although the model formulated above gives a quite good qualitative account of the salient features observed in slowly deforming, dense granular materials, it leaves much to be desired as a practical tool for performing quantitative calculations representing the behaviour of specific materials. Both the number and the difficulty of the experiments that would be needed to establish the functional relation (3.2), and hence the complete form of the nest of yield surfaces in principal stress space, are prohibitive without further assumptions about the shape of these surfaces. Experimental characterization of the yield conditions would become practicable only if the form of (3.2) were specified, except for the values of a small number of parameters, so that experiments could be limited to those needed to evaluate these parameters for a material of interest. However, even supposing this were possible so that an explicit form for the function $f(p, \rho)$ could be found, it is unlikely this would be of such simple form that the solution of (3.2), (3.3), and (3.4) for p, q, and γ could be expressed explicitly in terms of the derivatives of the velocity components, thereby allowing the stress components to be eliminated from the momentum equations using (3.1).

Apart from these manipulative difficulties, it must be borne in mind that the proposed constitutive relations are applicable only when the material is deforming, and doing so sufficiently slowly that collisional momentum transfer can be neglected completely. In many practical situations, parts of the material may not be deforming at all, at any given time, and the location of the boundaries separating deforming from nondeforming material then becomes part of the problem. A case in point is a granular material discharging from a conical hopper of wide angle. This is a situation where stationary shoulders of material adjacent to the hopper walls are separated from flowing material nearer the axis. In the upper part of the cone, this moves and deforms quite slowly, so it might be describable by a constitutive model of the type just outlined. Near the hole at the bottom where the material emerges, on the other hand, flow and deformation are both rapid, so momentum transfer by collisions and velocity fluctuations may be important. Between these locations, both contributions to the stress will be of comparable importance, and we do not know how they should be combined to provide constitutive relations for this regime Within the stationary shoulders, yet another constitutive model is applicable, namely that of an elastic solid. Finally, the boundary separating the shoulders from the flowing material must be located, using appropriate jump conditions, and its position might be expected to vary as the hopper discharges. Problems of this complexity, though not uncommon in engineering and geophysical applications, have not been solved in detail. What progress has been made has resulted from sweeping, though not unreasonable approximations.

Real fluidized beds may well contain regions where the material, while deforming, is dense enough to exhibit the phenomenon of yield, so a fully satisfactory treatment of their dynamical behaviour should include some features of the type described above in its model for S^p. Although some success has, nevertheless, attended treatments that neglect yield in "fully fluidized" suspensions, inclusion of the yield stress is imperative for any discussion of the processes of fluidization and defluidization, by which a packed bed becomes a fluidized bed on increasing the fluid flow rate and vice versa on decreasing the flow. We shall now address this type of problem.

3.4 The Processes of Fluidization and Defluidization

3.4.1 Theory

As a preliminary to examining fluidization and defluidization we need to extract the special form taken by the constitutive ideas of Section 3.3 when the deformation of the particle assembly is constrained to be one-dimensional. Referring to Figure 3.8, only the points of the yield locus with normals $\mathbf{n}_1$ and $\mathbf{n}_4$ correspond to one-dimensional deformations, with dilation at the former and consolidation at the latter. For the consolidating material, represented by the point with normal $\mathbf{n}_4$, the major principal stress is aligned with the axis of consolidation and its magnitude is the abscissa of this point. Dilation, on the other hand, is possible only at the origin where both principal stresses vanish. The magnitude of the major principal stress for consolidation depends on the particular yield locus considered, that is, on the bulk density of the material, and it increases with increasing bulk density. Thus, for either dilation or consolidation in one dimension, the normal stress p_{yy} acting on planes orthogonal to the direction of motion can be related uniquely to the bulk density or particle volume fraction, and the form of this relation is sketched in Figure 3.10. Here ϕ_m is the maximum value of the volume fraction that can be achieved by one-dimensional consolidation. The normal stress p_c responsible for consolidation increases without bound as $\phi \rightarrow \phi_m$ from below. ϕ_0 denotes the lowest concentration at which the assembly of particles is capable of supporting stress through a structure of sustained interparticle contacts, so $p_c = 0$ when $\phi = \phi_0$. The material can dilate only when $p_{yy} = p_d = 0$, and this value for the dilational yield stress is also shown in Figure 3.10. For values of p_{yy} lying between p_d and p_c, the assembly behaves as an elastic solid with a modulus of the same order of magnitude as that of the material from which the particles are

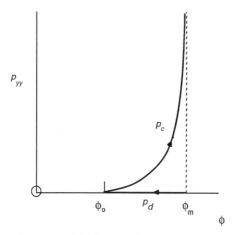

Figure 3.10. Normal stress at yield for one-dimensional dilation and consolidation.

formed. For typical mineral particles, therefore, the assembly can be regarded as essentially incompressible when $p_d < p_{yy} < p_c$.

Let us now examine the bearing of these considerations on the processes of fluidization and defluidization. Consider a bed of particles supported above a horizontal porous plate in a vertical tube of circular cross section. The interstitial space is filled with a fluid that can be regarded as incompressible and Newtonian. Then, when the fluid is at rest, the volume fraction of solids is close to ϕ_m and the bed is supported against the force of gravity by a combination of forces exerted at the porous plate, frictional forces between the walls of the tube and particles in contact with them, and buoyancy. If an upward flow of fluid is established, the pressure drop between the fluid at the bottom and top of the bed is increased and part of the weight of the particles is then supported by the drag forces exerted on them by the fluid. When the fluid flow is large enough, it is found that the bed expands, eventually reaching and passing the point where $\phi = \phi_0$ (see Figure 3.10). Then, the particle assembly is no longer capable of forming a structure of contacts capable of supporting a compressive load. If the flow is subsequently decreased progressively, the bed contracts and ϕ increases, eventually returning to a value near ϕ_m when the flow reaches zero. What has been described is, of course, the process of fluidization, followed by defluidization and we will now analyse these in more detail. Our treatment follows closely that of Tsinontides & Jackson (1993).

At each stage of fluidization or defluidization, we envisage the particle assembly to be at rest on average, in the sense that the local average value of the particle velocity vanishes everywhere. Thus, the momentum equation for the particle phase reduces to a force balance, found by setting the left hand side of Equation (2.28′) equal to zero. The stress tensor S^p in this equation is the resultant of contributions from contact forces transmitted through an extended structure together with momentum transfer and collisional impulses resulting from fluctuations of individual particle velocities about the local average. The former have been discussed in the earlier part of this chapter, the latter in Chapter 2. The contributions from velocity fluctuations are expected to be related uniquely to ϕ for the assembly of particles supported in equilibrium and to approach zero both as $\phi \to 0$ and as $\phi \to \phi_m$. The contribution from contact forces, on the other hand, is not uniquely related to ϕ. For a given value of ϕ, its value depends on whether the assembly is at the point of yielding in dilation, yielding in consolidation, or somewhere in between, as indicated in Figure 3.10.

Furthermore, when the assembly of particles can transmit stress through sustained contacts (i.e., for $\phi > \phi_0$) frictional forces may be exerted at contacts between particles and the walls of the confining tube. These are directed vertically and are proportional in magnitude to the normal forces at the contact points. Thus, in their presence, the principal axes of S^p cannot be aligned vertically and horizontally at the wall. It follows, therefore, that the processes of fluidization and defluidization are not merely examples of one-dimensional dilation or consolidation; when $\phi > \phi_0$ the stress field must be at least two-dimensional. Detailed solutions have been found computationally for similar stress fields in granular materials confined within cylindrical bins (Horne & Nedderman, 1976), but a simple and much earlier approach (Janssen, 1895) has proved adequate to give a good picture of the build up of stress with increasing depth in bins. A modification of this method will be adopted here to treat fluidization and defluidization.

The two-dimensional nature of the stress and displacement fields during dilation or consolidation will be neglected, so each of these processes will be regarded as examples of one-dimensional yielding. Then, as discussed earlier, the principal axes of S^p will be vertical and horizontal, and the values of the principal stresses will be independent of position in the cross section. If they take nonzero values, there will be a normal force exerted between the particles and the bounding wall and this may be associated with a frictional force that is directed upward or downward. (As already noted, the existence of this tangential force means that the stress field within the bed cannot really be one-dimensional, but this we will ignore.) The normal force on the wall will be found from the one-dimensional treatment of stress within the bed, and the bounding values for the frictional force will be calculated from this using a coefficient of limiting friction, in the usual way. These values are achieved only when the material is sliding, or is about to slide, up or down the wall. Then, if a coordinate z is measured down from the upper surface of the particulate material, a vertical force balance on the slice of the bed between z and $z + dz$ gives

$$\frac{\pi D^2}{4}\left[\frac{dp_{zz}}{dz} - \rho_s \phi g - nf_z\right] \pm \pi D\tau_f = 0,$$

where D is the diameter of the tube, nf_z is the fluid-particle interaction force and τ_f is the friction force per unit area between the particles and the tube wall. The positive sign before the last term certainly applies to consolidation where the particles are slipping or are about to slip down the wall, whereas the negative sign must apply for a dilating bed. When the bed is neither at the point of consolidation nor dilation, the last term can have either sign and may take any value between the limits of friction. The interaction force $n\mathbf{f}$ was discussed in detail in Chapter 2. We shall assume for simplicity that the Reynolds number, based on the velocity of the fluid, is always small enough to justify using a Richardson-Zaki form for the drag force (see Eq. (2.80)), when

$$nf_z = -\frac{(\rho_s - \rho_f)\phi g}{v_t \varepsilon^n}\bar{u} - \rho_f \phi g,$$

where $\bar{u}$ is the superficial velocity of the fluid (i.e., the volume flow rate divided by the cross sectional area, $\pi D^2/4$) and v_t is the terminal velocity of fall of an isolated particle in the fluid. Combining the last two equations gives

$$\frac{dp_{zz}}{dz} \pm \frac{4}{D}\tau_f = (\rho_s - \rho_f)\phi g\left[1 - \frac{1}{\varepsilon^n}\frac{\bar{u}}{v_t}\right]. \tag{3.5}$$

The contribution to $\mathbf{S}^p$ from particle velocity fluctuations is given by (2.91) so, since $\mathbf{v} = 0$, it reduces to an isotropic pressure $p^p(\phi)$, which is also its contribution to p_{zz}. During yield the contributions to p_{zz} and p_{rr} from contact forces are the coordinates of points such as $\mathbf{n}_4$ or $\mathbf{n}_1$ on a yield locus (see Figure 3.8), for consolidation and dilation, respectively. These are functions of ϕ and their contributions to p_{zz} are given by the curves sketched in Figure 3.10. We will assume, with Janssen, that they make proportional contributions to p_{rr}, so that their contributions to p_{zz} and p_{rr} during consolidation are $p_c(\phi)$ and $jp_c(\phi)$, respectively, where j is the so-called Janssen coefficient. During dilation the contributions of contact forces to both p_{zz} and p_{rr} vanish. Assembling these results, in Equation (3.5) we must substitute

$$p_{zz} = \begin{cases} p^p(\phi) + p_c(\phi), & \tau_f = \mu j p_c(\phi) & \text{during consolidation,} \\ p^p(\phi), & \tau_f = 0 & \text{during dilation.} \end{cases} \tag{3.6}$$

Note that only the part of p_{rr} resulting from contact forces, namely $jp_c(\phi)$, generates a corresponding frictional traction $\mu j p_c(\phi)$, where μ is the coefficient of friction between the particles and the tube wall. With the substitutions (3.6) Equation (3.5) becomes a first order differential equation for ϕ as a function of z, to be integrated from a suitable initial value at the bed surface, $z = 0$. Its solution provides the axial solids concentration profile during consolidation or dilation.

A force balance on the fluid in a horizontal slice of the bed relates the fluid pressure gradient to its superficial velocity:

$$\frac{dp}{dz} = \frac{(\rho_s - \rho_f)\phi g}{v_t} \frac{\bar{u}}{\varepsilon^n}. \tag{3.7}$$

This does not include the gravitational contribution $\rho_f g$.

It is now useful to introduce dimensionless variables, scaling z with the height H_m that the bed would occupy with the particles packed at volume fraction ϕ_m, and scaling the fluid pressure and stresses in the particle phase by $H_m(\rho_s - \rho_f)g\phi_m$, the buoyant weight of the bed per unit cross sectional area. Also, v_t can be related to $\bar{u}_m$, the fluid superficial velocity at which the drag force on a bed of particles with $\phi = \phi_m$ matches their buoyant weight, with the result $v_t/\bar{u}_m = 1/\varepsilon_m^n$. ($\bar{u}_m$ is simply the conventional 'minimum fluidization velocity' for a bed with concentration ϕ_m when defluidized.) Then for a consolidating bed Equation (3.5) becomes

$$\frac{d\tilde{p}^p}{d\zeta} + \frac{d\tilde{p}_c}{d\zeta} + J\tilde{p}_c = \frac{\phi}{\phi_m}\left[1 - \frac{\bar{u}}{\bar{u}_m}\left(\frac{\varepsilon_m}{\varepsilon}\right)^n\right], \tag{3.8}$$

where $J = 4\mu j H_m/D$, $\zeta = z/H_m$ and $\tilde{p}^p$ and $\tilde{p}_c$ are dimensionless stresses, scaled as above. The corresponding equation for a dilating bed is obtained from (3.8) by setting $\tilde{p}_c = 0$. With the same scaling Equation (3.7) becomes

$$\frac{d\tilde{p}}{d\zeta} = \frac{\phi}{\phi_m}\frac{\bar{u}}{\bar{u}_m}\left(\frac{\varepsilon_m}{\varepsilon}\right)^n, \tag{3.9}$$

where $\tilde{p}$ is the dimensionless fluid pressure. Once the dependence of p^p and p_c on ϕ is specified Equations (3.8) and (3.9), as we shall see, permit the bed height and the fluid pressure drop to be predicted, as functions of $\bar{u}$, during defluidization. At each value of $\bar{u}$, they also provide profiles of ϕ and p_{zz} within the bed. Similar results are obtainable for dilation simply by setting $\tilde{p}_c = 0$ in (3.8).

3.4.2 Predictions

First consider a simple model, in which yield stresses are neglected ($\tilde{p}_c = 0$) and the reversible particle pressure $\tilde{p}^p$ is taken to be a function of ϕ only. Then (3.8) reduces to

$$\frac{d\tilde{p}^p}{d\phi}\frac{d\phi}{d\zeta} = \frac{\phi}{\phi_m}\left[1 - \frac{\bar{u}}{\bar{u}_m}\left(\frac{\varepsilon_m}{\varepsilon}\right)^n\right]. \tag{3.10}$$

The value of ϕ at the bed surface, $\zeta = 0$, is such that $\tilde{p}^p = 0$, so Equations (3.9) and (3.10) can be integrated down the bed, starting from this value of ϕ and an arbitrary value of $\tilde{p}$. The bottom of the bed, $\zeta = H/H_m$, is located by equating the total volume of solids to $(\pi D^2/4)\phi_m H_m$. In dimensionless terms this condition takes the form

$$\int_0^{H/H_m} \phi \, d\zeta = \phi_m. \tag{3.11}$$

The dimensionless pressure drop across the bed is then given by $\Delta \tilde{p} = \tilde{p}(H/H_m) - \tilde{p}(0)$. Repeating this calculation for a sequence of values of $\bar{u}/\bar{u}_m$ generates curves of bed height and pressure drop as functions of the fluid flow rate.

The result of these computations will be presented for a specific form of the dependence of particle phase pressure on concentration, namely

$$\tilde{p}^p = \begin{cases} P\left(\dfrac{\phi - \phi_0}{\phi_m - \phi}\right) & \text{for } \phi_0 < \phi < \phi_m, \\ 0 & \text{for } \phi < \phi_0, \end{cases} \tag{3.12}$$

where P is a dimensionless constant. Then integration starts at $\zeta = 0$ with $\phi = \phi_0$. We shall take $\phi_0 = 0.5$ and $\phi_m = 0.65$ in (3.12) and set $n = 4$ in (3.10).

Results of the computations with these parameter values are shown in Figure 3.11 for $P = 0.05$, $P = 0.01$, and in the limit $P \to 0$. The limiting case corresponds to the naive picture of the process of fluidization. For $\bar{u} < \bar{u}_m$ the bed height takes the constant value H_m and the pressure drop increases linearly with $\bar{u}$ until it just reaches the buoyant weight of the particles when $\bar{u} = \bar{u}_m$. As $\bar{u}$ increases beyond $\bar{u}_m$ the pressure drop remains constant and the bed height increases monotonically. The "minimum fluidization" condition, $\bar{u} = \bar{u}_m$, is marked by discontinuous changes in the slopes of both the pressure drop and the bed height curves. For the curves with nonzero values of P, in contrast, these breaks in slope are missing and the transition from fixed to fluidized bed becomes more gradual as P is increased.

The type of behaviour illustrated is not limited to the particular functional form (3.12) for $\tilde{p}^p$. Any smooth function that starts from zero, increases monotonically, and diverges as $\phi \to \phi_m$ will generate smooth pressure drop and bed height curves, qualitatively similar to those in Figure 3.11 for $P = 0.01$ and $P = 0.05$, and will therefore fail to predict a sharp condition of minimum fluidization. In the limiting case $P \to 0$, for which there is a sharply defined condition of minimum fluidization, $\tilde{p}^p$ vanishes for $\phi < \phi_m$ and jumps discontinuously to an unbounded value when $\phi = \phi_m$. Physically this means that the

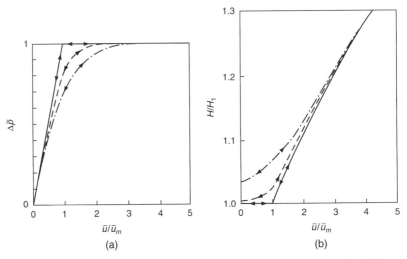

Figure 3.11. (a) (Pressure drop curves. No yield stress. — $P \to 0$; - - - $P = 0.01$; ---
$P = 0.05$. (b) Bed height curves. No yield stress. - - - $P = 0.01$; --- $P = 0.05$.

particle assembly offers no resistance to consolidation until ϕ reaches ϕ_m, at
which point it becomes perfectly rigid.

Real fluidized beds do however exhibit sharp discontinuities in the slopes
of the pressure drop and bed height curves as $\bar{u}$ is increased; clearly then, any
model of stress in the particle assembly that takes account only of $\tilde{p}^p$, and that
represents this as a monotone increasing function of ϕ that diverges smoothly as
$\phi \to \phi_m$, cannot represent this observed behaviour. To generate pressure drop
and bed height curves that have the observed sort of jumps in slope, $\tilde{p}^p$ must
approach a finite limit as $\phi \to \phi_m$ but should then jump discontinuously to an
unbounded value when $\phi = \phi_m$.

When the only contributor to stress in the particle phase is the pressure p^p
the pressure drop and bed height curves, such as those exhibited in Figure 3.11,
are the same for both fluidization and defluidization. We will now examine how
these curves are affected by the presence of an irreversible yield stress of the sort
discussed above. For simplicity let us assume that this is the only form of stress
in the particle phase and thus set $\tilde{p}^p = 0$ in Equation (3.8). For a qualitatively
correct representation of $\tilde{p}_c$ it is not unreasonable to postulate a dependence on
ϕ with the same algebraic form as (3.12), namely

$$\tilde{p}_c = \begin{cases} C\left(\dfrac{\phi - \phi_0}{\phi_m - \phi}\right) & \text{for } \phi_0 < \phi < \phi_m, \\ 0 & \text{for } \phi < \phi_0. \end{cases} \qquad (3.13)$$

Then, during the consolidation of the bed that occurs as the fluid flow is

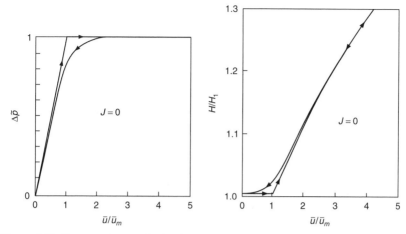

Figure 3.12. Pressure drop and bed height curves. Yield stress alone. $C = 0.01$. Infinite bed diameter ($J = 0$).

decreased,

$$\frac{d\tilde{p}_c}{d\phi}\frac{d\phi}{d\zeta} + J\tilde{p}_c = \frac{\phi}{\phi_m}\left[1 - \frac{\bar{u}}{\bar{u}_m}\left(\frac{\varepsilon_m}{\varepsilon}\right)^n\right]. \qquad (3.14)$$

The curves relating bed height and pressure drop to $\bar{u}$ are generated by integrating (3.14) and (3.9) downward from the bed surface and using the condition (3.11) to terminate the integration. The initial value of $\tilde{p}$ is arbitrary but ϕ must start from ϕ_0 at $\zeta = 0$.

The parameter value $J = 0$ corresponds to a bed of infinite diameter and for this case the results are the same as those shown in Figure 3.11. They are reproduced as the defluidization curves in Figure 3.12 for $C = 0.01$ and in Figure 3.14 for $C = 0.05$. The corresponding curves for a bed of bounded diameter ($J = 10$) are shown in Figure 3.13 for $C = 0.01$ and in Figure 3.15 for $C = 0.05$. For both values of C the height of the fully collapsed bed ($\bar{u} = 0$) is larger when $J = 10$ than when $J = 0$, as would be expected, since friction at the walls contributes to the support of the particles in the bed of bounded diameter. Correspondingly, for each value of $\bar{u}$, the fluid pressure drop is smaller in the bed of bounded diameter than in the infinite bed since it does not have to support the full buoyant weight of the particles.

Following complete defluidization, with $\bar{u}$ decreased to zero, we must consider what happens when the process is reversed by increasing $\bar{u}$ progressively. To understand this we must examine the profile of $\tilde{p}_{zz}$ in the fully collapsed bed when $\bar{u}$ has just been reduced to zero. Then $\tilde{p}_{zz} = \tilde{p}_c(\phi)$, which follows when ϕ

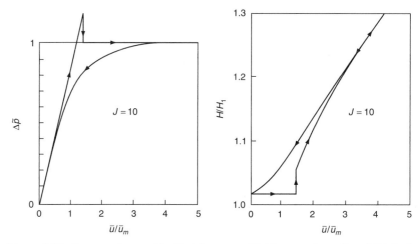

Figure 3.13. Pressure drop and bed height curves. Yield stress alone. $C = 0.01$. Bed of bounded diameter ($J = 10$).

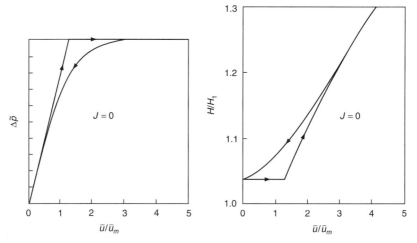

Figure 3.14. Pressure drop and bed height curves. Yield stress alone. $C = 0.05$. Infinite bed diameter ($J = 0$).

is found by integrating (3.14) with $\bar{u} = 0$. Both $\phi(\zeta)$ and $\tilde{p}_c(\zeta)$, obtained in this way, are exhibited in Figure 3.16 for parameter values $C = 0.05$ and $J = 10$. Now consider what happens when $\bar{u}$ is increased from zero. Since the fluid then exerts an upward drag force on the particles $\tilde{p}_{zz}$ increases less rapidly with increasing ζ, so it lies between the yield limits (zero and $\tilde{p}_c$) and the particle assembly is neither on the verge of consolidating nor of dilating. Thus $\tilde{p}_{zz}$ is no longer related algebraically to ϕ but is an independent variable. For a small, but nonvanishing value of $\bar{u}$ the shape of the relation between $\tilde{p}_{zz}$ and ζ is not

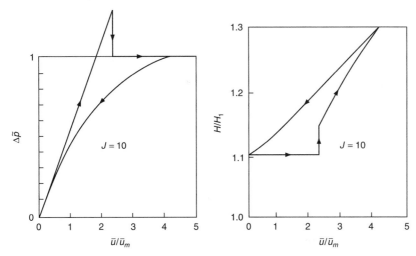

Figure 3.15. Pressure drop and bed height curves. Yield stress alone. $C = 0.05$. Bed of bounded diameter ($J = 10$).

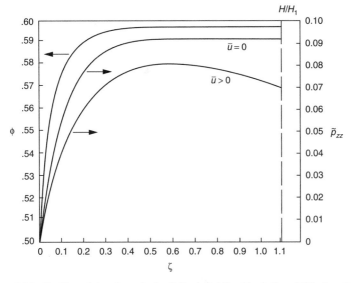

Figure 3.16. Profiles of ϕ and p_{zz} in the fully defluidized bed. $C = 0.05$; $J = 10$. Also shown is a sketched profile of p_{zz} for small $\bar{u} > 0$.

hard to see. When $\bar{u} = 0$ Figure 3.16 shows that $\tilde{p}_{zz}$ is almost constant in the lower part of the bed. For a small positive value for $\bar{u}$ it then follows that $\tilde{p}_{zz}$ must decrease on moving down this part of the bed. Consequently, $\tilde{p}_{zz}$ must pass through a maximum at some intermediate value of ζ and the profile of $\tilde{p}_{zz}(\zeta)$ must be as sketched in Figure 3.16 when $\bar{u} > 0$.

With further increases in $\bar{u}$ the value of $\tilde{p}_{zz}$ at the lower surface of the particle bed will continue to decrease until it reaches zero, which is the condition for dilational yield. At this point, however, $\tilde{p}_{zz}$ remains positive for all other values of $\zeta > 0$, so the bed cannot dilate at any other level. Consequently, we expect it to move up *en bloc*. The unsupported lower surface will then become unstable and will shed material, which will descend to form a new fluidized bed, resting on the distributor plate, but with a smaller volume fraction of solids than the original bed. The new bed would be expected to be at the dilational yield limit and will therefore transmit no stress.

The point at which bed expansion begins is characterized by the condition that $\tilde{p}_{zz} = 0$ at $\zeta = H(0)/H_m$, where $H(0)$ is the height of the fully defluidized bed with $\bar{u} = 0$. To find $\tilde{p}_{zz}(\zeta)$ at this point we must integrate the vertical force balance, noting that the frictional force at the bounding wall will now act downward, since the bed is on the point of lifting, and it will be related to the normal stress on the wall by the condition of limiting friction. Thus the force balance is

$$\frac{d\tilde{p}_{zz}}{d\zeta} - J\tilde{p}_{zz} = \frac{\phi}{\phi_m}\left[1 - \frac{\bar{u}}{\bar{u}_m}\left(\frac{1 - \phi_m}{1 - \phi}\right)^n\right]. \tag{3.15}$$

Also, since the bed has not yet expanded, $\phi(\zeta)$ is the same as the known particle concentration profile in the bed when $\bar{u} = 0$ at the end of the defluidization process. Thus $\tilde{p}_{zz}$ is the only unknown in (3.15), which can be solved explicitly, giving

$$p_{zz}(H(0)/H_m)e^{-JH(0)/H_m} - p_{zz}(0) = \int_0^{H(0)/H_m} e^{-J\zeta}A(\zeta)\,d\zeta$$

$$-\frac{\bar{u}}{\bar{u}_m}\int_0^{H(0)/H_m} e^{-J\zeta}B(\zeta)\,d\zeta,$$

where

$$A(\zeta) = \frac{\phi}{\phi_m} \quad \text{and} \quad B(\zeta) = \frac{\phi}{\phi_m}\left(\frac{1 - \phi_m}{1 - \phi}\right)^n.$$

The value of the fluid velocity at which the bed begins to expand is then obtained by setting $\tilde{p}_{zz}(H(0)/H_m) = \tilde{p}_{zz}(0) = 0$, giving

$$\frac{\bar{u}}{\bar{u}_m} = \frac{\int_0^{H(0)/H_m} e^{-J\zeta}A(\zeta)\,d\zeta}{\int_0^{H(0)/H_m} e^{-J\zeta}B(\zeta)\,d\zeta}, \tag{3.16}$$

and the corresponding pressure drop is found by integrating (3.9), with the result

$$\Delta\tilde{p} = \frac{\bar{u}}{\bar{u}_m}\int_0^{H(0)/H_m} B(\zeta)\,d\zeta. \tag{3.17}$$

As $\bar{u}$ increases from zero to the value given by (3.16) the bed height remains fixed at $H(0)$ and the pressure drop is proportional to $\bar{u}$. This determines those parts of the pressure drop and bed height curves in Figures 3.12 to 3.15 up to the point of initial expansion. For beds of bounded diameter Figures 3.13 and 3.15 show that the pressure drop overshoots to $\Delta\tilde{p} > 1$ before the bed begins to expand. Thus the pressure drop exceeds the buoyant weight of the particles before expansion is initiated. The bed can remain at rest and unexpanded in these circumstances only because its weight is supplemented by downward frictional forces exerted at the wall.

Beyond the point of initial expansion the pressure drop and bed height curves are easily found. As noted above the bed is then stress free, so we set $\tilde{p}_c = 0$ in the force balance (3.14), with the result

$$\phi = 1 - (1 - \phi_m)\left(\frac{\bar{u}}{\bar{u}_m}\right)^{1/n}.$$

Also, since the volume of solid material is the same in both the expanded and unexpanded beds, $\phi H = \phi_m H_m$. Then ϕ can be eliminated between these results, giving

$$\frac{H}{H_m} = \frac{\phi_m}{1 - (1 - \phi_m)(\bar{u}/\bar{u}_m)^{1/n}}, \tag{3.18}$$

which defines the bed height curve. The dimensionless pressure drop is now unity since the pressure drop must balance the buoyant weight of the particles. The parameters C and J do not appear in (3.18), so the curves of bed height $v.\ \bar{u}$ during fluidization are common to Figures 3.12, 3.13, 3.14, and 3.15.

Although the particle assembly transmits no stress after the fluid velocity is increased beyond the point of initial expansion, it is wrong to regard this expanded bed as being fluidized. Until the bed is expanded beyond $\phi = \phi_0$, where the hysteresis loop closes, the particle assembly is still capable of supporting a nonvanishing compressive load without consolidating. Indeed, if the fluid flow is decreased at any time, while ϕ is still larger than ϕ_0, it will do just that, and the bed will not contract until the condition for compressive yield is reached (see Tsinontides & Jackson (1993)).

The fluidization and defluidization curves presented in Figures 3.11 to 3.15 indicate how the nature of stresses transmitted within the particle assembly is reflected in the curves relating pressure drop and bed height to fluid flow. Conversely, examination of these curves should allow one to discriminate between the different types of stress, and even to estimate these stresses quantitatively. For example, if measured fluidization and defluidization curves coincide accurately, over the whole range of values of $\bar{u}$, we can conclude that there are no yield limits for dilation and consolidation. Furthermore, if the observed curves closely resemble those for the limiting case $P \to 0$ in Figure 3.11, there must be negligible stress transmission in the bed for any volume fraction of particles smaller than ϕ_m, with a sudden transition to complete incompressibility at $\phi = \phi_m$. The characteristic signature of this physical behaviour is vanishing bed expansion and pressure drop proportional to $\bar{u}$, when $\bar{u} < \bar{u}_m$, and pressure drop constant and exactly balancing the buoyant weight of the particles, when $\bar{u} > \bar{u}_m$. The symptoms of the existence of a significant particle phase pressure p^p for $\phi < \phi_m$, in contrast, are a pressure drop that decreases gradually below the buoyant weight of the particles as the fluid flow is decreased and a bed height that begins to increase gradually as soon as there is a nonvanishing fluid flow. This is seen in the curves for $P = 0.01$ and 0.05 in Figure 3.11. Notice also that the bed height for $\bar{u} = 0$ is larger than H_m when $P \neq 0$ and it increases with P.

In contrast to the above, if the particle assembly has yield limits for consolidation and dilation that differ from each other over an extended interval of ϕ below ϕ_m, the fluidization and defluidization curves will differ in this interval of ϕ. This will be observed even in beds of large diameter in relation to their height, and the magnitude of the hysteresis provides a measure of the size of the compressive yield stress. (See Figures 3.12 and 3.14.) In beds of smaller diameter the hysteresis loop opens wider and a new phenomenon appears at the point of initial bed expansion on the fluidization curves. (See Figures 3.13 and 3.15.) The pressure drop overshoots beyond the buoyant weight of the particles before the bed begins to expand; then it falls back to balance the buoyant weight once expansion is initiated. This is mirrored by a discontinuous jump in the bed height at the point of initial expansion. Both the size of the overshoot in pressure drop and that of the jump in bed height increase as the bed diameter is decreased. With a bed of given diameter the magnitudes of both discontinuities increase as the compressive yield stress increases, as can be seen by comparing Figures 3.13 and 3.15. Finally, the height of the bed when fully collapsed, with $\bar{u} = 0$, is larger for the bed of bounded diameter than for the infinite bed. This is to be expected because of the support provided by frictional forces at the wall.

An overshoot in pressure drop and a jump in bed height at initial expansion can also be a result of the particle assembly being cohesive, so that it is able to sustain a nonvanishing tensile stress before it dilates. However, discontinuities arising from this cause will persist unchanged however large the bed diameter, in contrast to the behaviour described above.

3.4.3 Experiments

Surprisingly, while there have been many measurements of pressure drop and bed height during the process of fluidizing the bed by increasing the flow rate,

careful measurement of these quantities during both fluidization and defluidization, for the same particle bed and in tubes of differing diameters, appears to
have been limited to a single study (Tsinontides, 1992; Tsinontides & Jackson,
1993). Figures 3.17 and 3.18, taken from Tsinontides (1992), show the pressure drop and the bed height, as functions of the fluid superficial velocity, for
fluidized cracking catalyst (FCC) in tubes with internal diameters of 2.53 cm
and 4.99 cm, respectively. The fluidizing medium was dry air and the mass of
particulate material loaded, per unit cross-sectional area of the bed, was the
same in both cases. The mean particle diameter was approximately 60 μm,
with a broad size distribution between 20 μm and 100 μm and a much smaller
secondary peak in the size distribution around 1–2 μm. Note that the bed height
was not scaled by H_m, as in the above theory, since this is not easy to measure
without a bed of very large cross section. Instead the total volume of the partic-
ulates was calculated as m/ρ_s, where m is the mass of catalyst charged and ρ_s
the intrinsic density of the solid material. Dividing this by the cross-sectional
area of the bed gave the scaling height, H_1, and the quantity H/H_1 was then
used as the indicator of bed expansion. Of course ρ_s, the density of the porous
material from which the particles are formed, is not easily measurable so the
value used to determine H_1 is an estimate. Thus, although the scaling of H is
defined unambiguously it is, to some extent, arbitrary.

The observations are seen to conform well with the theoretical predictions
shown in Figures 3.12 to 3.15, which are based on a model in which stress
associated with random fluctuations in particle velocities is neglected. The
hysteresis between the curves for fluidization and defluidization is prominent in
both Figure 3.17 and Figure 3.18, and it is of just the sort predicted theoretically.
Also as predicted the loops are more open, the overshoot in $\Delta\tilde{p}$ is larger, and
the bed height at zero fluid flow is greater in the bed of smaller diameter. In
both figures the hysteresis loops close at approximately the same scaled bed
height, $H/H_1 \approx 1.9$, so this should correspond to the particle volume fraction
ϕ_0 at which the compressive yield stress is reduced to zero. The observations
differ from the theoretical predictions in only one significant respect: for large
values of the gas flow the dimensionless pressure drop $\Delta\tilde{p}$ does not appear to
approach the value unity but remains appreciably smaller, so the gas pressure
drop is not supporting the full weight of the solid particles. The reason for this
is not clear. It may be indicative of the existence of a significant contribution
p^p to the particle phase stress, associated with fluctuations in velocity of
individual particles, which would be expected to survive to bed expansions
beyond ϕ_0 where the particle assembly no longer exhibits yield stresses.

As was pointed out earlier, the part of the fluidization curves (increasing gas
flow) between the point of initial expansion and the point where the hysteresis

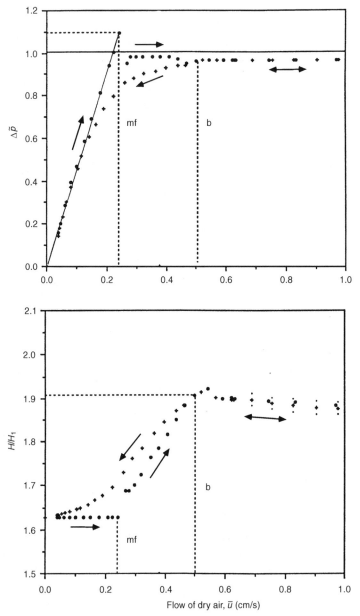

Figure 3.17. Complete fluidization–defluidization cycle for FCC in a tube of 2.53 cm internal diameter. (From Tsinontides & Jackson, 1993.)

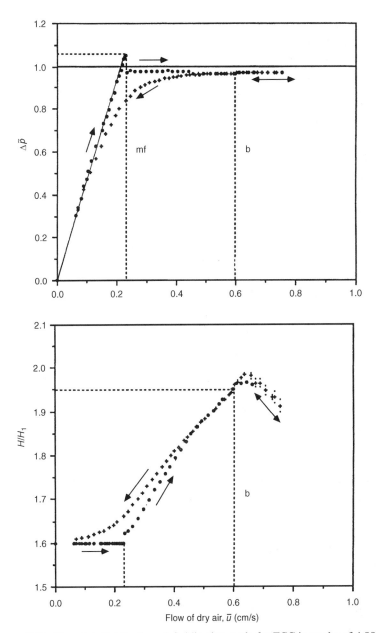

Figure 3.18. Complete fluidization–defluidization cycle for FCC in a tube of 4.99 cm internal diameter. (From Tsinontides & Jackson, 1993.)

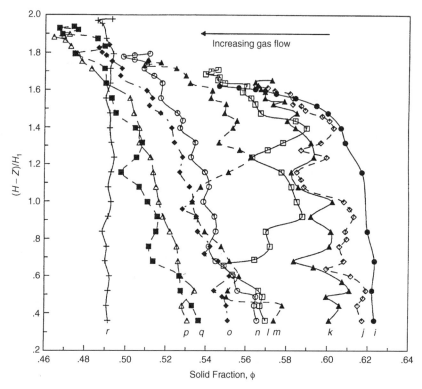

Figure 3.19. Experimentally determined profiles of solids fraction as a function of height in a bed of FCC, fluidized by air, in a tube of 4.99 cm internal diameter. (From Tsinontides & Jackson, 1993.)

loop closes represents a bed that is expanded, but not fluidized, since it is still capable of sustaining a nonvanishing compressive stress before it consolidates. Striking confirmation of this is provided by gamma ray absorption measurements of the bulk density of the bed, as a function of height on the centre line (Tsinontides & Jackson, 1993). These are reproduced in Figure 3.19. The curve labeled (i) corresponds to the fully defluidized bed, with zero air flow. The curve labeled (r) corresponds to a gas flow considerably larger than that at which the hysteresis loop closes. Between these two extremes we see large-amplitude swings in the bulk density on moving down the bed, and scans perpendicular to the axis have shown that these extend across the complete section of the bed. Thus, the bed is built of layers of alternating high and low bulk density material. Furthermore, this is not a transient phenomenon. Repeat measurements indicate that the structures present in profiles such as those labeled (j), (k), and (l) persist unchanged for days if the air flow is held constant, and they can even survive a limited amount of mechanical disturbance induced by tapping the tube. Structures of this sort, with long-term stability, could not possibly survive if the bed behaved like a fluid, without yield limits.

The large hysteresis loops seen in Figures 3.17 and 3.18 result from the ability of FCC to form structures capable of supporting a compressive load over an extended interval of particle volume fraction. This property accounts, in part, for the desirable behaviour of FCC in particle circulation loops, as we shall see. Other materials, such as smooth glass beads, do not have this property. Expanded beds of glass beads can support virtually zero stress until the particle volume fraction is increased to a value very close to ϕ_m, though they become completely rigid when $\phi = \phi_m$. Correspondingly, their fluidization and defluidization curves almost coincide, with any hysteresis confined to a very narrow interval of ϕ adjacent to ϕ_m.

Careful measurements of fluidization and defluidization curves, such as those reported by Tsinontides (1992), have the potential to shed useful light on the nature and magnitude of stresses transmitted within the particle phase. If these measurements were repeated for a range of bed depths and bed diameters they would be able to distinguish effects of wall friction from effects of cohesion between the particles and to show how the compressive yield strength of the particle assembly depends on its concentration. It would be particularly valuable to exploit measurements of this sort in combination with one of the techniques now becoming available for direct measurement of the particle temperature (Cody et al., 1996; Menon & Durian, 1997), since the latter provide a direct indication of the existence of a contribution to the stress from p^p. Figure 3.20 shows results of Cody & Goldfarb (1996) who measured both bed expansion and shot noise due to particle impacts on the wall of a bed of FCC. Figure 3.20(a) shows the bed height as a function of a variable U_s / U_{mf}^*. In our notation U_s is $\bar{u}$, the superficial fluid velocity, while U_{mf}^* denotes, not the superficial velocity $\bar{u}_{mf}$ at which expansion begins, but that at which noise is first detected, and it is seen that $U_{mf}^* = 4.57\bar{u}_{mf}$. Figure 3.20(b) shows v_n, the root mean square fluctuation in the particle velocity component normal to the wall, scaled as indicated, as a function of U_s / U_{mf}^*. It is clear that the bed remains completely silent until $U_s = U_{mf}^*$, despite the fact that it has already experienced substantial expansion when the fluid flow reaches this value. These observations are entirely consistent with the conclusion, reached on the basis of the measurements of Tsinontides & Jackson, that a bed of particles of this type expands for some distance before it is released from yield stresses and becomes fully fluidized. Over a large part (possibly all) of this expansion the measurements of Cody & Goldfarb show that p^p is negligibly small.

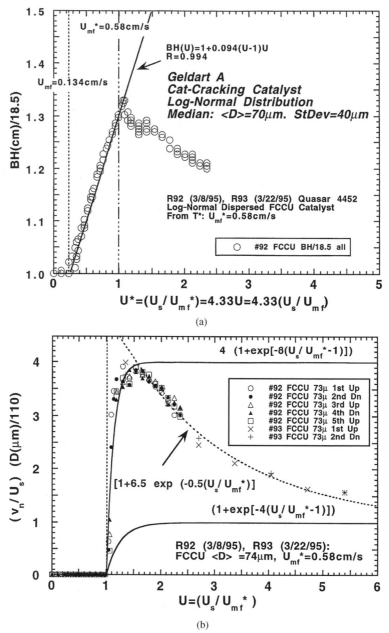

Figure 3.20. Bed height and normal component of velocity fluctuations for FCC. (From Cody & Goldfarb, 1996.)

The existence of a broad interval of expansion in which yield stresses persist, but p^p is negligible, may be peculiar to certain materials such as FCC. There are other particulate materials, such as small glass beads, that also expand smoothly well beyond minimum fluidization but for which Cody & Goldfarb found evidence of a significant level of p^p.

References

Christofferson, J., Mehrabadi, M. M., & Nemat-Nasser, S. 1981. A micromechanical description of granular material behaviour. *J. Appl. Mech.* **48**, 339–344.

Cody, G. D. & Goldfarb, D. J. 1996. Discontinuity in particle granular temperature of gas fluidized beds across Geldart A/B boundary. Materials Research Soc. Fall Meeting, Boston, Dec. 2–6.

Cody, G. D., Goldfarb, D. J., Storch, G. V., & Norris, A. N. 1996. Particle granular temperature in gas fluidized beds. *Powder Technol.* **87**, 211–232.

Coulomb, C. A. 1776. Essai sur une application des règles de maximis et minimis à quelques problèmes de statique, relatifs à l'architechture. *Acad. R. Sci. Mem. Math. Phys. Divers Savants* **7**, 343–382.

Drucker, D. C. & Prager, W. 1952. Soil mechanics and plastic analysis or limit design. *Q. Appl. Math.* **10**, 157–165.

Goddard, J. D. & Didwania, A. K. 1998. Computations of dilatancy and yield surfaces for assemblies of rigid frictional spheres. *Q. J. Mech. Appl. Math.* **51**, 15–43.

Horne, M. R. 1965. The behaviour of an assembly of rotund, rigid, cohesionless particles. Parts I & II. *Proc. Roy. Soc. London.* **A286**, 62–97.

Horne, M. R. 1969. The behaviour of an assembly of rotund, rigid, cohesionless particles. Part III. *Proc. Roy. Soc. London.* **A310**, 21–34.

Horne, R. M. & Nedderman, R. M. 1976. Analysis of the stress distribution in two-dimensional bins by the method of characteristics. *Powder Technol.* **14**, 93–102.

Janssen, H. A. 1895. Versuche über Getreidedruck in Silozellen. *Z. Ver. Deutsch. Ing.* **39**, 1045–1049.

Jenike, A. W. & Shield, R. T. 1959. On the plastic flow of Coulomb solids beyond failure. *J. Appl. Mech.* **26**, 599–602.

Johnson, P. C. & Jackson, R. 1987. Frictional–collisional constitutive relations for granular materials, with application to plane shearing. *J. Fluid Mech.* **176**, 67–93.

Johnson, P. C., Nott, P., & Jackson, R. 1990. Frictional–collisional equations of motion for particulate flows and their application to chutes. *J. Fluid Mech.* **210**, 501–535.

Kanatani, K-I. 1981. A theory of contact force distribution in granular materials. *Powder Technol.* **28**, 167–172.

Menon, N. & Durian, D. J. 1997. Diffusing wave spectroscopy of dynamics in a three-dimensional granular flow. *Science* **275**, 1920–1922.

Roscoe, K. H. 1970. The influence of strains in soil mechanics. *Geotech.* **20**, 129–170.

Roscoe, K. H., Schofield, A. N., & Wroth, C. P. 1958. On the yielding of soils. *Geotech.* **8**, 22–33.

Rowe, P. W. 1962. The stress–dilatancy relation for static equilibrium of an assembly of particles in contact. *Proc. Roy. Soc. London.* **A269**, 500–527.

Savage, S. B. 1982. Granular flows down rough inclines – review and extension. In *Proc. of U.S. – Japan Seminar on New Models and Constitutive Relations in the Mechanics of Granular Materials*, ed. J. T. Jenkins & M. Satake, pp. 261–282. Elsevier.

Schofield, A. N. & Wroth, C. P. 1968. *Critical State Soil Mechanics.* McGraw-Hill.

Shield, R. T. 1955. On Coulomb's law of failure in soils. *J. Mech. Phys. Solids* **4**, 10–16.

Thornton, C. 1979. The conditions for failure of a face-centered cubic array of uniform rigid spheres. *Geotech.* **29**, 441–459.

Tsinontides, S. C. 1992. A theoretical and experimental investigation of the mechanics of fluidized gas–particle systems. Ph.D. dissertation, Princeton University.

Tsinontides, S. C. & Jackson, R. 1993. The mechanics of gas fluidized beds with an interval of stable fluidization. *J. Fluid Mech.* **255**, 237–274.

4

Stability of the uniformly fluidized state

4.1 Introduction

From the earliest applications of fluidized beds it was noticed that most liquid fluidized beds present a smooth appearance, expanding progressively as the fluid flow rate is increased, while most gas fluidized beds are markedly heterogeneous, containing numerous rising pockets of gas essentially free of particles. These gas pockets were referred to as "bubbles" and the terms "particulate" and "aggregative" were introduced to describe the two types of behaviour. Quite early Wilhelm & Kwauk (1948) proposed an empirical criterion to distinguish between them, suggesting that the bed behaves aggregatively if $Fr = \bar{u}_m^2/(g d_p) > 1$ and particulately otherwise, where $\bar{u}_m$ denotes the minimum fluidization velocity and d_p is the particle diameter. Though the mechanics of motion around bubbles was puzzling at first, a simple but elegant analysis by Davidson (1961) revealed its salient features. Davidson & Harrison (1963) went on to show that large enough bubbles might be expected to self-destruct by entrainment of solid particles from below, and hence they speculated that the distinction between particulate and aggregative behaviour might hinge on the maximum size of bubble that can survive.

At about the same time a quite different approach to this question was being explored independently by a number of workers (Jackson, 1963; Pigford & Baron, 1965; Murray, 1965). The hope was that a conventional hydrodynamic stability analysis of the uniformly fluidized state would yield a stability boundary separating particulate behaviour, on the stable side, from aggregative behaviour on the unstable side. Unfortunately this hope was not realized. The earliest stability analysis (Jackson, 1963) predicted that all fluidized suspensions should be unstable and, more disturbingly, that the rate of growth of the instability should increase monotonically as its wavelength tends to zero. The

99

only encouraging feature was a comparison of rates of growth for disturbances of similar wavelengths in gas and liquid fluidized beds. This was found to be larger by one to two orders of magnitude in typical gas fluidized beds than in typical liquid fluidized beds. Indeed, in gas fluidized beds disturbances of "reasonable" wavelength were found to increase in amplitude by a factor e while rising through a distance of the same order of magnitude as their own linear dimensions, suggesting that these beds are highly unstable and tend to separate rapidly into regions of high and low particle concentration.

The physically unacceptable prediction of most rapid growth at vanishingly short wavelength is clearly a consequence of the fact that the simple particle-phase momentum equation adopted by Jackson contains no terms representing energy dissipation that increases with increasing deformation rate; in other words, viscous dissipation. This type of term might be expected to result from physical mechanisms such as momentum transport associated with fluctuations in particle velocities or contact interactions between particles. When such terms are included (Pigford & Baron, 1965; Anderson & Jackson, 1967, 1968) there is a bounded value of the wavelength for which the disturbances are amplified most rapidly, but the suspension remains unstable in all cases. However, if momentum transport by the particles generates viscous stresses it must presumably also give rise to an isotropic contribution to the stress, that is, an effective pressure associated with the particle phase. This was recognized by Anderson & Jackson (1967), but it was estimated to be a quite small effect, with minimal influence on disturbance growth rates. However, Garg & Pritchett (1975) later pointed out that, if this pressure increases sufficiently rapidly with particle concentration, it is capable of stabilizing the bed. Thus, the prospect of finding a stability boundary that might distinguish between particulate and aggregative behaviour was resurrected.

To predict the existence and location of such a boundary an expression relating the effective particle-phase pressure to the concentration of the particles is needed, and in 1984 this was provided, in effect, by a simple and explicit stability criterion proposed by Foscolo & Gibilaro, though it was not until some time later that they interpreted their criterion in these terms (Foscolo & Gibilaro, 1987). The Foscolo–Gibilaro criterion appeared to have remarkable success in discriminating between systems observed to bubble and those that do not; for example, it predicted that water fluidized beds of glass beads of the order of 1 mm diameter should be stable and, indeed, no bubbles are observed in these beds, whereas air fluidized beds of the same particles, which certainly bubble, were predicted to be unstable. However, experiments to be discussed later in Section 4.6 (Anderson & Jackson, 1969; El Kaissy & Homsy, 1976; Nicolas et al., 1994, 1996) show unequivocally that water fluidized beds of small

(≈1 mm diameter) glass beads, though devoid of bubbles, are not uniform but are traversed by rising wave patterns of varying particle concentration whose initial development resembles closely the instability waves predicted theoretically. In view of this a stability criterion, such as that of Foscolo & Gibilaro, which judges these particulate beds to be stable, is actually at variance with the experimental evidence. This is not surprising since it is by no means justified to assume that nonbubbling beds are necessarily stable. Visible bubbles represent large-amplitude departures from the state of uniform fluidization, whereas stability theory examines the initial development of very small amplitude perturbations, so it is perfectly possible for a suspension to be linearly unstable without the resulting disturbances ever developing into recognizable bubbles.

Since bubbling beds are clearly unstable and even typically "particulate" beds (e.g., small glass beads) reveal themselves to be unstable on closer examination, we might ask whether there are any systems that can be expanded beyond the point of initial fluidization without showing symptoms of instability. There are, in fact, two possible candidates. First, it is well known that certain gas fluidized beds of small particles, such as the fluid cracking catalyst used in oil refining, can be expanded substantially without any trace of visible bubbling, or any other motion of the particles. Second, a much narrower range of stable expansion has been reported (Ham et al., 1990) for certain beds of small particles fluidized by water–glycerine mixtures. There is reason to suppose (see Section 3.4.3 above and Tsinontides & Jackson (1993)) that stabilization in the first of these cases is not simply a result of particle-phase pressure, but the second case could well provide evidence for stabilization by this mechanism. One is tempted to speculate further that beds of very small particles, with a density not much greater than that of the fluidizing medium, might be stable over extended ranges of expansion due to the influence of the particle-phase effective pressure. Unfortunately, however, there is no reliable experimental evidence on this point. If stabilization by the effective pressure occurs it is here that it will most likely be found, so a careful, systematic experimental exploration of this class of suspensions would be of great value.

On the theoretical side, credible methods for predicting the particle-phase pressure are clearly needed. Differing mechanisms have been proposed by Mutsers & Rietema (1977), Foscolo & Gibilaro (1987), Batchelor (1988), Koch (1990), and, in a series of papers, by Buyevich (1971a,b; 1972a,b,c) and Buyevich & Kapbasov (1994). These have provoked sharp controversies, which are not yet resolved.

Although linear stability analyses have accounted for the existence of slowly growing waves in liquid fluidized beds and explained why similar waves grow much faster in most gas fluidized beds, they have nothing to say about their ultimate fate and cannot explain why the waves do not develop into bubbles in sufficiently deep liquid fluidized beds or, indeed, whether they are the precursors of bubbles in the gas fluidized beds. This has led to a number of attempts to extend the theory of one-dimensional disturbances to take into account the nonlinearities in the equations of motion (Fanucci et al., 1979; Liu, 1982; Needham & Merkin, 1983, 1986; Ganser & Drew, 1990), but these have not revealed any qualitative distinction between the predicted behaviour of gas and liquid fluidized beds. The ultimate fate of the one-dimensional, vertically propagating waves that dominate, as we shall see, in linear stability theory can also be explored by bifurcation analysis of the full equations of motion (Danckworth & Sundaresan, 1991; Göz, 1992), but again this has failed to reveal any qualitative differences between the gas and liquid fluidized cases. Thus, if the continuum equations of motion contain the physical mechanism needed to discriminate between bubbling and nonbubbling behaviour, it must manifest itself in motions that are not one-dimensional. Indeed, secondary, two-dimensional instabilities of the primary one-dimensional waves have been observed experimentally in water fluidized beds that are wide in one horizontal dimension (Didwania & Homsy, 1981), and they have also been studied theoretically by Didwania & Homsy (1982), Needham & Merkin (1984), Batchelor & Nitsche (1991), and Batchelor (1993). These studies suggest that the secondary instabilities might play a crucial role in determining whether or not the primary instabilities develop into bubbles. As we shall see in Chapter 5 this question appears to have been resolved by a series of direct computational studies of solutions of the full equations of motion. By supplementing the fully developed primary waves with a small perturbation, then generating the subsequent evolution of the motion by direct numerical integration of the equations of motion, Anderson, Sundaresan & Jackson (1995) were able to show that a solution with all the major attributes of a rising bubble developed in a bed of 200 μm diameter glass beads fluidized by ambient air, while in a bed of 1 mm diameter glass beads fluidized by water the solution degenerated into a complex motion with only small-amplitude density variations. Subsequently, making use of both bifurcation analysis and numerical integration, Glasser et al. (1996, 1997) succeeded in accounting for most of the features that characterize the contrasting dynamical behaviour of these two systems.

In this chapter we shall confine our attention to linear stability analysis of the uniformly fluidized state, deriving the full dispersion relations for propagating density waves of small amplitude and examining the conditions under which they grow or decay. We shall also look at the results of experimental studies of wave propagation in liquid fluidized beds to see to what extent they support the predictions and hence the continuum equations on which they are based. Work that goes beyond this to look at perturbations with structure in two spatial dimensions, and introduces nonlinear effects associated with the full equations of motion, will be covered in the next chapter.

4.2 Small Perturbations of the Uniformly Fluidized State

We now consider a base state representing a uniformly fluidized suspension of infinite extent, with the particles suspended at rest by an upward flow of fluid sufficient to provide a drag force that exactly balances their buoyant weight. In particular we shall examine the stability of this solution against small perturbations using conventional linear stability analysis. To do this we must, of course, have an explicit form for the equations of motion. For the continuity equations we have (2.14) and (2.15), namely

$$\frac{\partial \varepsilon}{\partial t} + \nabla \cdot (\varepsilon \mathbf{u}) = 0 \qquad (4.1)$$

and

$$\frac{\partial \phi}{\partial t} + \nabla \cdot (\phi \mathbf{v}) = 0. \qquad (4.2)$$

The fluid and particle phase momentum equations, (2.33) and (2.34), will be closed by adopting the Newtonian expressions (2.90) and (2.91) for the effective stress tensors. For the drag force we take the form (2.80), which is appropriate for small values of a Reynolds number based on the particle diameter and the slip velocity. The virtual mass force is given by (2.87) and the lift force will be set equal to zero. Then

$$n\mathbf{f}_2 = \beta(\phi)(\mathbf{u} - \mathbf{v}) + \frac{\rho_f \phi C_v}{1 - \phi} \left(\frac{D_f \mathbf{u}}{Dt} - \frac{D_f \mathbf{v}}{Dt} \right),$$

with

$$\beta(\phi) = \frac{(\rho_s - \rho_f)\phi g}{v_t (1 - \phi)^{n-1}},$$

where n is the Richardson–Zaki exponent (see (2.79)). Thus, the momentum equations take the following explicit form:

$$\rho_f \frac{D_f \mathbf{u}}{Dt} = -\beta(\mathbf{u} - \mathbf{v}) - \frac{\rho_f \phi C_v}{1 - \phi} \left(\frac{D_f \mathbf{u}}{Dt} - \frac{D_p \mathbf{v}}{Dt} \right) + \rho_f \mathbf{g} - \nabla p^f$$
$$+ \nabla \cdot \left\{ \mu^f \left(\nabla \mathbf{u} + \nabla \mathbf{u}^T - \frac{2}{3}(\nabla \cdot \mathbf{u})\mathbf{I} \right) \right\} \qquad (4.3)$$

and

$$\rho_s \phi \frac{D_p \mathbf{v}}{Dt} = \beta(\mathbf{u} - \mathbf{v}) + \frac{\rho_f \phi C_v}{1 - \phi}\left(\frac{D_f \mathbf{u}}{Dt} - \frac{D_p \mathbf{v}}{Dt}\right) + (\rho_s - \rho_f)\phi\mathbf{g} + \rho_f \phi \frac{D_f \mathbf{u}}{Dt}$$

$$- \nabla p^p + \nabla \cdot \left\{\mu^p \left(\nabla \mathbf{v} + \nabla \mathbf{v}^T - \frac{2}{3}(\nabla \cdot \mathbf{v})\mathbf{I}\right)\right\}. \tag{4.4}$$

We are interested in the stability of the uniformly fluidized suspension, with $\mathbf{u} = \mathbf{u}_0 = \mathbf{i}\,u_0$ and $\mathbf{v} = 0$, where $\mathbf{i}$ denotes the unit vector in the upward vertical direction. In this state the momentum balances (4.3) and (4.4) reduce to

$$\beta(\phi_0)u_0 + \rho_f g + dp_0^f/dx = 0 \tag{4.5}$$

and

$$\beta(\phi_0)u_0 - (\rho_s - \rho_f)\phi_0 g = 0, \tag{4.6}$$

where x is a coordinate measured vertically upward. This uniform state is then perturbed slightly, writing $\mathbf{u} = \mathbf{i}u_0 + \mathbf{u}_1$, $\mathbf{v} = \mathbf{v}_1$, and $p^f = p_0^f + p_1^f$, and the equations are linearized in the perturbations, giving

$$-\frac{\partial \phi_1}{\partial t} + (1 - \phi_0)\nabla \cdot \mathbf{u}_1 - u_0 \frac{\partial \phi_1}{\partial x} = 0, \tag{4.7}$$

$$\frac{\partial \phi_1}{\partial t} + \phi_0 \nabla \cdot \mathbf{v}_1 = 0, \tag{4.8}$$

$$\left[\rho_f + \frac{\rho_f \phi_0 C_v}{1 - \phi_0}\right]\left(\frac{\partial \mathbf{u}_1}{\partial t} + u_0 \frac{\partial \mathbf{u}_1}{\partial x}\right) - \frac{\rho_f \phi_0 C_v}{1 - \phi_0}\frac{\partial \mathbf{v}_1}{\partial t}$$

$$= -\mathbf{i}\rho_f g - \beta_0(\mathbf{u}_1 - \mathbf{v}_1) - \mathbf{i}\beta_0' \phi_1 - \nabla p_1^f$$

$$+ \mu_0^f \left(\nabla^2 \mathbf{u}_1 + \frac{1}{3}\nabla(\nabla \cdot \mathbf{u}_1)\right), \tag{4.9}$$

$$\left[\rho_s \phi_0 + \frac{\rho_f \phi_0 C_v}{1 - \phi_0}\right]\frac{\partial \mathbf{v}_1}{\partial t} - \rho_f \phi_0 \left[1 + \frac{C_v}{1 - \phi_0}\right]\left(\frac{\partial \mathbf{u}_1}{\partial t} + u_0 \frac{\partial \mathbf{u}_1}{\partial x}\right)$$

$$= -\mathbf{i}(\rho_s - \rho_f)g\phi_1 + \beta_0(\mathbf{u}_1 - \mathbf{v}_1) + \mathbf{i}u_0 \beta_0' \phi_1 - p_0^{p'}\nabla \phi_1$$

$$+ \mu_0^p \left(\nabla^2 \mathbf{v}_1 + \frac{1}{3}\nabla(\nabla \cdot \mathbf{v}_1)\right). \tag{4.10}$$

Here a suffix zero indicates that the term is evaluated at conditions corresponding to the base state, and a prime denotes differentiation with respect to ϕ. Equations (4.7) to (4.10) then form a set of eight linear partial differential equations in the six components of the vectors $\mathbf{u}_1$ and $\mathbf{v}_1$, together with the scalars ϕ_1 and p_1^f.

Solutions can be found by writing

$$(\mathbf{u}_1, \mathbf{v}_1, \phi_1, p_1) = \left(\hat{\mathbf{u}}_1, \hat{\mathbf{v}}_1, \hat{\phi}_1, \hat{p}_1^f\right) \exp(st) \exp(i\mathbf{k} \cdot \mathbf{x}),$$

where $\hat{\phi}_1$, $\hat{p}_1^f$, and the components of $\hat{\mathbf{u}}_1$ and $\hat{\mathbf{v}}_1$ are constants. Equations (4.7) to (4.10) then reduce to a set of eight linear algebraic equations with these as unknowns. A nonvanishing solution exists only if the determinant of their coefficients vanishes, and this condition relates s to the vector $\mathbf{k}$. If the components of $\mathbf{k}$ are real these solutions represent travelling waves with wave vector $\mathbf{k}$. For each $\mathbf{k}$ the determinantal equation has eight roots, which will, in general, be complex numbers. Thus the waves grow or decay as they propagate depending on the sign of the real part of s.

It is possible to combine the above equations to give a single partial differential equation to be satisfied by ϕ_1, and for this purpose the fluid-phase momentum equation (4.9) is not needed. Taking the divergence of (4.10) and eliminating $\nabla \cdot \mathbf{u}_1$ and $\nabla \cdot \mathbf{v}_1$ from the result with the help of (4.7) and (4.8) we find that

$$\frac{1}{\tau_0}\left[\frac{\partial \phi_1}{\partial t} + V\frac{\partial \phi_1}{\partial x}\right] - \nu_e \frac{\partial}{\partial t}\nabla^2\phi_1 + \frac{\partial^2 \phi_1}{\partial t^2} + c_1 \frac{\partial^2 \phi_1}{\partial x \, \partial t}$$

$$+ \frac{1}{2}c_1 u_0 \frac{\partial^2 \phi_1}{\partial x^2} - \frac{p_0^{p'}}{A\rho_s}\nabla^2\phi_1 = 0, \tag{4.11}$$

where the new coefficients introduced are given by

$$A = 1 + \frac{\rho_f}{\rho_s}\frac{\phi_0}{1 - \phi_0}\left(1 + \frac{C_v}{\phi_0(1 - \phi_0)}\right), \tag{4.12}$$

$$\tau_0 = \frac{A\rho_s\phi_0(1 - \phi_0)}{\beta_0} = \frac{A(1 - \phi_0)u_0}{(1 - \rho_f/\rho_s)g}, \tag{4.13}$$

$$\nu_e = \frac{4\mu_0^p/3}{A\phi_0\rho_s}, \tag{4.14}$$

$$V = u_0\left[2\phi_0 - 1 + \frac{\phi_0(1 - \phi_0)\beta_0'}{\beta_0}\right] = n\phi_0 u_0, \tag{4.15}$$

$$c_1 = \frac{2u_0}{A}\frac{\rho_f}{\rho_s}\frac{\phi_0}{1 - \phi_0}\left(1 + \frac{C_v}{1 - \phi_0}\right). \tag{4.16}$$

The second of Equations (4.15) follows upon introducing the explicit expression for β given above.

Substituting a trial solution $\phi_1 = \hat{\phi}_1 \exp(st) \exp(i\mathbf{k} \cdot \mathbf{x})$ into (4.11) leads to a quadratic equation relating s to $\mathbf{k}$, since (4.11) is second order in time. The two roots of this equation (which are, of course, two of the eight roots of the determinantal equation referred to above) then yield nonvanishing solutions for ϕ_1. The remaining six roots of the determinantal equation must, therefore, belong to solutions for which $\phi_1 = 0$. But then it follows from Equations (4.7) and (4.8) that $\nabla \cdot \mathbf{u}_1 = \nabla \cdot \mathbf{v}_1 = 0$ and, consequently, $\mathbf{k} \cdot \mathbf{u}_1 = \mathbf{k} \cdot \mathbf{v}_1 = 0$. Thus these solutions are transverse waves that generate no disturbances of the particle concentration. We shall not study these but will confine our attention to the solutions of (4.11), which should be visible as waves of varying particle concentration.

Before going further it is useful to rewrite (4.11) in dimensionless form. There are several different choices for the length and time scales needed to do this. For the purpose of stability theory the equations are reduced to their simplest form if dimensionless Cartesian coordinates and a dimensionless time are defined by

$$(\chi, \eta, \zeta) = \frac{(x, y, z)}{\sqrt{\nu_e \tau_0}}, \qquad \tau = \frac{V}{\sqrt{\nu_e \tau_0}}. \tag{4.17}$$

Then (4.11) becomes

$$\frac{\partial \phi_1}{\partial \tau} + \frac{\partial \phi_1}{\partial \chi} - \frac{\partial}{\partial \tau} \nabla^2 \phi_1 + \frac{V}{\nu}$$

$$\times \left\{ \frac{\partial^2 \phi_1}{\partial \tau^2} + \frac{c_1}{V} \frac{\partial^2 \phi_1}{\partial \tau \partial \chi} + \frac{1}{2} \frac{c_1 u_0}{V^2} \frac{\partial^2 \phi_1}{\partial \chi^2} - \frac{p_0^{p'}}{A \rho_s V^2} \nabla^2 \phi_1 \right\} = 0, \tag{4.18}$$

where ∇^2 denotes the Laplacian in the scaled space coordinates and v is a parameter with the dimensions of velocity defined by

$$v = \sqrt{\frac{\nu_e}{\tau_0}} = \sqrt{\frac{\frac{4}{3} \mu_0^p (1 - \rho_f / \rho_s) g}{A^2 u_0 \rho_s \phi_0 (1 - \phi_0)}}. \tag{4.19}$$

We seek a solution of (4.18) in the form of plane waves with wave vector $\boldsymbol{\kappa}$:

$$\phi_1 = \hat{\phi}_1 \exp(\sigma \tau) \exp(i \boldsymbol{\kappa} \cdot \boldsymbol{\chi}), \tag{4.20}$$

where $\boldsymbol{\chi} = (\chi, \eta, \zeta)$. Then σ must satisfy the following quadratic equation:

$$\sigma^2 + \left[\frac{v}{V} (1 + \kappa^2) + i \kappa \frac{c_1}{V} \cos \theta \right] \sigma$$

$$+ \left[\frac{p_0^{p'}}{A \rho_s} - \frac{1}{2} c_1 u_0 \cos^2 \theta \right] \frac{\kappa^2}{V^2} + i \kappa \frac{v}{V} \cos \theta = 0, \tag{4.21}$$

where $\kappa = |\boldsymbol{\kappa}|$ and θ denotes the angle subtended between the vector $\boldsymbol{\kappa}$ and the upward vertical direction. Equation (4.21) has the form

$$\sigma^2 + (p + iq)\sigma + P + i Q = 0, \tag{4.21'}$$

where

$$p = (1 + \kappa^2)v/V, \qquad q = \cos\theta\, \kappa c_1/V,$$
$$P = (d - e\cos^2\theta)\kappa^2/V^2, \quad Q = \cos\theta\, \kappa v/V, \tag{4.22}$$

with $d = p_0^{p'}/A\rho_s$ and $e = \frac{1}{2}c_1 u_0$.

The roots of (4.21′) are given by

$$2\sigma = -(p + iq) \pm \sqrt{(p^2 - q^2 - 4P) + i(2pq - 4Q)} \tag{4.23}$$

and the root of interest here is that with the larger real part. Its real and imaginary parts are given by

$$\alpha = \mathrm{Re}(\sigma) = \frac{1}{2}[M - p], \qquad \mathrm{Im}(\sigma) = \frac{1}{2}[N - q], \tag{4.24}$$

where

$$M = \sqrt{\left(\frac{\sqrt{a^2 + b^2} + a}{2}\right)}, \quad N = \mathrm{sgn}(b)\sqrt{\left(\frac{\sqrt{a^2 + b^2} - a}{2}\right)}, \tag{4.25}$$

with the positive value taken for each square root, and

$$a = p^2 - q^2 - 4P, \quad b = 2pq - 4Q.$$

Using (4.22) these expressions become

$$a = \left(\frac{v}{V}\right)^2 (1 + \kappa^2)^2 + \frac{\kappa^2}{V^2}\left\{\left(2c_1 u_0 - c_1^2\right)\cos^2\theta - 4d\right\}, \tag{4.26}$$

$$b = 2\kappa\frac{v}{V}\left\{(1 + \kappa^2)\frac{c_1}{V} - 2\right\}\cos\theta. \tag{4.27}$$

Equations (4.24) to (4.27) provide a prescription for computing the dimensionless growth rate α (=Re(σ)) and velocity of propagation v (= $-$Im(σ)/κ) for a small-amplitude disturbance wave with any specified wave vector κ. The corresponding functions $\alpha(\kappa, \theta)$ and $v(\kappa, \theta)$ are referred to as the *dispersion relations* for small disturbances, and we shall now examine their form in some detail.

4.3 The Dispersion Relations

First consider the dependence of the growth rate α on the angle θ, for given values of κ and the other parameters of the system. Since p is independent of θ the first of Equations (4.24) shows that α is largest when M is largest. But M is positive, so M is largest when M^2 is largest. Thus α is largest at the value of θ that maximizes M^2. But from (4.25)

$$2\frac{d(M^2)}{d\theta} = \left(1 + \frac{a}{\sqrt{a^2+b^2}}\right)\frac{da}{d\theta} + \frac{b}{\sqrt{a^2+b^2}}\frac{db}{d\theta}. \qquad (4.28)$$

Now from (4.27) it is seen immediately that $b\, db/d\theta < 0$ for $0 < \theta < \pi/2$, so the second term on the right-hand side of (4.28) is negative throughout this interval of θ. Also, from (4.26)

$$\frac{da}{d\theta} = -\frac{\kappa^2}{V^2}c_1(2u_0 - c_1)\sin 2\theta,$$

so $da/d\theta$ is opposite in sign to $(2u_0 - c_1)$ in $0 < \theta < \pi/2$. However, from the definition of c_1 in (4.16)

$$2u_0 - c_1 = \frac{2u_0}{A}\left[1 + \frac{\rho_f}{\rho_s}\frac{C_v}{1-\phi_0}\right] > 0.$$

Hence it follows that $da/d\theta < 0$ throughout this interval. Furthermore, the expression in brackets that multiplies $da/d\theta$ on the right-hand side of (4.28) is clearly positive for all values of a; so the first term on the right-hand side of (4.28) is also negative. Consequently, $d(M^2)/d\theta < 0$ for $0 < \theta < \pi/2$, showing that M, and hence α, is largest when $\theta = 0$. Thus, for any given value of κ, the growth rate is biggest for a disturbance that propagates in the upward vertical direction. We may, therefore, limit attention to these disturbances, replacing (4.26) and (4.27) by

$$a = \left(\frac{v}{V}\right)^2(1+\kappa^2)^2 + \frac{\kappa^2}{V^2}\{c_1(2u_0 - c_1) - 4d\} \qquad (4.26°)$$

and

$$b = \frac{2\kappa v}{V}\left\{(1+\kappa^2)\frac{c_1}{V} - 2\right\}. \qquad (4.27°)$$

Now from (4.24), $\alpha > 0$ and the disturbance grows exponentially provided $M > p$. But since M and p are both positive quantities this condition is equivalent to $M^2 > p^2$, or

$$\frac{\sqrt{a^2+b^2}+a}{2} > \left(\frac{v}{V}\right)^2(1+\kappa^2)^2.$$

Let us first suppose that

$$4\frac{p_0^{p'}}{A\rho_s} = 4d < c_1(2u_0 - c_1)$$

or, equivalently,

$$d - \frac{1}{2}u_0 c_1 < -c_1^2/4. \tag{4.29}$$

Then from (4.26°), $a > 0$ for all values of κ, and the above condition for α to be positive may be written as

$$a\left[\sqrt{1 + \frac{b^2}{a^2}} + 1\right] > \left(\frac{v}{V}\right)^2 (1 + \kappa^2)^2$$

or, putting in the explicit expression (4.26°) for a,

$$\left(\left(\frac{v}{V}\right)^2 (1 + \kappa^2)^2 + \frac{\kappa^2}{V^2}\{c_1(2u_0 - c_1) - 4d\}\right)$$

$$\times \left(\frac{\sqrt{1 + (b^2/a^2)} + 1}{2}\right) > \left(\frac{v}{V}\right)^2 (1 + \kappa^2)^2.$$

Whenever (4.29) is satisfied the first factor on the left-hand side is itself larger than the right-hand side, while the second factor is positive and larger than one. Thus this inequality is satisfied, showing that (4.29) is a *sufficient* condition for α to be positive for all values of κ, in other words, for disturbances of all wavelengths to grow.

Now turn to cases for which $p_0^{p'}$ is large enough that (4.29) is violated. Since we are limiting attention to $\theta = 0$ the disturbances are independent of the coordinates η and ζ so (4.18) reduces to

$$\frac{\partial \phi_1}{\partial \tau} + \frac{\partial \phi_1}{\partial \chi} - \frac{\partial}{\partial \tau}\frac{\partial^2 \phi_1}{\partial \chi^2} + \frac{V}{v}\left(\frac{\partial^2 \phi_1}{\partial \tau^2} + \frac{c_1}{V}\frac{\partial^2 \phi_1}{\partial \tau \partial \chi} - \frac{d - \frac{1}{2}c_1 u_0}{V^2}\frac{\partial^2 \phi_1}{\partial \chi^2}\right) = 0$$

or

$$\frac{\partial \phi_1}{\partial \tau} + \frac{\partial \phi_1}{\partial \chi} - \frac{\partial}{\partial \tau}\frac{\partial^2 \phi_1}{\partial \chi^2} + \frac{V}{v}\left(\frac{\partial}{\partial \tau} + \frac{c_u}{V}\frac{\partial}{\partial \chi}\right)\left(\frac{\partial}{\partial \tau} - \frac{c_d}{V}\frac{\partial}{\partial \chi}\right)\phi_1 = 0, \tag{4.30}$$

where

$$c_u = \frac{1}{2}\left\{\sqrt{c_1^2 - 2(u_0 c_1 - 2d)} + c_1\right\},$$

$$c_d = \frac{1}{2}\left\{\sqrt{c_1^2 - 2(u_0 c_1 - 2d)} - c_1\right\}. \tag{4.31}$$

In view of our assumption that (4.29) is violated, both c_u and c_d are real and c_u is positive.

Note that an equation obtained by erasing all but the first two terms of (4.30), namely

$$\frac{\partial \phi_1}{\partial \tau} + \frac{\partial \phi_1}{\partial \chi} = 0,$$

represents a wave motion travelling upward with unit dimensionless velocity (i.e., with dimensional velocity V). Hypothetical waves of this sort are called *continuity waves*; they are analogous to the waves of concentration that develop in traffic on a highway. In contrast, an equation obtained by retaining only the last term of (4.30), namely

$$\left(\frac{\partial}{\partial \tau} + \frac{c_u}{V} \frac{\partial}{\partial \chi} \right) \left(\frac{\partial}{\partial \tau} - \frac{c_d}{V} \frac{\partial}{\partial \chi} \right) \phi_1 = 0,$$

represents a pair of waves, with dimensionless velocities c_u/V and $-c_d/V$, respectively (i.e., with dimensional velocities c_u and $-c_d$). These are called *dynamical waves* and they are analogous to elastic waves in a solid. However, it must be emphasized that the concept of dynamical waves, and their velocities, appears in the context of fluidized beds only when $p_0^{p'}$ is large enough that the inequality (4.29) is violated.

Now if we seek plane wave solutions of (4.30) in the form (4.20), with $\kappa = (\kappa, 0, 0)$, the secular equation (4.21) can be expressed as

$$\sigma(1 + \kappa^2) + i\kappa + \frac{V}{\upsilon}\left(\sigma + i\kappa \frac{c_u}{V} \right)\left(\sigma - i\kappa \frac{c_d}{V} \right) = 0. \qquad (4.32)$$

Writing $\sigma = \alpha - i\kappa\upsilon$ allows us to separate the real and imaginary parts of (4.32), giving

$$\alpha(1 + \kappa^2) = \frac{V}{\upsilon}\left\{ \kappa^2\left(\upsilon - \frac{c_u}{V} \right)\left(\upsilon + \frac{c_d}{V} \right) - \alpha^2 \right\} \qquad (4.33)$$

and

$$\upsilon\left(1 + \kappa^2 + 2\frac{V}{\upsilon}\alpha \right) = 1 + \frac{V}{\upsilon}\left(\frac{c_u - c_d}{V} \right)\alpha. \qquad (4.34)$$

Since (4.29) is violated we can no longer conclude that α is positive for all values of κ, so we seek values of κ for which α vanishes. Setting $\alpha = 0$ in (4.33) and (4.34) and eliminating υ between the resulting equations we find that

$$\kappa^2\left(\frac{1}{1 + \kappa^2} - \frac{c_u}{V} \right)\left(\frac{1}{1 + \kappa^2} + \frac{c_d}{V} \right) = 0,$$

which has roots $\kappa = 0, \kappa_1, \kappa_2$, where

$$\kappa_1, \kappa_2 = \sqrt{\frac{V}{c_u} - 1}, \quad \sqrt{-\frac{V}{c_d} - 1}. \qquad (4.35)$$

When (4.29) is satisfied there are no real values of c_u and c_d and consequently, as we know already, α is positive for all values of $\kappa > 0$ and there are no real roots κ_1 and κ_2. These roots first appear as a coincident pair when $d - \frac{1}{2}u_0c_1 = -c_1^2/4$. Then $c_u = -c_d = c_1/2$ and consequently $\kappa_1 = \kappa_2 = (2V/c_1 - 1)^{1/2}$. If d is increased further the roots separate, κ_1 decreases and κ_2 increases, and $\kappa_2 \to \infty$ when $d - \frac{1}{2}u_0c_1 \to 0$. For still larger values of d the quantity $d - \frac{1}{2}u_0c_1$ becomes positive, c_d becomes positive, and κ_2 is no longer real. When $d - \frac{1}{2}u_0c_1$ increases to $V(V - c_1)$ the root $\kappa_1 \to 0$ and, finally, for still larger values of d, $\alpha(\kappa)$ once more has no real zeros in $\kappa > 0$, but now it is negative for all nonvanishing values of κ. (When κ_1 and κ_2 are both real and positive, α is negative for $\kappa_1 < \kappa < \kappa_2$ and is positive for other values of $\kappa > 0$.)

To complete the qualitative picture of the function $\alpha(\kappa)$ we need only examine its limiting behaviour as $\kappa \to 0$ and as $\kappa \to \infty$. This is easily obtained from the explicit expressions provided by (4.24)–(4.27), and the limiting values of the dimensionless propagation velocity can also be found at the same time. The results are

$$\alpha \approx \kappa^2 \left[\frac{V(V - c_1) + \frac{1}{2}u_0c_1 - d}{vV} \right], \quad v \to 1 \quad \text{as } \kappa \to 0 \qquad (4.36)$$

and

$$\alpha \to \frac{\frac{1}{2}u_0c_1 - d}{vV}, \quad v \to \frac{c_1}{2V} \quad \text{as } \kappa \to \infty. \qquad (4.37)$$

All the above results can now be organized into a systematic description of the form of the function $\alpha(\kappa)$ and how this changes as d is increased progressively from zero. There are four qualitatively distinct intervals, as follows.

(a) $d - \frac{1}{2}u_0c_1 < -c_1^2/4$ α vanishes for $\kappa = 0$, is positive for all other values of κ, and approaches the positive limit given by (4.37) as $\kappa \to \infty$.

(b) $-c_1^2/4 < d - \frac{1}{2}u_0c_1 < 0$ There exist positive wavenumbers κ_1 and κ_2 ($\kappa_1 < \kappa_2$) such that α vanishes for $\kappa = 0, \kappa_1$, and κ_2. κ_1 decreases and κ_2 increases as d is increased, and they lie within the following intervals:

$$\sqrt{\frac{2V}{c_1} - 1} > \kappa_1 > \sqrt{\frac{V}{c_1} - 1}, \quad \sqrt{\frac{2V}{c_1} - 1} < \kappa_2 < \infty.$$

Real values of c_u and c_d also exist and lie within the intervals

$$c_1/2 < c_u < c_1, \quad -c_1/2 < c_d < 0.$$

In each of these inequalities the left-hand limits are taken when $d - \frac{1}{2}u_0c_1 = -c_1^2/4$ and the right-hand limits when $d - \frac{1}{2}u_0c_1 = 0$.

α takes positive values for $0 < \kappa < \kappa_1$ and for $\kappa > \kappa_2$, while it takes negative values for $\kappa_1 < \kappa < \kappa_2$. It approaches the positive limit, given by (4.37), as $\kappa \to \infty$.

(c) $0 < d - \frac{1}{2}u_0c_1 < V(V - c_1)$ There exists one positive number κ_1 such that α vanishes for $\kappa = 0$ and κ_1. κ_1 decreases as d is increased, and it lies within the following interval:

$$\sqrt{\frac{V}{c_1} - 1} > \kappa_1 > 0.$$

Real values of c_u and c_d also exist and lie within the intervals

$$c_1 < c_u < V, \quad 0 < c_d < V - c_1.$$

In each of these inequalities the left-hand limit is taken when $d - \frac{1}{2}u_0c_1 = 0$ and the right-hand limit when $d - \frac{1}{2}u_0c_1 = V(V - c_1)$.

α takes positive values for $0 < \kappa < \kappa_1$ and negative values for $\kappa > \kappa_1$. It approaches a negative limit, given by (4.37), as $\kappa \to \infty$.

(d) $V(V - c_1) < d - \frac{1}{2}u_0c_1$ α vanishes for $\kappa = 0$, is negative for all other values of κ, and approaches the negative limit given by (4.37) as $\kappa \to \infty$.

The behaviour of the function $\upsilon(\kappa)$ is less interesting. As κ is increased υ changes monotonically between the limits given in (4.36) and (4.37).

Note that (4.36) shows that the dimensional speed of propagation of the disturbance approaches V as the wavenumber tends to zero; in other words the disturbance becomes a continuity wave when its wavelength becomes very long. It can also be shown that the speed of propagation of the neutrally stable disturbance with $\kappa = \kappa_1$ is c_u. Thus it corresponds to the upwardly propagating dynamic wave.

The behaviour summarized in (a) to (d) above is illustrated by Figures 4.1(i) to 4.1(vii). These show α as a function of κ for a system with $c_1/V = 0.5$, $v/V = 2$, and successively increasing values of $(d - \frac{1}{2}u_0c_1)/V^2$. Figures 4.1(i), (ii), and (iii) belong to case (a), 4.1(iv) belongs to case (b), 4.1(v) and (vi) belong to case (c), and 4.1(vii) belongs to case (d). Figure 4.1(vii) represents a stable case, where α decreases monotonically from zero as κ increases. In all the other figures there exist values of κ for which α is positive, so the bed is unstable. At the smallest values of $(d - \frac{1}{2}u_0c_1)/V^2$, α increases monotonically from zero as κ increases, so the most rapidly amplified disturbances are those with vanishingly short wavelength. This behaviour is illustrated by Figure 4.1(i). For larger values of $(d - \frac{1}{2}u_0c_1)/V^2$ the growth rate has a stationary maximum and, except for Figure 4.1(ii) and (iii), this is also the largest value attained by α. In Figures 4.1(i), (ii), and (iii) α assumes its largest value in the limit $\kappa \to \infty$. However, for most systems this limit is not of practical interest. When there is a stationary maximum the corresponding value of α can usually be exceeded only for values of κ so large as to be unrealistic, in the sense that the

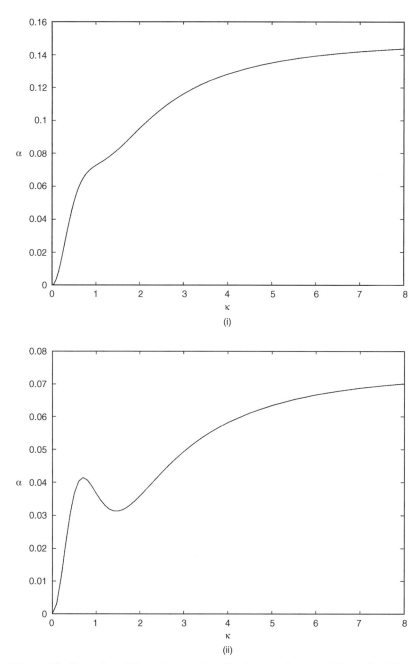

Figure 4.1. Examples of dispersion relations for the growth rate. $c_1/V = 0.5$, $v/V = 2.0$ in all cases. (i) $(d - \frac{1}{2}u_0c_1)/V^2 = -0.3$. (ii) $(d - \frac{1}{2}u_0c_1)/V^2 = -0.15$. (iii) $(d - \frac{1}{2}u_0c_1)/V^2 = -0.065$. (iv) $(d - \frac{1}{2}u_0c_1)/V^2 = -0.03$. (v) $(d - \frac{1}{2}u_0c_1)/V^2 = 0.01$. (vi) $(d - \frac{1}{2}u_0c_1)/V^2 = 0.475$. (vii) $(d - \frac{1}{2}u_0c_1)/V^2 = 0.525$.

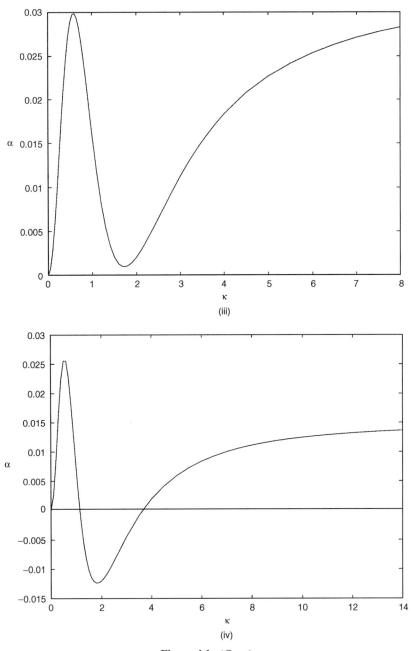

Figure 4.1. (*Cont.*)

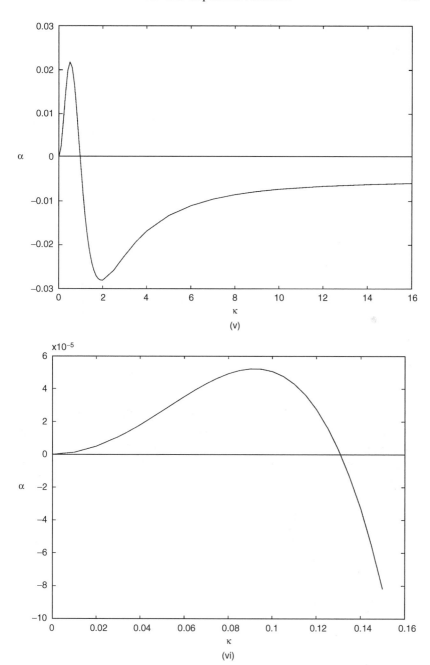

Figure 4.1. (*Cont.*)

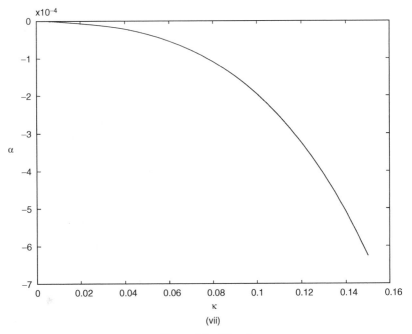

Figure 4.1. (*Cont.*)

corresponding physical wavelength is of the same order as the particle size. For such disturbances the continuum equations are clearly inappropriate.

Some treatments of small disturbances, for example, those of Jackson (1963) and Foscolo & Gibilaro (1984, 1987), are based on an inviscid form of the particle momentum equation, so we should also examine the form of the dispersion relations when $v_e = 0$. In this case our choice of dimensionless variables is clearly inappropriate. However, if we return to the dimensional form (4.11) the differential equation satisfied by ϕ_1 in the inviscid case is

$$\frac{1}{\tau_0}\left[\frac{\partial\phi_1}{\partial t} + V\frac{\partial\phi_1}{\partial x}\right] + \frac{\partial^2\phi_1}{\partial t^2} + c_1\frac{\partial^2\phi_1}{\partial x\partial t} - \left(d - \frac{1}{2}u_0c_1\right)\frac{\partial^2\phi_1}{\partial x^2} = 0,$$

and a solution of the form $\phi_1 = \hat{\phi}_1 e^{st} e^{ikx}$ exists only when s satisfies the following quadratic equation:

$$s^2 + \left(\frac{1}{\tau_0} + ikc_1\right)s + \left(d - \frac{1}{2}u_0c_1\right)k^2 + \frac{ikV}{\tau_0} = 0.$$

If $s = m + in$, the real and imaginary parts of this equation are

$$m^2 - n^2 + \frac{m}{\tau_0} - nkc_1 + \left(d - \frac{1}{2}u_0c_1\right)k^2 = 0$$

and

$$2mn + \frac{n}{\tau_0} + kmc_1 + \frac{kV}{\tau_0} = 0.$$

We first examine the conditions for neutral stability. Setting $m = 0$ and eliminating n between the real and imaginary parts of the above secular equation then gives the condition

$$k^2 \left[d - \frac{1}{2}u_0c_1 - V(V - c_1) \right] = 0.$$

This is satisfied for $k = 0$, of course, whatever the value of d. It is also satisfied for all values of k when $d - \frac{1}{2}u_0c_1 = V(V - c_1)$. Thus, when this condition is satisfied, the dispersion relation for $\mathrm{Re}(s)$ is represented geometrically in the $(k, \mathrm{Re}(s))$-plane by the k axis. Unlike the more general case, where $\mu^p \neq 0$, there are no values of $k > 0$ at which $\mathrm{Re}(s) = 0$ when $d - \frac{1}{2}u_0c_1 \neq V(V - c_1)$. Thus $\mathrm{Re}(s)$ retains the same sign for all positive values of k.

Careful examination of the limiting behaviour of $\mathrm{Re}(s)$ for large k reveals that

$$\mathrm{Re}(s) \approx k \left(\frac{u_0c_1 - 2d - \frac{1}{2}c_1^2}{2} \right)^{1/2} \quad \text{as } k \to \infty \quad \text{when } d - \frac{1}{2}u_0c_1 < -c_1^2/4$$

and

$$\mathrm{Re}(s) \to \frac{1}{2\tau_0} \left\{ \frac{2V - c_1}{\left(c_1^2 - 2u_0c_1 + 4d \right)^{1/2}} - 1 \right\} \quad \text{as } k \to \infty \quad \text{when}$$

$$d - \frac{1}{2}u_0c_1 > -c_1^2/4.$$

Thus $\mathrm{Re}(s)$ approaches a negative asymptote as $k \to \infty$ when $d - \frac{1}{2}u_0c_1 > V(V - c_1)$. It approaches a positive asymptote as $k \to \infty$ when $-c_1^2/4 < d - \frac{1}{2}u_0c_1 < V(V - c_1)$, and the value of this asymptote increases without bound as $d - \frac{1}{2}u_0c_1 \to -c_1^2/4$. For $d - \frac{1}{2}u_0c_1 < -c_1^2/4$, $\mathrm{Re}(s)$ becomes proportional to k when k is large and it therefore increases without bound. This behaviour is sketched in Figure 4.2, which should be contrasted with Figure 4.1.

For this inviscid case note that there is no bounded value of k for which the disturbances grow most rapidly. A nonvanishing value for the viscosity μ^p is therefore necessary to establish a characteristic length scale for the disturbances.

4.4 The Stability of the Uniformly Fluidized State

The above discussion of the dispersion relations has shown that disturbances with positive growth rates ($\alpha > 0$) always exist for some values of the

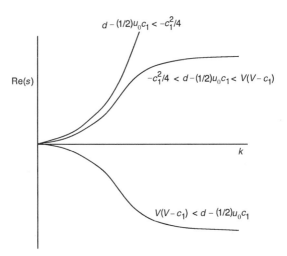

Figure 4.2. Dispersion relations for the inviscid case.

wavenumber unless

$$d - \frac{1}{2}u_0 c_1 > V(V - c_1) \tag{4.38}$$

or, inserting the value of d,

$$\frac{p_0^{p'}}{A\rho_s} > V(V - c_1) + \frac{1}{2}u_0 c_1. \tag{4.39}$$

This inequality is a necessary condition for stability so the value of $p_0^{p'}$ must be sufficiently large if the bed is to be stable. The possibility of securing a stable bed in this way was first pointed out by Garg & Pritchett (1975). Putting in the explicit expressions for V and c_1, we can rewrite (4.39) in the following form:

$$\frac{p_0^{p'}}{\rho_s u_0^2} > n^2 \phi_0^2 + \frac{\rho_f}{\rho_s} \frac{\phi_0}{1 - \phi_0} \left\{ (n\phi_0 - 1)^2 + C_v \left[1 + \frac{(n-1)^2 \phi_0}{1 - \phi_0} \right] \right\}.$$

Furthermore, u_0 can be related to ϕ_0 using the Richardson–Zaki relation (see (2.63) and (2.64)), with the result $u_0 = v_t (1 - \phi_0)^{n-1}$. Thus the above becomes

$$\frac{p_0^{p'}}{\rho_s v_t^2} > n^2 \phi_0^2 (1 - \phi_0)^{2(n-1)}$$

$$\times \left[1 + \frac{\rho_f}{\rho_s} \frac{\phi_0}{(1 - \phi_0)} \left\{ \left(1 - \frac{1}{n\phi_0} \right)^2 + \frac{C_v}{n^2 \phi_0^2} \left(1 + \frac{(n-1)^2 \phi_0}{1 - \phi_0} \right) \right\} \right].$$
$$\tag{4.40}$$

Note that the value of the effective viscosity μ_0^p has no effect on the stability criterion. Indeed for an unstable bed, with $V > c_u$, (4.35) shows that the value of κ_1, where the dispersion curve crosses the axis $\alpha = 0$, is independent of the effective viscosity. However, changing the value of μ_0^p will change the maximum value attained by the dimensionless growth rate, α. It is important to emphasize that these statements refer to the *dimensionless* propagation parameters. The dimensional critical wavenumber is given by $k_1 = \kappa_1/(\nu_e \tau_0)^{1/2}$ so, since $\nu_e \propto \mu_0^p$, the value of k_1 increases without bound as the effective viscosity is reduced to zero. Thus, in the limit of zero effective viscosity, the growth rate retains the same sign for all bounded values of the (dimensional) wavenumber, as we have seen in the previous section.

In order to make (4.40) into a completely explicit stability criterion all that is needed is an expression for $p_0^{p'}$. Only two publications give expressions for this quantity as an explicit function of ϕ. A remarkably simple form was proposed by Foscolo & Gibilaro (1987), namely

$$p_0^{p'} = 3.2 g d_p \phi (\rho_s - \rho_f), \tag{4.41}$$

where d_p is here the particle diameter. Then, neglecting all but the first term on the right-hand side of (4.40), and using (4.41) for $p_0^{p'}$, we recover the stability condition in the form posed by the above authors, namely

$$\frac{(g d_p)^{1/2}}{v_t} \left(\frac{\rho_s - \rho_f}{\rho_s} \right)^{1/2} > 0.56 n \phi_0^{1/2} (1 - \phi_0)^{n-1}. \tag{4.42}$$

The terms in (4.40) that have been neglected to obtain (4.42) arise from differences between our particle phase momentum equation, (4.10), and that of Foscolo & Gibilaro, given as Equation (2.77). For example, the latter authors do not include the effect of virtual mass.

Batchelor, however, suggests the following, much more complicated expression for $p_0^{p'}$:

$$p_0^{p'} = \rho_s v_t^2 \left\{ \frac{d}{d\phi} \left[\frac{\phi^2}{\phi_m} \left(1 - \frac{\phi}{\phi_m} \right) (1 - \phi)^{2p} \right] + \frac{\alpha \gamma a g}{v_t^2} \left(1 - \frac{\rho_f}{\rho_s} \right) \right\}, \tag{4.43}$$

where α, γ, and p are parameters that take the values $\alpha = \gamma = 1$ and $p = 5.5$ when the Reynolds number is small for an isolated particle falling at its terminal velocity. Since (4.40) is itself based on an expression for V that is correct only for small Reynolds number, (4.43) can be used consistently in (4.40) with this choice of parameter values. (Batchelor's expression (4.43) is not limited to small values of the Reynolds number. He tabulates values of α, γ, and p for Reynolds numbers up to 2,000.) The first term in (4.43) arises from the transport of momentum by fluctuations in particle velocities and the second from hydrodynamic diffusion. The form of each depends on some arbitrary, but reasonable assumptions. To obtain the first term it has to be assumed that the mean square fluctuation in particle velocities is given by an expression of the form

$$\langle v'^2 \rangle = H(\phi)U^2 \quad \text{with} \quad U = v_t(1 - \phi)^p.$$

Batchelor then argues further that the fluctuations in velocity must vanish when $\phi \to 0$, since each particle is then unaffected by the presence of neighbours, and also when $\phi \to \phi_m$, since the particles are then locked into a structure of mutual contacts. The simplest form for $H(\phi)$ consistent with these limits is $H = (\phi/\phi_m)(1 - \phi/\phi_m)$. The second term in (4.43) is formulated in terms of the particle diffusion coefficient D as

$$\frac{\gamma g}{U} \left(1 - \frac{\rho_f}{\rho_s} \right) D,$$

and the diffusion coefficient is postulated to be given by $D = \alpha a v_t (1 - \phi)^p$, leading to the form shown explicitly in (4.43).

The simplicity of the Foscolo–Gibilaro expression (4.41) for $p_0^{p'}$ is appealing, compared with the more cumbersome form of (4.43). However, the validity of its physical basis is questionable; as Batchelor writes,"all these calculations (*of Foscolo & Gibilaro*) are flawed by the misconception that the elasticity of the particle configuration is related to the dependence of the mean fluid drag force on the particle concentration". In contrast to the approaches of Foscolo & Gibilaro, and of Batchelor, who postulate a fluid mechanical origin for p_0^p, Rietema argues that the effective elasticity of the particle phase is primarily a result of solid–solid forces, particularly in gas fluidized beds. This view is expounded at length in a number of papers and a monograph (Rietema, 1991), and we will return to it in Section 4.7.

Using (4.31) the condition (4.38) for stability can be written $c_u > V$, where c_u is the speed of the upward-travelling dynamic wave and V is the speed of the continuity wave, that is, the wave of concentration and velocity fluctuations that can propagate in a suspension without inertia, where the particle velocity is simply a function of concentration (Kynch, 1952). This form for the stability criterion, which originated with Wallis (1969), has since been quoted widely. However, it has tempted a number of authors to the quite unjustified conclusion that $c_u < V$ is a necessary condition for instability. This is far from true since we have seen that the bed is always unstable when the inequality (4.29) is satisfied, and in that case c_u does not even exist! Indeed, the early analysis of Jackson (1963), which demonstrated the existence of the instability, was based on dynamical equations for which there is no dynamic wave. As pointed out by Batchelor (1988) all that is needed to generate an instability is inertia and the

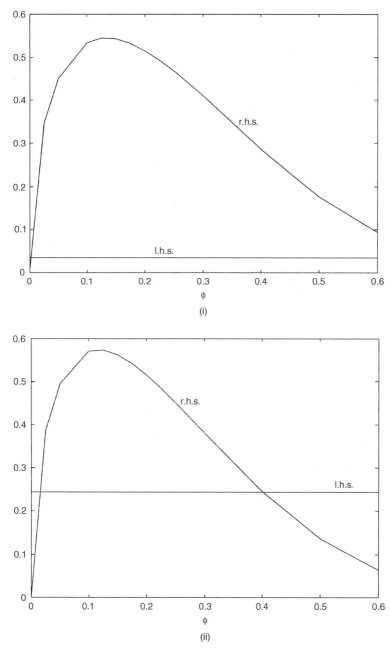

Figure 4.3. Left- and right-hand sides of the Foscolo–Gibilaro criterion. (i) 200 μm diameter glass beads in air. (ii) Fluid cracking catalyst in air. Mean diameter of 60 μm. (iii) 1 mm diameter glass beads in water.

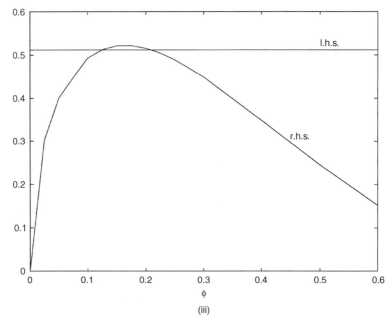

(iii)

Figure 4.3. (*Cont.*)

existence of a continuity wave. In the absence of inertia the relative velocity of the two phases is simply a function of the void fraction, and the form of this function determines the continuity wave speed. However, inertia causes the fluctuations in particle velocity to be out of phase with the fluctuations in void fraction, and it is this that leads to their amplification. In view of these remarks the frequently drawn analogies among fluid bed instabilities, sound waves in gases, and waves in traffic flow should be viewed with caution.

Let us now examine the predictions of the above stability criteria for one or two specific cases. Figures 4.3(i), (ii), and (iii) show the left- and right-hand sides of the Foscolo–Gibilaro criterion (4.42) for 200 μm diameter glass beads fluidized by air at ambient conditions, 60 μm mean diameter FCC particles fluidized by air at ambient conditions, and 1 mm diameter glass beads fluidized by water, respectively. The criterion indicates that the air fluidized glass beads should be unstable for all but extremely small values of ϕ, and this is consistent with the observation that such beds bubble freely as soon as they are fluidized. For the bed of FCC particles, in contrast, there is an interval of stability extending down to $\phi \approx 0.4$ and a second interval of stability for $\phi < 0.01$. This appears to be consistent with the well-established fact that beds of this type of particle can be expanded significantly before bubbling becomes visible, though they do bubble when expanded sufficiently. Finally, the water fluidized bed is predicted to be stable everywhere, except for a short interval $0.12 < \phi < 0.21$, and beds of this type are, indeed, observed not to bubble. This success in discriminating between bubbling and nonbubbling behaviour was cited by the authors as a triumph for their stability criterion. The stability criterion of Batchelor also shows a gas fluidized bed of $200 \, \mu$m diameter glass beads to be unstable, except at very small values of ϕ, and it indicates the existence of an interval of stable fluidization, just beyond minimum fluidization, for particles such as FCC fluidized by air. For water fluidized beds, such as the bed of 1 mm diameter glass beads, the nature of its prediction depends entirely on the value assumed for the parameter α.

However, though the criteria of Foscolo & Gibilaro, and of Batchelor, give conditions for growth or decay of *very small amplitude* disturbances of the state of uniform fluidization, their success has been judged above on the basis of their agreement with observations of bubbling, which is a *very large amplitude* phenomenon. It cannot be emphasized too strongly that instability, implying that small disturbances will begin to grow, does *not* imply that such disturbances will necessarily continue to grow into anything resembling bubbles. Indeed, the prediction of stability for the water fluidized bed of 1 mm glass beads by the Foscolo–Gibilaro criterion is quite contrary to what is observed experimentally. Liquid fluidized beds of this sort actually exhibit an observable instability of exactly the sort found above, consisting of plane waves with horizontal wavefronts, originating near the bottom of the bed and propagating upward with increasing amplitude but no change in geometry for a limited distance. Beyond this they develop into more complicated structures, as we will see in Chapter 5, but not into bubbles of the sort routinely seen in gas fluidized beds (Anderson & Jackson, 1969; El-Kaissy & Homsy, 1976; Homsy et al., 1980; Ham et al., 1990; Nicolas et al., 1996). Thus, a prediction of stability for such beds is actually a serious failure for a criterion of linear stability.

The prediction of an interval of no bubbling beyond minimum fluidization for gas fluidized FCC beds is in accord with observations but there is a reason for this that is unrelated to linear stability theory. This will be discussed in Section 4.7 below.

Both the Foscolo–Gibilaro criterion and Batchelor's criterion predict that all fluidized beds become stable for sufficiently small values of ϕ, that is, when sufficiently expanded. This is not a universal feature of the linear stability criterion but depends on how quickly $p_0^{p'} \to 0$ as $\phi \to 0$. Specifically, as can be seen from (4.40), if $p_0^{p'} \propto \phi^y$ for small values of ϕ, the bed becomes stable in the limit of small ϕ when $y < 2$, while it remains unstable when $y > 2$. The experimental evidence is inconclusive, though clearly recognizable structures are not observed at very high bed expansions.

4.5 Behaviour of the Dominant Disturbance in Unstable Beds

As pointed out in Section 4.3, when the bed is unstable the dominant disturbance of practical interest corresponds to the stationary maximum of α. This value, α_m, the corresponding dimensionless wavenumber κ_m, and the dimensionless velocity of propagation υ_m are functions of three dimensionless parameters, namely $(d - \frac{1}{2}u_0c_1)/V^2$, c_1/V, and v/V, as can be seen from Equations (4.22)–(4.27). From the condition of stability, (4.38), unstable disturbances exist only when

$$(d - \tfrac{1}{2}u_0c_1)/V^2 < 1 - \frac{c_1}{V},$$

thereby limiting the ranges of the parameters to be explored.

Figures 4.4(i) and (ii) show α_m as a function of $(d - \frac{1}{2}u_0c_1)/V^2$, for each of three values of c_1/V. Figure 4.4(i) corresponds to $v/V = 0.1$, representing rather a small value for μ_0^p, while $v/V = 2$ for Figure 4.4(ii). The value of the parameter c_1/V is strongly dependent on the density ratio ρ_f/ρ_s. $c_1/V = 0.0003$ is representative of a typical gas fluidized bed, while $c_1/V = 0.5$ corresponds to a typical liquid fluidized bed. Each curve starts from the axis $\alpha_m = 0$ at a point determined by the condition of limiting stability, such that $d = d_l$, where $(d_l - \frac{1}{2}u_0c_1)/V^2 = 1 - (c_1/V)$. α_m then increases monotonically as d is decreased below d_l, and the curve terminates when d becomes so small that a stationary maximum of α no longer exists (see Section 4.3 and, in particular, Figure 4.1(i)). Comparison of Figures 4.4(i) and 4.4(ii) shows that, for given values of the other two parameters, the maximum growth rate is always smaller when $v/V = 2$ than when $v/V = 0.1$. This is not surprising since increasing the effective particle-phase viscosity reduces the rate of growth of the dominant disturbance.

The form of Figures 4.4 suggests immediately that it would be useful to replot the results with the abscissa replaced by $(d_l - d)/V^2$, since d_l is easily found from the limiting stability condition. The results are presented in this form in Figures 4.5(i) and 4.6(i). Also of interest are the dimensionless wavelength λ_m $(= 2\pi/\kappa_m)$ of the dominant disturbance and its dimensionless velocity of propagation υ_m, and these are shown in Figures 4.5(ii) and (iii) and 4.6(ii) and (iii), for $v/V = 0.1$ and $v/V = 2$ respectively. With plots in this form the abscissa represents, in an obvious sense, the "distance" from the stability limit. At the stability limit it is seen that $\lambda_m \to \infty$ and $\upsilon_m \to 1$, showing that the wavelength of the dominant disturbance increases without bound and its velocity of propagation approaches that of the continuity wave. This illustrates the real difficulty of choosing a suitable scaling length for this type of dynamical model. The natural scale is the wavelength of the dominant instability, but this varies very rapidly with distance from the stability boundary when the

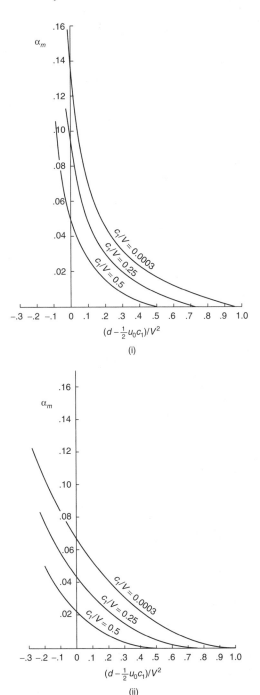

Figure 4.4. Dependence of the growth rate of the dominant mode on the bed parameters. (i) $v/V = 0.1$. (ii) $v/V = 2$.

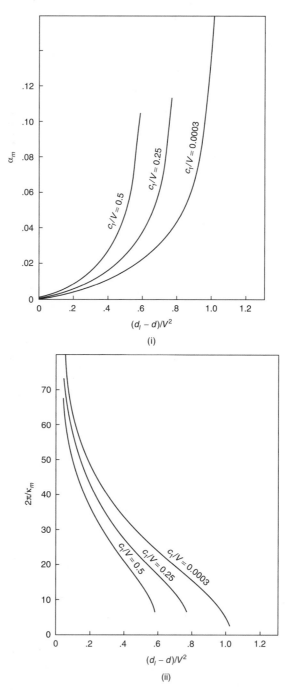

Figure 4.5. Propagation properties of the dominant disturbance with $v/V = 0.1$. (i) Dimensionless maximum growth rate. (ii) Dimensionless wavelength. (iii) Dimensionless velocity.

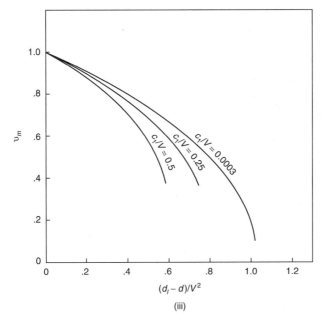

(iii)

Figure 4.5. (*Cont.*)

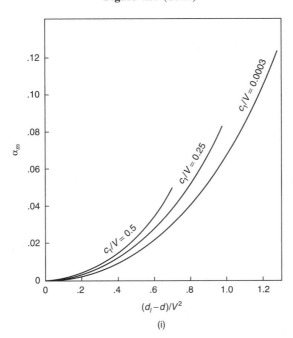

(i)

Figure 4.6. Propagation properties of the dominant disturbance with $v/V = 2.0$. (i) Dimensionless maximum growth rate. (ii) Dimensionless wavelength. (iii) Dimensionless velocity.

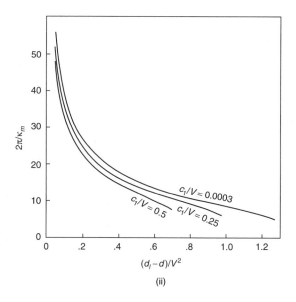

(ii)

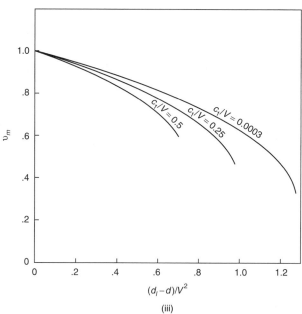

(iii)

Figure 4.6. (*Cont.*)

Table 4.1. *Typical parameter values*

200 μm glass beads/air	1 mm glass beads/water
$\rho_s = 2.2$ g/cm^3	$\rho_s = 2.2$ g/cm^3
$\rho_f = 0.0013$ g/cm^3	$\rho_f = 1$ g/cm^3
$\phi_0 = 0.57$	$\phi_0 = 0.57$
$\mu_0^p = 7.60$ g/(cm·s)	$\mu_0^p = 7.60$ g/(cm·s)
n (Richardson–Zaki) $= 4.25$	n (Richardson–Zaki) $= 3.65$
$v_t = 142$ cm/s	$v_t = 14.3$ cm/s
C_V (virtual mass coeff.) $= 0.5$	C_V (virtual mass coeff.) $= 0.5$
$p_0^{p'} = 605$ dyn/cm^2	$p_0^{p'} = 14.9$ dyn/cm^2
$u_0 = 9.14$ cm/s	$u_0 = 1.53$ cm/s
$A = 1.002$	$A = 2.832$
$\tau_0 = 0.00402$ s	$\tau_0 = 0.00348$ s
$V = 22.2$ cm/s	$V = 3.18$ cm/s
$c_1 = 0.031$ cm/s	$c_1 = 1.41$ cm/s
$v = 44.8$ cm/s	$v = 28.6$ cm/s
$d\ (=p_0^{p'}/A\rho_s) = 274$ cm^2/s^2	$d\ (=p_0^{p'}/A\rho_s) = 2.40$ cm^2/s^2
$c_1/V = 0.0014$	$c_1/V = 0.443$
$v/V = 2.02$	$v/V = 8.99$
$c_u = 16.6$ cm/s	$c_u = 2.1$ cm/s

bed operates near this boundary, so there is no universally appropriate choice of scale. However, in practice this problem may not arise. As pointed out in Chapter 3, the onset of fluidization is associated with the loss of the yield stress associated with networks of particles in sustained contact. For many beds instabilities of bounded linear dimensions appear as soon as the velocity of the fluidizing fluid becomes large enough to relieve the contacts responsible for this yield stress. This implies that any source of reversible pressure p^p is sufficiently weak that the corresponding value of d is already significantly smaller than d_l at the point where fluidization is initiated. Thus, the bed is always unstable when it is fluidized, and the wavelength of the dominant instability at the point of minimum fluidization sets a reasonable length scale.

Figure 4.6, for which v/V has a value consistent with the order of magnitude of measurements of bed viscosity, indicates that our choice of scaling has been reasonably successful. Though the curves for different values of c_1/V do not coincide, they differ by far less than corresponding plots of the dimensional growth rate, wavelength, and propagation velocity.

To give a feeling for the values of the various parameters in typical fluidized beds Table 4.1 lists them for a beds of 200 μm diameter glass beads fluidized by air at ambient conditions, and for 1 mm diameter glass beads fluidized by water.

The value of μ_0^p is taken to be the same for both liquid and gas fluidized beds, since rheological measurements of the effective viscosity do not reveal much dependence on the nature of the fluidizing fluid. This value is also typical of the order of magnitude found in experimental measurements. The values of $p_0^{p'}$ are chosen so that each bed is unstable and the "distance" from the stability boundary, represented by $(d_l - d)/V^2$, is essentially the same in both cases, namely 0.44 for the air fluidized bed and 0.43 for the water fluidized bed.

The dispersion relations for the dimensionless growth rates are shown in Figures 4.7(i) and (ii) for these air fluidized and water fluidized beds to give an idea of what can be expected in practical cases. Though this cannot be seen on the figures, each curve approaches a negative asymptote as $\kappa \to \infty$. The dimensionless growth rate takes its largest value α_m at a stationary maximum where $\kappa = \kappa_m$. The values of κ_m for the two cases are similar, whereas α_m for the air fluidized bed is about three times the corresponding value for the water fluidized bed. However, this does not reveal the true difference between the two beds, which is partly concealed by the scaling. The *dimensional* growth rates, $\mathrm{Re}(s)$, for the dominant disturbances are 1.45 s^{-1} for the air fluidized bed and 0.115 s^{-1} for the water fluidized bed, so the dominant disturbance grows almost thirteen times faster in the air fluidized bed. However, the dimensional wavelengths for the dominant disturbances are similar, namely 3.06 cm for the air fluidized bed and 1.43 cm for the water fluidized bed. These comparisons are quite typical for gas and liquid fluidized systems.

4.6 Experimental Evidence

The predicted growth of disturbances in gas fluidized beds is, on the whole, so rapid that it would be difficult to observe its course in detail. Nevertheless, the fact that bubbles form almost instantly in such beds is certainly not inconsistent with this prediction. In liquid fluidized beds, in contrast, where the growth is much slower, detailed observation of the properties of propagating disturbances is practicable. The first objective of such observations is to check whether plane density waves propagate upward through the bed, as predicted by the theory, whether the waves grow in amplitude in the direction of their motion, and whether there is a dominant wavelength that grows most quickly. If these qualitative predictions are supported by the observations, experiments should then be directed to measuring the wavelengths, velocities, and growth rates of the waves, over ranges of bed expansions and with different types of particle. Some quantitative comparisons with the predictions of the theory should then be possible.

Some informal reports of density waves in water fluidized beds of glass beads circulated in the early 1960s and a systematic experimental study of the

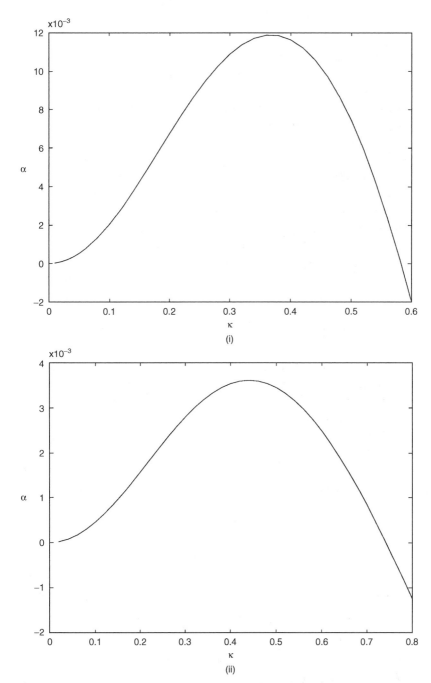

Figure 4.7. Growth rates in typical air and water fluidized beds. (i) 200 μm glass beads in air. (ii) 1 mm glass beads in water.

phenomenon was first published at the end of that decade (Anderson & Jackson, 1969). Figure 4.8 reproduces an early photograph of wave motion in a bed of glass beads, with a nominal diameter of 2 mm, fluidized by water in a vertical tube of 1.54 in nominal internal diameter (Anderson, 1967). The left-hand image shows the lower part of the bed, with the flange for the distributor visible at its lowest point, while the upper part of the bed is shown on the right. The lowest point of the right-hand image coincides with the highest point of the left-hand image, so the development of the waves can be followed upward from the bottom of the bed without interruption. The apparatus is viewed against the background of a diffusing screen strongly illuminated from behind so that regions of lower particle concentration, which transmit more light, appear as light bands. These are seen to be spaced at fairly regular intervals and the contrast between the light and dark bands, which is minimal at the foot of the bed, increases with increasing height. The whole pattern moves upward with a characteristic velocity. Thus, the visual impression is in accord with the theoretical predictions.

The dominant period of the waves and their speed of propagation can be determined directly from visual observation of these patterns. To examine their growth in amplitude Anderson (1967) used a simple light transmission technique that has also been adopted by subsequent investigators. A collimated beam of light, with diameter significantly larger than the particle size but small compared to the diameter of the bed, is transmitted diametrically through a horizontal section of the bed. The intensity of the emergent beam then depends on the fraction of the light scattered at particle surfaces, and hence on the mean concentration of the particles along the path of the beam. The relation between this concentration and the signal generated by a photocell intercepting the emergent beam is established by calibration experiments with a known mass of particles in a short bed, where waves cannot grow to a significant amplitude. Figure 4.9, taken from Anderson & Jackson (1969), shows the signal from the photocell as a function of time, at various heights above the distributor, for a bed of 0.127 cm diameter glass beads in a tube of 1 in nominal internal diameter. At the lowest station there is little evidence of organised behaviour but, on moving up the bed, a strong periodicity emerges. This is most striking and regular at a height of 36 in, above which there appears to be some coalescence of adjacent peaks, leading to an increase in the period.

Power spectra at various heights, constructed from these records, are shown in Figure 4.10 taken from Anderson (1969). At the lowest station the spectral density is a monotone decreasing function of frequency, showing no evidence of periodic behaviour, but on moving up the bed a dominant and quite narrow peak

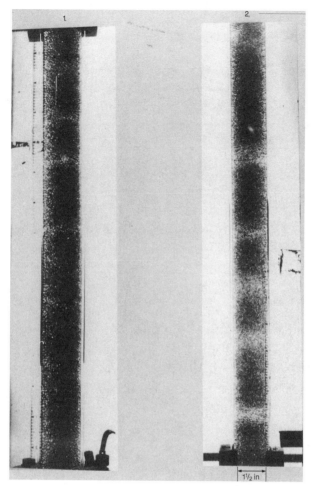

Figure 4.8. Photograph of instability waves. 2 mm glass beads in 1.5 in ID tube. (From Anderson, 1967.)

emerges at a frequency rather less than 1 Hz. The amplitude of this component, as measured by the square root of the spectral density at the frequency of the peak, is plotted on a logarithmic scale as a function of height in Figure 4.11. The linearity of this plot for small amplitudes indicates that the growth is exponential, as predicted theoretically. Observations of this sort, using essentially the same technique, have also been made by El-Kaissy & Homsy (1976) and, most recently, by Nicolas et al. (1996), and spectra of density fluctuations at a sequence of increasing heights can also be found in these publications. Like

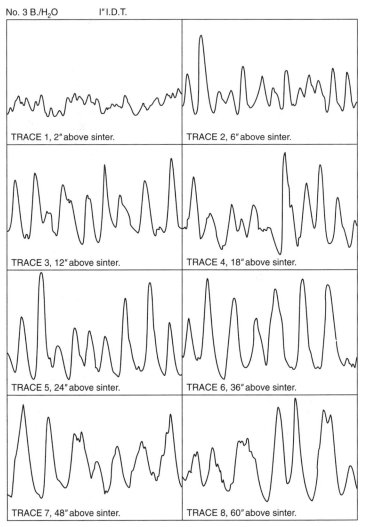

Figure 4.9. Light transmission as a function of time at a sequence of heights. (Reprinted by permission from Anderson & Jackson (1969). Copyright 1969 American Chemical Society.)

Figure 4.10 they show a dominant periodic component that grows with increasing height and, typically, has a frequency of the order of 1 Hz. The latter authors also observed the propagation of forced periodic perturbations, induced by a sinusoidal vertical displacement of the porous distributor. Their frequency is then at the disposal of the experimenter thus offering, in principle, the possibility of exploring the complete form of the dispersion relation.

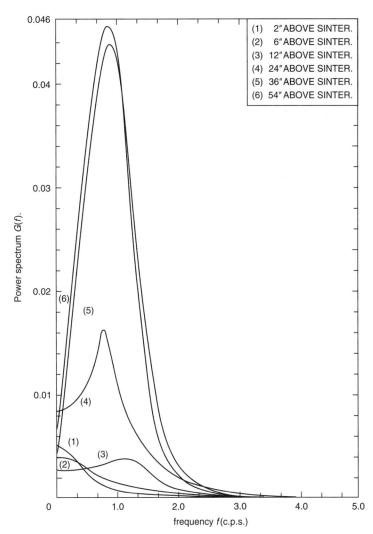

Figure 4.10. Power spectra of transmitted light signals from Figure 4.9. (From Anderson, 1967.)

Both Anderson & Jackson (1969) and El-Kaissy & Homsy (1976) made measurements for beds of glass beads of various sizes, ranging from 0.59 mm to 2.07 mm diameter and, in the case of Anderson & Jackson, for three bed diameters of 1.27 cm, 2.54 cm, and 3.81 cm. In each case propagating waves with the behaviour described above were seen over a substantial range of bed expansions. Thus the qualitative features of small disturbances predicted by the equations of motion appear to be borne out by all these observations.

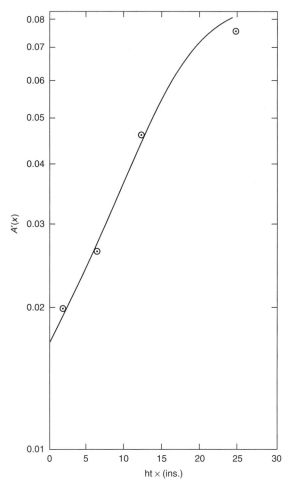

Figure 4.11. Wave amplitude as a function of height. (From Anderson, 1967.)

In each of the above studies the wave motion appeared first at a bed expansion not very different from minimum fluidization and no attempt was made to identify the precise point at which this occurred. However, Ham et al. (1990), who made a careful study of the point of initiation of instabilities in liquid fluidized beds, found a nonvanishing difference between the fluid velocity u_m at which bed expansion first occurs (the "minimum fluidization velocity", in the conventional sense) and the velocity u_c at which large-scale organized motion can first be observed, indicative of the onset of instability. Correspondingly the void fraction ε_c is slightly larger than the value ε_m at minimum fluidization. Measurements were made for particles of different sizes and densities fluidized

Table 4.2. *Delayed onset of instability*

ρ_s (g/cm^3)	d_p (cm)	Fluid[a]	u_c/u_m	$\varepsilon_c/\varepsilon_m$
2.42	0.0655	1	1.18	1.05
2.47	0.0460	1	1.14	1.04
2.49	0.0325	1	1.21	1.04
2.47	0.0230	1	1.15	1.03
2.47	0.0165	1	1.35	1.07
1.19	0.0655	1	1.42	1.04
1.19	0.0285	1	1.20	1.01
4.14	0.0325	1	1.18	1.02
2.49	0.0325	2	1.20	1.02
2.49	0.0325	3	1.23	1.04
2.49	0.0325	4	1.54	1.10
2.49	0.0325	5	1.82	1.12

[a]Fluid 1: Water; $v = 1.00$ cS, $\rho_f = 1.000$ g/cm^3.
Fluid 2: 18.30% glycerol; $v = 1.68$ cS, $\rho_f = 1.041$ g/cm^3.
Fluid 3: 31.96% glycerol; $v = 2.45$ cS, $\rho_f = 1.077$ g/cm^3.
Fluid 4: 43.80% glycerol; $v = 4.03$ cS, $\rho_f = 1.108$ g/cm^3.
Fluid 5: 60.02% glycerol; $v = 8.50$ cS, $\rho_f = 1.146$ g/cm^3.

by water and by water–glycerol mixtures of increasing density and kinematic viscosity. The measured ratios u_c/u_m and $\varepsilon_c/\varepsilon_m$ are given in Table 4.2 and they do not reveal any obvious dependence on particle size or density, though both seem to increase monotonically with the concentration of glycerol in the fluid, for particles of given size and density.

A particularly interesting result of this work is the observed variation of the frequency of the dominant wave on reducing the fluidization velocity. The peak in the power spectrum was found to move to smaller values of the frequency as u was decreased. This peak remained discernable down to a frequency as low as 0.1 Hz but, when u reached the value u_c, it disappeared. Since the velocity of propagation retains a finite value this behaviour is entirely consistent with the theoretical prediction that the dominant wavelength tends to infinity on approaching the stability limit.

It now remains to examine the extent to which it has been possible to test the quantitative predictions of the theory. Measurement of the growth rate and velocity of small-amplitude waves as functions of the wavenumber, and comparison of the results with the dispersion relations predicted theoretically, would provide a searching test of the proposed equations of motion. The experiments on forced oscillations proposed by Nicolas et al. (1996) are directed to this end, but they face difficulties because of masking by the rapid growth of the dominant mode. Existing experimental evidence is therefore restricted to measurements of the wavelength growth rate, and speed of propagation of this mode. These can then be compared with theoretical predictions, such as those presented in Figures 4.5 and 4.6. However, the theory contains certain parameters characteristic of the particular system and base state, so it can predict propagation properties of small disturbances only if the values of these parameters have been determined independently. Let us examine this point further, since it is the main obstacle faced by attempts to validate the equations.

We have already seen that the wavelength, growth rate, and velocity of propagation of the dominant wave can be predicted in terms of the three dimensionless parameters, c_1/V, v/V, and $(d - \frac{1}{2}u_0c_1)/V^2$. From the definitions of d, v, V, and c_1 these, in turn, are expressible in terms of certain dimensionless combinations of physical properties of the bed, namely ϕ_0, ρ_f/ρ_s, $\phi_0\beta_0'/\beta_0$, $C_v(\phi_0)$, $g\mu_0^p/(\rho_s u_0^3)$ and $p_0^{p'}/(\rho_s u_0^2)$. The first two of these are known and the third is easily deducible from measurements of bed expansion versus fluid flow rate, since the force balance equation (4.6) in the unperturbed bed can be differentiated with respect to ϕ_0 to give

$$\frac{\phi_0\beta_0'}{\beta_0} = 1 - \frac{\phi_0u_0'}{u_0},$$

where primes denote differentiation with respect to ϕ_0. However, the last three quantities are much more difficult to evaluate. There are some ideas about the virtual mass coefficient, and these have been discussed in Chapter 2. There are also some experimental estimates of an "effective viscosity" for fluidized beds, based largely on rheological measurements with Couette or torsional pendulum viscometers. These have been discussed by Schugerl (1971); they indicate values of the order of several poises in both gas and liquid fluidized systems. Since this is much larger than the viscosity of the fluidizing fluid it has been customary to interpret the measurements as indicating the value of μ^p. Regarding the value of $p_0^{p'}$ there is no reliable experimental information; however, as pointed out in Chapter 3, something could be learned, in principle, from careful studies of bed expansion and axial density profiles in *stable* fluidized beds. The difficulty, of course, is that the beds tend to become unstable once they are expanded significantly beyond the point of minimum fluidization. We are, therefore, left with three dimensionless parameters whose precise values are not known, though we have some idea of the orders of magnitude of two of them. In addition we have some idea how they ought to vary as the bed is expanded: we expect all three to be increasing functions of ϕ_0. In this situation, with three quantities to be predicted and three groups whose precise values are unknown, it is conceivable that experimental measurements of the wavelength, growth rate, and speed of propagation of the dominant wave could be fitted exactly, for any base state, simply by treating the values of these three groups as available quantities. Then agreement between predictions and measurements would be valueless as a test of the theory. However, this misrepresents the situation to some degree. The question is not whether a match between theory and experiments can be obtained with a free choice of values for C_v, μ_0^p, and $p_0^{p'}$, but whether such a match can be obtained over a range of bed expansions, using values of these quantities that are consistent with what we know about their orders of magnitude and their dependence on ϕ_0. If it can, then this provides some support for the theory.

An attempt along the above lines to compare measurement and prediction for the properties of the dominant disturbance was made quite early by Anderson & Jackson (1969), who observed the development of waves in cylindrical beds confined within tubes of three different diameters, D, as indicated in Table 4.3. Their measurements of growth rate were hampered by lack of modern data reduction facilities, so that routine spectral analyses of density fluctuations were not practicable, and the wave amplitude had to be estimated from the average of a number of measurements of the difference between successive peaks and troughs in the density versus time records. In selecting the parameter values to be used for the theoretical predictions these authors took $C_v = 1/2$ throughout, so they were left with only μ_0^p and $p_0^{p'}$ as adjustable parameters. Their results are recorded in Table 4.3, where the wavenumber and growth rate are expressed in the dimensionless forms $\tilde{\kappa} = kd_p$ and $\tilde{\alpha} = \mathrm{Re}(s)d_p/u_0$, which differ from those used previously. With one or two exceptions the agreement between measured and predicted quantities is fair. Furthermore, it is obtained using values of the effective viscosity with credible orders of magnitude, and with values for both μ_0^p and $p_0^{p'}$ that decrease monotonically as ε_0 is increased, as they should.

A similar comparison of theory and experiment has been reported by Homsy et al. (1980), with the results recorded in Table 4.4. The density variations were measured using light transmission, as in the earlier work of Anderson & Jackson, but with access to more modern data reduction facilities these authors could deduce the propagation characteristics of the dominant disturbance from the observed density variations with greater confidence. Scrutiny of Table 4.4 reveals rather clearly the difficulty encountered in attempting to use the observed properties of the dominant wave to test the theory and evaluate its parameters. For most sets of measured values of $\tilde{\kappa}$, $\tilde{\alpha}$, and V/u_0 the table shows two or even three possible choices for the parameters of the theory, all of which yield predictions of these three quantities in acceptable agreement with the measurements.

Table 4.3. Results of Anderson & Jackson

d_p (cm)	D (cm)	u_0 (cm/s)	ε_0	$\tilde{\kappa}$	$\tilde{\alpha} \times 10^3$	V/u_0	$\tilde{\kappa}$	$\tilde{\alpha} \times 10^3$	V/u_0	$\frac{4}{3}\dfrac{\mu_0^p \times 10^{-3}}{\mu}$	$\dfrac{\phi_0 p_0''}{\beta_3 u_0^2}$	$\dfrac{\phi_0 C_v}{\varepsilon_0}$
					Observed propagation properties			Calculated propagation properties			Parameter values	
0.207	3.81	8.90	0.418	0.31	3.5	0.74	0.19	3.3	0.94	11.2	0.31	0.70
		9.08	0.433	0.26	6.2	0.88	0.23	5.5	0.87	8.7	0.23	0.65
		9.49	0.451	0.25	8.7	0.94	0.25	8.2	0.83	6.8	0.16	0.61
	2.54	8.84	0.457	0.29	—	0.66	0.26	4.4	0.79	7.8	0.21	0.59
0.127	2.54	5.31	0.463	0.22	4.9	0.99	0.22	5.0	1.09	4.5	0.31	0.58
		5.43	0.470	0.24	6.2	0.99	0.24	6.6	1.04	3.8	0.25	0.56
		5.63	0.483	0.25	9.5	1.04	0.27	9.0	1.01	3.1	0.18	0.54
		5.83	0.496	0.25	10.8	1.05	0.28	10.9	0.99	2.7	0.15	0.51
		5.92	0.500	0.26	—	1.09	0.28	11.5	0.99	2.6	0.13	0.50
	1.27	5.03	0.503	0.47	—	0.78	0.27	1.7	0.92	3.9	0.28	0.49
		5.24	0.515	0.44	—	0.84	0.29	4.2	0.90	3.2	0.20	0.47
		5.44	0.528	0.44	—	0.87	0.32	6.2	0.88	2.8	0.16	0.45
0.086	2.54	2.86	0.469	0.22	3.9	1.30	0.23	3.7	1.30	2.4	0.57	0.57
		3.03	0.485	0.20	5.0	1.42	0.26	5.6	1.22	2.1	0.42	0.53
	1.27	2.82	0.504	0.31	2.7	1.23	0.26	2.7	1.23	2.1	0.50	0.49
		3.10	0.529	0.31	3.2	1.26	0.26	3.4	1.14	1.9	0.35	0.45
		3.27	0.541	0.31	—	1.29	0.24	2.6	1.07	1.9	0.29	0.42
0.064	1.27	1.74	0.495	0.28	1.9	1.46	0.26	1.9	1.41	1.3	0.78	0.51
		1.93	0.507	0.27	3.4	1.43	0.29	3.4	1.32	1.2	0.54	0.49
		2.00	0.511	0.25	—	1.48	0.29	3.7	1.30	1.2	0.49	0.48
		2.12	0.519	0.26	3.9	1.44	0.29	4.0	1.26	1.1	0.42	0.46
		2.26	0.530	0.26	—	1.44	0.28	4.5	1.22	1.1	0.35	0.44

Table 4.4. Results of Homsy, El-Kaissy & Didwania

	Experimental conditions			Observed propagation properties			Calculated propagation properties			Parameter values		
d_p (cm)	u_0 (cm/s)	ε_0		$\bar{\kappa}$	$\bar{\alpha} \times 10^3$	V/u_0	$\bar{\kappa}$	$\bar{\alpha} \times 10^3$	V/u_0	$\frac{4}{3}\frac{\mu_0^p \times 10^{-3}}{\mu}$	$\frac{\phi_0 p_0''}{\rho_s u_0^2}$	$\frac{\phi_{h1} C_v}{\varepsilon_0}$
0.059	2.37	0.401		0.07–0.157	0.4	1.48	0.125	0.417	1.46	5.5	7	3
							0.10	0.40	1.44	9.5	8	4
							0.125	0.422	1.41	7	5	3
	2.46	0.422		0.2	2.7	1.5	0.2	2.7	1.32	2.5	9	6
							0.2	2.6	1.32	3.5	5	3
	2.62	0.427		0.17–0.25	4.4	1.45–1.72	0.2	4.6	1.39	1.5	17	11
							0.225	4.35	1.36	1.5	8	5
0.083	3.63	0.410		0.062–0.115	1.0	1.2–1.32	0.125	1.04	1.20	7.5	8	6
							0.10	1.03	1.27	6.5	7	5
							0.10	1.04	1.22	9.5	9	7
	3.95	0.425		0.186–0.279	1.83	1.26–1.35	0.20	1.93	1.23	1.5	4	2
	4.22	0.441		0.296–0.31	7.9	1.25	0.225	7.8	1.23	1.5	8	7
							0.25	7.95	1.20	1.0	9	6
0.110	5.80	0.450		0.17–0.32	1.87	0.97–1.03	0.15	1.89	0.968	2.0	13	14
							0.15	1.79	0.983	1.5	16	17
							0.175	1.89	0.991	1.0	10	10
	5.88	0.456		0.17–0.323	6.6	1.0	0.225	6.81	0.957	0.5	11	12
	5.99	0.461					0.25	6.6	0.96	0.4	10	8
0.156	8.46	0.441		0.075–0.32	3.6	0.92	0.2	3.56	0.914	1.5	4	4
							0.175	3.56	0.908	2.0	7	8
							0.15	3.50	0.921	1.5	15	18
	8.93	0.459		0.25–0.34	13.7	0.91	0.275	13.8	0.88	1.0	6	8
	9.45	0.473		0.31–0.39	23.0	0.94	0.325	23.9	0.85	0.4	3	3
							0.375	22.1	0.86	0.6	7	5

In comparing these alternatives note that those with a larger value for $\phi_0 C_v / \varepsilon_0$ tend to have a larger value of $\phi_0 p_0^{p'} / (\rho_s u_0^2)$. This is not surprising since increasing the value of C_v increases the predicted growth rate of the dominant disturbance, while increasing the value of $p_0^{p'}$ has the opposite effect. Consequently, when both are increased these effects tend to cancel. A more striking illustration of this is provided by comparing the values for these parameters in Table 4.4 with those in Table 4.3. As a consequence of setting $C_v = 1/2$ the values of the parameter $\phi_0 C_v / \varepsilon_0$ used in Anderson & Jackson's predictions are much smaller than those found in Table 4.4. Correspondingly, to secure a reasonable match with the measurements, the values of $\phi_0 p_0^{p'} / (\rho_s u_0^2)$ selected by Anderson & Jackson are also much smaller than the corresponding values in Table 4.4. The values of μ_0^p / μ, however, are of the same order of magnitude in both tables.

In summary, observations of spontaneously occurring disturbances propagating in liquid fluidized beds reveal features in qualitative agreement with the predictions of the linearized equations of motion, even though these equations reflect the most naive assumptions used to achieve closure. Attempts to devise a more quantitative test by comparing observed and predicted values of the wavelength, growth rate, and speed of the dominant wave show some encouraging trends but are frustrated by the lack of reliable, independently determined values for the parameters that appear in the equations of motion. The presence of bubbles in most gas fluidized beds is not inconsistent with the prediction of very rapidly growing instabilities, but we again caution against linking the appearance, or nonappearance, of bubbles directly to the speed of growth of small disturbances. Bubbling is a large-amplitude phenomenon not to be explained by linearized equations of motion.

4.7 Some Other Aspects of Stability

We conclude this chapter with brief discussions of three unrelated topics: consequences of including an equation of balance for pseudothermal energy in the stability analysis, the existence of a wide interval of stable expansion for some gas fluidized beds of small particles, and instabilities that could initiate circulation patterns in beds of bounded dimensions.

4.7.1 Stability Analyses Including a Balance of Pseudothermal Energy

The stability analysis described above has been based on equations of motion resulting from simple empirical closure assumptions of the sort described in Section 2.4 of Chapter 2. However, Section 2.3 described some closures with a more fundamental basis, in particular two closures due to Koch and to Buyevich, respectively, limited to cases where the Stokes number is large. In each of these the continuity and momentum equations for the two phases were supplemented by a balance equation for pseudothermal energy, represented by a "particle temperature", $T = \frac{1}{3} \langle \mathbf{u'} \cdot \mathbf{u'} \rangle^p$. Koch (1990) developed a stability analysis based on the linearized form of his equations and concluded that the uniform suspension is always unstable when $St \gg \phi^{-3/2}$, as required for the validity of his closure. This was the case both when the balance equation for T was included in the analysis, and when it was omitted (by simply assigning a constant value to T). However, Koch's closure is appropriate only for particle concentrations so small that terms of $O(\phi^2)$ in the equations of motion can be neglected in comparison to terms of $O(\phi)$, so his conclusion should not be extrapolated to denser fluidized beds.

Koch & Sangani (1999) have analysed the stability of the uniformly fluidized state using equations based on their explicit closure for the momentum balances described above in Section 2.3.2. They show examples of dispersion relations for perfectly elastic particles with various combinations of the Stokes number and the solids volume fraction, and in all but the most dilute suspensions they find the bed to be unstable for each value of the Stokes number investigated. Figure 4.12 shows their computed value of the dimesionless growth rate for the fastest growing disturbance, as a function of the solids volume fraction, for Stokes numbers between 5 and 100. Each curve shows a narrow interval

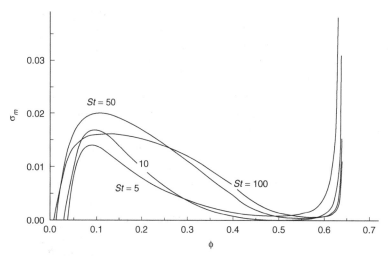

Figure 4.12. Maximum growth rate as a function of ϕ for various values of the Stokes number, from the theory of Koch & Sangani (1999).

of stability for small values of ϕ. As ϕ decreases the growth rate becomes positive and passes through a maximum, before decreasing again to a relatively small value near $\phi = 0.5$. Finally, the growth rate increases very rapidly as ϕ increases towards its maximum value of 0.65. In practice, however, it seems likely that this sharp increase will be prevented by the growing importance of forces exerted between particles at points of sustained contact, which are not taken into account in the equations. In interpreting Figure 4.12 note that σ_m in the diagram is a dimensionless growth rate obtained by dividing the dimensional form by v_t/a.

4.7.2 Beds with an Extended Interval of Stable Expansion

With certain gas fluidized beds of fine particles there is an extended interval of apparently stable expansion beyond the point of minimum fluidization. This is much wider than the narrow interval of stability observed by Ham et al. (1990) in liquid fluidized beds (see Section 4.6) and it has attracted a good deal of attention as a result of its technical importance in the case of fluid cracking catalyst, which provides an outstanding example of this phenomenon. However, there is compelling evidence that attempts to account for this behaviour in terms of stabilization by a hydrodynamically generated particle-phase pressure are misguided, at least in the case of cracking catalyst. Quite early Rietema & Mutsers (1973) demonstrated that, within the interval of stable expansion, such beds exhibit a nonvanishing angle of repose. When the bed is tilted its surface does not remain level but behaves like a granular material, tilting with the container until it reaches some limiting angle of slope. This indicates that sustained particle–particle contact forces are still playing a significant role. Other evidence of the importance and nature of contact forces, and the way in which they stabilize the expanded bed, has been accumulated by Rietema and coworkers and is now summarized in a monograph (Rietema, 1991). Experiments of Tsinontides & Jackson (1993), discussed in Section 3.4.3 above, also demonstrate beyond doubt that fluid cracking catalyst retains a compressive yield strength over a substantial range of expansion beyond minimum fluidization. In this interval, which appears to extend to the point where bubbling is first seen, the particles are not actually suspended in gas and the linear stability analysis of Sections 4.2 to 4.5 is inapplicable. Tsinontides & Jackson worked only with this particular material, so it should not necessarily be concluded that yield stress is involved whenever there is an interval of smooth expansion. For example, the relatively narrow intervals of this sort found by Ham et al. (1990) in liquid fluidized beds may be true examples of a linear stability criterion at work, of the type derived in the present chapter. This would merit further investigation, extending the observations to smaller and less dense particles.

4.7.3 Circulatory Instabilities

All the work so far described in this chapter has addressed the stability of beds of unbounded spatial extent, for which the length scale of importance is the wavelength of the most rapidly growing instability, which is an intrinsic property of the bed in its unperturbed state. In addition there is another type of motion that is suggestive of an instability of a different sort. Beds of all but the smallest diameters share a common feature, namely a persistent and frustrating tendency to develop circulatory motion, with particles ascending in one or more regions of the cross section and descending elsewhere. Such an effect can clearly result from fluid maldistribution if the flow resistance of the porous bed support varies from place to place, but convective motion can persist despite the most careful attention to ensuring a uniform distributor resistance. This suggests the presence of some mechanism capable of generating circulation patterns even in the presence of a perfectly uniform distributor; in other words, a circulatory instability of the uniformly expanded bed. This possibility has been investigated by Medlin et al. (1974).

A mechanism by which circulation might be generated, even with a uniform distributor, is not difficult to visualize. Consider a uniform bed of bounded depth. Then the fluid pressure drop across the bed is the same at all points of the cross section, just balancing the weight of the suspended particles per unit cross-sectional area, and the particles are at rest everywhere. But now suppose there is some nonuniformity in the distribution of particles, so that the concentration of particles is lower over one part of the cross section and higher over the remainder. Then in the former part the pressure drop exceeds the weight of the particles, and consequently they begin to move upward, while in the latter part the pressure drop is less than the weight of the particles, so they start to move down, thus initiating circulation. Whether or not this incipient circulation subsequently grows might be expected to depend on the properties of the bed and its linear dimensions, while the nature of the distributor should also be implicated, since a uniform and large pressure drop in the distributor will clearly discourage circulation by this mechanism.

To analyse this possibility Medlin et al. (1974) confined attention to two-dimensional motions in a bed of bounded depth D, as sketched in Figure 4.13. The bed is visualized as contained in a channel of infinite height, bounded by parallel, vertical walls. The x axis is directed vertically up and the distributor is a horizontal porous plate located in the plane $x = 0$. Motion within the bed is described by the same equations as those used throughout this chapter and the side walls are assumed to be perfectly smooth, so that they constrain only the y component of velocity. Boundary conditions for the motion of both particles and fluid must be specified above and below the bed. Those associated with

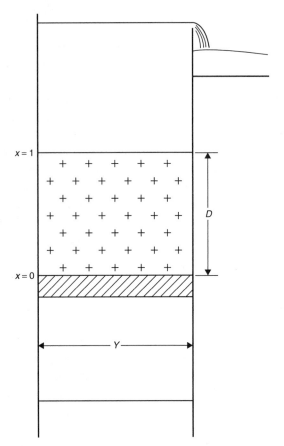

Figure 4.13. Bounded fluidized bed.

the particles are relatively straightforward. At $x = 0$ the distributor constrains the x component of the particle velocity to vanish, while the roughness of the distributor plate determines its influence on the y component. For a perfectly rough plate $v_y = 0$, and for a perfectly smooth plate the tangential component of the particle phase stress must vanish at $x = 0$. Both these possibilities were considered. At $x = D$ it is simplest to assume that the upper surface of the bed remains horizontal, but there is no constraint on lateral movement of the particles in this plane. More realistically the surface could be treated as a free (i.e., stress free) surface and allowed to deform, but this raises some questions of consistency with the base state. When fluid inertia is negligible, as in the case of a gas fluidized bed, the required set of boundary conditions can then be completed simply by assuming that the fluid pressure remains unperturbed below the plane of the distributor and above the plane of the bed surface. However, for liquid fluidized beds the inertia of the fluid prevents sudden transitions from unperturbed motions above the bed and below the distributor, to perturbed motion within the bed, and it is necessary to analyse the perturbed motion throughout all three regions, with appropriate matching conditions as the fluid enters the distributor and as it leaves the upper surface of the bed. The boundary conditions on the fluid are then imposed effectively at $x = -\infty$, where the flow is assumed to be distributed uniformly over the cross section of the channel, and at $x = \infty$, where the fluid pressure is assumed to be uniform over the cross section.

Small perturbations of the uniform bed are postulated of the form

$$(\phi', u', v', p^{f'}) = (\tilde{\phi}(x), \tilde{u}(x), \tilde{v}(x), \tilde{p}^f(x))e^{iky}e^{st},$$

where k is real, and the four functions of x on the right-hand side are to be determined as solutions of the boundary value problem, with eigenvalue parameter s. The stability of the uniform bed against these perturbations then depends on the sign of the real part of s. As a result of the spatial uniformity of the unperturbed state, the linear, ordinary differential equations to be satisfied by $\tilde{\phi}$, $\tilde{u}$, $\tilde{v}$, and $\tilde{p}^f$ have constant coefficients, so their general solutions can be written down immediately. These contain seven arbitrary constants, and the boundary conditions provide seven homogeneous, linear equations, to be satisfied by these constants, which possess a nontrivial solution only if the determinant of their coefficients vanishes, thus determining the eigenvalues of s. This description of the eigenvalue problem, while correct in its essentials, is somewhat oversimplified for the general case in which fluid inertia can not be neglected. Then, as noted above, the boundary conditions on the fluid must be moved to $x = \pm\infty$, and perturbations in the clear fluid regions above and below the bed must be matched to the perturbations within the bed. Though this adds significantly to the complexity of the solution it introduces no new difficulties of principle. Details of the work can be found in the original publication.

For given values of the bed depth and the parameters in the equations of motion it is found that a stability boundary exists as a relation between a critical wavelength λ_c ($\lambda = 2\pi/k$) and the ratio $\Delta p_d/\Delta p_b$ of the distributor pressure drop to the bed pressure drop in the unperturbed state. This stability boundary is exhibited in Figure 4.14 for two different sets of bed parameters, listed in Table 4.5, characteristic of liquid and gas fluidized beds of the same particles, respectively. (The no-slip condition is imposed on the particles at the surface of the distributor in both cases. Replacing this by a condition of free slip does not change the form of the results, though there are significant quantitative changes.) The domains of stability lie below and to the right of the boundary curves.

Table 4.5. *Parameter values for Figure 4.14*

Water fluidized bed	Air fluidized bed
$D = 86$ cm	$D = 86$ cm
$d_p = 0.086$ cm	$d_p = 0.086$ cm
$\rho_s = 2.86$ g/cm^3	$\rho_s = 2.86$ g/cm^3
$\rho_f = 1.0$ g/cm^3	$\rho_f = 0.0012$ g/cm^3
$\phi_0 = 0.58$	$\phi_0 = 0.58$
$\mu_0^p = 23$ poise	$\mu_0^p = 23$ poise
$\mu^f = 0.01$ poise	$\mu^f = 0.00022$ poise
$p_0^{p\prime} = 20$ dyn/cm^2	$p_0^{p\prime} = 0$
$C_v = 0.5$	$C_v = 0.5$

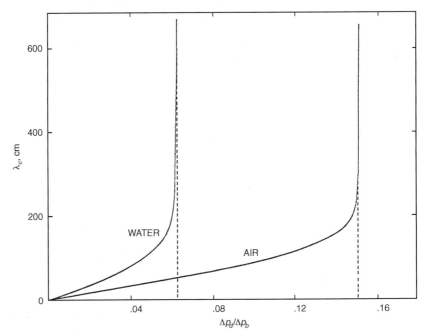

Figure 4.14. Stability boundaries for circulatory instabilities. (Reprinted by permission from Medlin et al. (1974). Copyright 1974 American Chemical Society.)

In each case, there exists a value of $\Delta p_d / \Delta p_b$, beyond which the bed is stable for all values of λ. For smaller values of $\Delta p_d / \Delta p_b$ the bed is stable for $\lambda < \lambda_c$ and unstable otherwise. Furthermore there is a relation between possible values of λ and the width W of the bed, arising from the requirement that $u_y = v_y = 0$ on each of the vertical bounding walls. This mandates that the separation between the walls must be an integer number of half wavelengths,

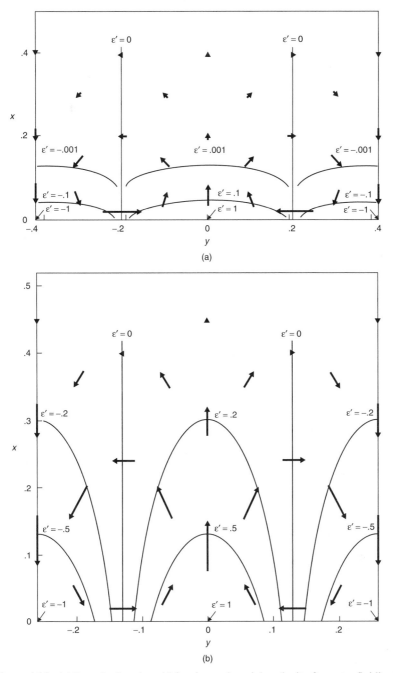

Figure 4.15. (a) Perturbations in void fraction and particle velocity for water fluidized bed. $\Delta p_d / \Delta p_b = 0.002$. (b) Perturbations in void fraction and particle velocity for air fluidized bed. $\Delta p_d / \Delta p_b = 0.01$. (Reprinted by permission from Medlin et al. (1974). Copyright 1974 American Chemical Society.)

or $\lambda = 2W/n$ ($n = 1, 2, \ldots$). If all these values of λ are smaller than λ_c, then the bed is stable against this kind of perturbation. Thus the ordinate axis in Figure 4.14 can be relabelled "$2W$" and the figure will then identify the regions of stability and instability in the plane of bed width and distributor pressure drop. For a given value of Δp_d the bed can always be stabilized by reducing its width sufficiently. For a given value of the width, it can always be stabilized by increasing the distributor pressure drop sufficiently.

The nature of the motion that results when the uniform bed is unstable can be seen in Figures 4.15(a) and (b). These show the particle velocity field (indicated in direction and magnitude by arrows) and the perturbation in void fraction, for the water and air fluidized beds, respectively. Each is presented for that value of λ at which the disturbance grows most rapidly, and the space coordinates, x and y, are both scaled by the bed depth. We see that the particles circulate upward, then outward in that part of the bed where ε' is positive (i.e., where the particle concentration is reduced). In the regions of increased particle concentration, in contrast, they move downward, then inward. In both cases the circulation is confined largely to the lower half of the bed.

Entirely as a result of the extraordinary difficulty of fabricating porous distributors that have completely uniform resistance to fluid flow it has proved very difficult to compare the predictions of this stability theory with experiment. This is a problem in all work on wide fluidized beds, but it is particularly acute in the present case since minor nonuniformities in the distributor resistance can generate circulation patterns in the bed that mask the effects analysed above. Nevertheless Agarwal et al. (1980) were able to explore the stability of water fluidized beds of 0.076 cm diameter glass beads in "two-dimensional" fluidized beds whose front and back faces were separated by only 1.9 cm. Beds of widths between 5 cm and 100 cm, with depths between 10 cm and 70 cm, were studied by varying the distributor pressure drop and observing whether or not circulatory motion was present. Elaborate precautions were taken to avoid circulation due to inhomogeneity of the distributor, and it was concluded that the observations were compatible with predictions, using credible values for the unknown parameters in the theory.

References

Agarwal, G. P., Hudson, J. L., & Jackson, R. 1980. Fluid mechanical description of fluidized beds. Experimental investigation of convective instabilities in bounded beds. *Ind. Eng. Chem. Fundam.* **19**, 59–66.
Anderson, T. B. 1967. The dynamics of fluidized beds, with particular reference to the stability of the fluidized state. Ph.D. thesis, University of Edinburgh.
Anderson, T. B. & Jackson, R. 1967. Fluid mechanical description of fluidized beds. Equations of motion. *Ind. Eng. Chem. Fundam.* **6**, 527–539.

Anderson, T. B. & Jackson, R. 1968. Fluid mechanical description of fluidized beds. Stability of the state of uniform fluidization. *Ind. Eng. Chem. Fundam.* **7**, 12–21.

Anderson, T. B. & Jackson, R. 1969. Fluid mechanical description of fluidized beds. Comparison of theory and experiment. *Ind. Eng. Chem. Fundam.* **8**, 137–144.

Anderson, K., Sundaresan, S., & Jackson, R. 1995. Instabilities and the formation of bubbles in fluidized beds. *J. Fluid Mech.* **303**, 327–366.

Batchelor, G. K. 1988. A new theory of the instability of a uniform fluidized bed. *J. Fluid Mech.* **193**, 75–110.

Batchelor, G. K. 1993. Secondary instability of a gas fluidized bed. *J. Fluid Mech.* **257**, 359–371.

Batchelor, G. K. & Nitsche, J. M. 1991. Instability of stationary unbounded stratified fluid. *J. Fluid Mech.* **227**, 357–391.

Buyevich, Yu. A. 1971a. On the fluctuations of concentration in disperse systems. The random number of particles in a fixed volume. *Chem. Eng. Sci.* **26**, 1195–1201.

Buyevich, Yu. A. 1971b. Statistical hydrodynamics of disperse systems. Part 1. Physical background and general equations. *J. Fluid Mech.* **49**, 489–507.

Buyevich, Yu. A. 1972a. Statistical hydrodynamics of disperse systems. Part 2. Solutions of the kinetic equations for suspended particles. *J. Fluid Mech.* **52**, 345–355.

Buyevich, Yu. A. 1972b. Statistical hydrodynamics of disperse systems. Part 3. Pseudo-turbulent structure of homogeneous suspensions. *J. Fluid Mech.* **56**, 313–336.

Buyevich, Yu. A. 1972c. On the fluctuations in concentration in disperse systems II. *Chem. Eng. Sci.* **27**, 1699–1708.

Buyevich, Yu. A. & Kapbasov, Sh. K. 1994. Random fluctuations in a fluidized bed. *Chem. Eng. Sci.* **49**, 1229–1243.

Danckworth, D. & Sundaresan, S. 1991. Time-dependent flow patterns arising from the instability of uniform fluidization. Presentation at *Symposium on the Mechanics of Fluidized Beds, 1–14 July 1991*, Stanford University.

Davidson, J. F. 1961. Discussion at Symposium on Fluidization. *Trans. Inst. Chem. Eng.* **39**, 230–232.

Davidson, J. F. & Harrison, D. 1963. *Fluidized Particles*. Cambridge University Press.

Didwania, A. K. & Homsy, G. M. 1981. Flow regimes and flow transitions in liquid fluidized beds. *Int. J. Multiphase Flow* **7**, 563–580.

Didwania, A. K. & Homsy, G. M. 1982. Resonant side-band instabilities in wave propagation in fluidized beds. *J. Fluid Mech.* **122**, 433–438.

El-Kaissy, M. M. & Homsy, G. M. 1976. Instability waves and the origin of bubbles in fluidized beds. Part 1: experiments. *Int. J. Multiphase Flow* **2**, 379–395.

Fanucci, J. B., Ness, N., & Yen, R-H. 1979. On the formation of bubbles in gas–particulate fluidized beds. *J. Fluid Mech.* **94**, 353–367.

Foscolo, P. U. & Gibilaro, L. G. 1984. A fully predictive criterion for the transition between aggregative and particulate fluidization. *Chem. Eng. Sci.* **39**, 1667–1675.

Foscolo, P. U. & Gibilaro, L. G. 1987. Fluid dynamic stability of fluidized suspensions: the particulate bed model. *Chem. Eng. Sci.* **42**, 1489–1500.

Ganser, G. H. & Drew, D. A. 1990. Nonlinear stability analysis of a uniform fluidized bed. *Int. J. Multiphase Flow* **16**, 447–460.

Garg, S. K. & Pritchett, J. W. 1975. Dynamics of gas fluidized beds. *J. Appl. Phys.* **46**, 4493–4500.

Glasser, B. J., Kevrekidis, I. G., & Sundaresan, S. 1996. One- and two-dimensional travelling wave solutions in gas fluidized beds. *J. Fluid Mech.* **306**, 183–221.

Glasser, B. J., Kevrekidis, I. G., & Sundaresan, S. 1997. Fully developed travelling wave solutions and bubble formation in fluidized beds. *J. Fluid Mech.* **334**, 157–188.

Göz, M. 1992. On the origin of wave patterns in fluidized beds. *J. Fluid Mech.* **240**, 379–404.

Ham, J. M., Thomas, E., Guazzelli, E., Homsy, G. M., & Anselmet, M-C. 1990. An experimental study of the stability of liquid fluidized beds. *Int. J. Multiphase Flow* **16**, 171–185.

Homsy, G. M., El-Kaissy, M. M., & Didwania, A. K. 1980. Instability waves and the origin of bubbles in fluidized beds. Part 2: comparison with theory. *Int. J. Multiphase Flow* **6**, 305–318.

Jackson, R. 1963. The mechanics of fluidized beds. I: the stability of the state of uniform fluidization. *Trans. Inst. Chem. Eng.* **41**, 13–21.

Koch, D. L. 1990. Kinetic theory for a monodisperse gas–solid suspension. *Phys. Fluids* **A2**, 1711–1723.

Koch, D. L. & Sangani, A. S. 1999. Particle pressure and marginal stability limits for a homogeneous monodisperse gas fluidized bed: kinetic theory and numerical simulation. *J. Fluid Mech.* **400**, 229–263.

Kynch, G. J. 1952. A theory of sedimentation. *Trans. Faraday Soc.* **48**, 166–176.

Liu, J. T. C. 1982. Note on a wave-hierarchy interpretation of fluidized bed instabilities. *Proc. R. Soc. London* **A380**, 229–239.

Medlin, J., Wong, H-W., & Jackson, R. 1974. Fluid mechanical description of fluidized beds. Convective instabilities in bounded beds. *Ind. Eng. Chem. Fundam.* **13**, 247–259.

Murray, J. D. 1965. On the mathematics of fluidization. Part I. Fundamental equations and wave propagation. *J. Fluid Mech.* **21**, 465–493.

Mutsers, S. M. P. & Rietema, K. 1977. The effect of interparticle forces on the expansion of a homogeneous gas-fluidized bed. *Powder Technol.* **18**, 239–248.

Needham, D. J. & Merkin, J. H. 1983. The propagation of voidage disturbances in a uniform fluidized bed. *J. Fluid Mech.* **131**, 427–454.

Needham, D. J. & Merkin, J. H. 1984. The evolution of two-dimensional small-amplitude voidage disturbances in a uniformly fluidized bed. *J. Eng. Math.* **18**, 119–132.

Needham, D. J. & Merkin, J. H. 1986. The existence and stability of quasi-steady periodic voidage waves in a fluidized bed. *Z. Angew. Math. Phys.* **37**, 322–339.

Nicolas, M., Chomaz, J-M., & Guazzelli, E. 1994. Absolute and convective instabilities of fluidized beds. *Phys. Fluids* **6**, 3936–3944.

Nicolas, M., Chomaz, J-M., Vallet, D., & Guazzelli, E. 1996. Experimental investigations on the nature of the first wavy instability in liquid fluidized beds. *Phys. Fluids* **8**, 1987–1989.

Pigford, R. L. & Baron, T. 1965. Hydrodynamic stability of a fluidized bed. *Ind. Eng. Chem. Fundam.* **4**, 81–87.

Rietema, K. 1991. *The Dynamics of Fine Powders*. Elsevier Applied Science.

Rietema, K. & Mutsers, S. M. P. 1973. The effect of interparticle forces on the expansion of a homogeneous gas-fluidized bed. *Proc. International Symposium on Fluidization and Its Applications*, Toulouse.

Schugerl, K. 1971. Rheological behaviour of fluidized systems. In *Fluidization*, ed. J. F. Davidson & D. Harrison, pp. 261–291. Academic Press.

Tsinontides, S. & Jackson, R. 1993. The mechanics of gas fluidized beds with an interval of stable fluidization. *J. Fluid Mech.* **255**, 237–274.

Wallis, G. B. 1969. *One-Dimensional Two-Phase Flow*. McGraw-Hill.

Wilhelm, R. H. & Kwauk, M. 1948. Fluidization of solid particles. *Chem. Eng. Progr.* **44**, 201–218.

5

Bubbles and other structures
in fluidized beds

5.1 Introduction

In the last chapter a distinction was made between "particulate" and "aggregative" fluidization. Beds of the latter type are characterized by the presence of large-amplitude departures from the ideal condition of uniform expansion, whereas the former appear uniform on casual inspection. Most gas fluidized beds are aggregative; most liquid fluidized beds are particulate. However, there are exceptions; for example, water fluidized beds of large dense particles, such as lead shot, exhibit inhomogeneities that resemble somewhat those seen in gas fluidized beds. Also, gas fluidized beds of very fine particles can often be expanded significantly before the characteristic inhomogeneities appear.

The unspecific term "inhomogeneities" has been used because the nature of the motions observed depends not only on the particular fluid and particles involved but also on the extent to which the bed is expanded. In dense aggregative beds that have not been expanded much beyond the point where inhomogeneities first appear, these take the form of bubbles, or sharply defined regions of very low particle concentration that rise through the bed with a speed that depends on their size. Figure 5.1 is a photograph by P. N. Rowe of a bubble rising in a "two-dimensional" gas fluidized bed, that is, a bed of rectangular cross section with one pair of its plane faces separated by only a centimeter or two. The interior of the bubble is a region almost free of particles that spans the entire distance between the front and back bounding walls of the bed. The photograph was taken with a camera moving up with the bubble, and the motion of the particles relative to the bubble can be pictured from the traces they leave during the 1/100 s exposure time. The overall impression is strikingly similar to that of a large gas bubble rising through a pool of liquid, both as regards the shape of the bubble itself and the nature of the dense phase motion in its vicinity.

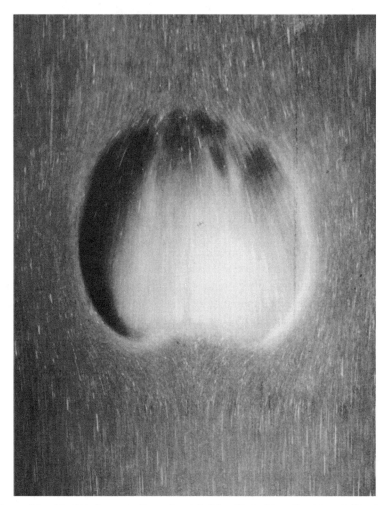

Figure 5.1. A bubble in a two-dimensional fluidized bed. (From Davidson & Harrison, 1971.)

Though "two-dimensional" beds are atypical, bubbles with the same general features are found in the interior of three-dimensional beds, as revealed by Figure 5.2, which is an X-ray image of such a bubble, also obtained by Rowe. A more detailed discussion of observations of the flow fields of both particles and fluid in the neighbourhood of the bubble will be deferred until some theoretical work on bubble motion has been described, since this is a case where the direction of the experimentation was suggested by the results of the theory.

As mentioned above and discussed in Chapter 4, air fluidized beds of small, light particles can be expanded considerably before visible bubbles appear, in

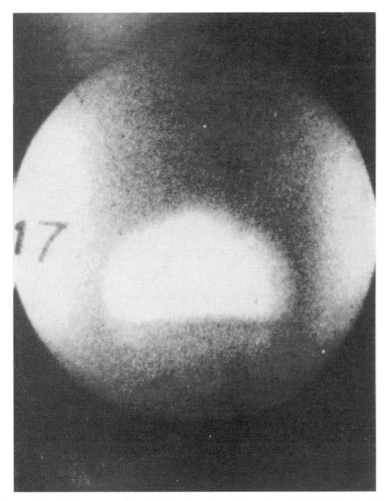

Figure 5.2. X-ray image of a bubble in a three-dimensional fluidized bed. (From Davidson & Harrison, 1963.)

contrast to materials such as coarse sand, or glass beads, that bubble vigorously as soon as they begin to expand. Which to expect can be predicted from an empirical classification due to Geldart (1973), who studied many fluid–particle combinations and concluded that a fairly sharp boundary separating the two types of behaviour can be identified in the plane of particle diameter and solid–fluid density difference. Geldart's classification is shown in Figure 5.3 as the line separating the regions labelled "A" and "B". Systems falling in region A can be expanded significantly before they bubble and are said to be "aeratable"; those in region B bubble immediately on expansion and are

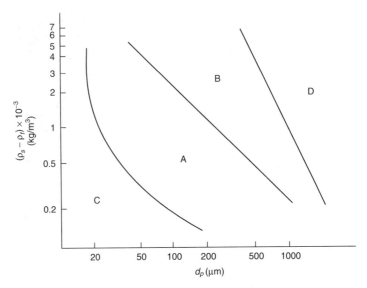

Figure 5.3. Geldart's classification.

referred to as "sand-like". Two other regions, of less concern to us, appear on the Geldart diagram. Very small particles, in region C, are classified as "cohesive". Van der Waals attractive forces between such particles are comparable with, or larger than, the forces exerted by gravity and fluid drag, with the result that the particles tend to stick together and fluidization is difficult. The region labelled "D" has to do with the behaviour of the particles in spouted beds and will not concern us.

The distinction between bubbling and nonbubbling behaviour is not a sharp one. Early studies of transitional systems were reported by Davidson & Harrison (1963), who experimented with lead, steel, glass, and ion-exchange resin particles fluidized by water, kerosene, and aqueous solutions of glycerol. Quite vigorous bubbling was observed with the particles of highest density (lead) fluidized by the liquid of lowest density (kerosene), and recognizable bubbles were also seen when lead shot was fluidized by water. However, large water bubbles injected into the fluidized bed of lead shot were broken up by their lower surface rising relative to the upper surface until the two met. (We shall see later that this observation is important.) There was some evidence of significant inhomogeneity in beds of steel spheres, but not with beds of glass or resin particles. Some later observations of beds of tungsten particles (which are denser than lead shot) fluidized by water showed that they bubble vigorously.

The above work on liquid fluidized beds is supplemented by a number of observations of gas fluidized beds operating at elevated pressures. These are reviewed by Jacob & Weimer (1987), who themselves report measurements on carbon powders of 44 μm and 112 μm diameters, over a very wide range of pressures. While all workers share a qualitative impression of "smoother" fluidization, with smaller average bubble sizes, as the pressure is increased, those who worked with small particles belonging to Geldart's class A were also able to measure the gas velocities for minimum bubbling and minimum fluidization, as functions of the pressure. The former was found to decrease and the latter to increase with increasing pressure, but both effects were rather small. For example, in the work of Jacob & Weimer u_{mf} decreased by about 12% while measurements of u_{mb}, though subject to considerable scatter, indicated an increase of the same order of magnitude. However, much of the reported work on the influence of pressure has used gases well above their critical temperatures, so that their behaviour is almost ideal over the range of pressures explored. For example, Jacob & Weimer worked with a synthesis gas consisting of 80% hydrogen and 20% carbon monoxide at ambient temperature, for which the ratio of density to pressure is almost exactly constant, indicating ideal behaviour, over the whole range of pressures extending to over 12,000 kPa. If the intermediate range between typical liquid and gas fluidized beds is to be explored it is important to work with a fluidizing medium that is just above its critical temperature at the experimental conditions, and to extend the measurements to pressures higher than the critical pressure. Then it should be possible to explore fluidization through the narrow interval over which the properties of the fluid change continuously from those of a gas to those of a liquid. An obvious choice is carbon dioxide, with a critical temperature of 31°C and a critical pressure of approximately 73 atmospheres, and this has been used with a variety of particles by Salatino et al. (1991). More studies using systems of this sort would be very welcome.

In this chapter we shall first review solutions of the equations of motion representing the motion of "adult" bubbles, whose interiors are devoid of particles. Then we go on to examine attempts to find the link between these and their origins in small-amplitude instabilities of the sort dealt with in Chapter 4. The objective of this is to show how the rapidly growing, plane wave instabilities grow into bubbles in typical gas fluidized beds and to explain why the corresponding instabilities in typical liquid fluidized beds do not. This is a puzzle of long standing; its resolution has had to await modern developments in nonlinear differential equations and computing. We shall also examine other types of structure whose existence can be predicted as a consequence of the equations of motion.

5.2 Davidson's Analysis of the Motion of Fully Developed Bubbles

Though the resemblance between bubbles in liquids and those observed in fluidized beds was obvious from the beginning it also posed a problem. The upper surface of a bubble rising in a liquid is stabilized by surface tension, but there is no analogue of this in the fluidized bed. This led to speculations by Reuter (1963a,b) that the particles forming the roof of the bubble were packed closely together to provide mutual support, rather like the stones in an arch, but this speculation proved to be sterile; no detailed picture of the motion of the fluid and the particles throughout the neighbourhood of the bubble could be developed from it. The essential breakthrough in our understanding of bubble motion came from a brief but elegant analysis by Davidson (1961). Subsequent treatments of bubble motion can be regarded as embellishments of this.

Davidson starts from the assumption that there exists an isolated, fully developed bubble, which rises through an infinite bed at a constant speed U_b and retains a constant size and shape. The system is then viewed from the rest frame of this bubble, in which the motion is independent of time. He further assumes that the bubble is spherical in shape and that the concentration of the particles is uniform everywhere outside, corresponding to a constant void fraction ε_0. Finally, the velocity field of the particles is simply taken to be the same as that of an inviscid fluid in potential flow round a spherical object (coinciding with the bubble) immersed in a uniform stream. Then if spherical coordinates (r, θ, ϕ) are set up, with origin at the centre of the bubble and polar axis pointing vertically upward, the solids velocity field is given by

$$\mathbf{v} = -\nabla V_p, \quad \text{where } V_p = U_b \cos\theta \left(r + \frac{r_b^3}{2r^2} \right), \tag{5.1}$$

in which r_b denotes the radius of the bubble and V_p is a velocity potential for the particle motion. Finally, the compressibility of the fluid is neglected and it is assumed to permeate through the particle assembly under a pressure gradient proportional to the relative velocity of the fluid and the particles, as in Darcy's Law. Thus

$$\nabla p^f + \beta_0(\mathbf{u} - \mathbf{v}) = 0, \tag{5.2}$$

where p^f is the fluid pressure and $\mathbf{u}$ and $\mathbf{v}$ are the local average velocities of fluid and particles, respectively. Also, since the fluid is assumed to be incompressible and the void fraction to be constant we have

$$\nabla \cdot \mathbf{u} = \nabla \cdot \mathbf{v} = 0. \tag{5.3}$$

Then, eliminating $\mathbf{u}$ and $\mathbf{v}$ between (5.2) and (5.3) yields

$$\nabla^2 p^f = 0. \tag{5.4}$$

Within the bubble the pressure is uniform, $p^f = p_b^f$, if we neglect both inertial and viscous contributions, which is reasonable for a gas. Also, at a large distance from the bubble, $\mathbf{u} - \mathbf{v} = \mathbf{i} u_0$, where $\mathbf{i}$ denotes the unit vector in the upward vertical direction and u_0 is the interstitial fluid velocity in the undisturbed bed. Then (5.2) shows that $\nabla p^f = -\mathbf{i}\beta_0 u_0$ when $r \gg r_b$. The solution of (5.4) that satisfies these conditions at $r = r_b$ and for $r \gg r_b$ is

$$p^f = p_b^f - \beta_0 u_0 \cos\theta \left(r - \frac{r_b^3}{r^2} \right). \tag{5.5}$$

Using (5.5) for p^f and (5.1) for $\mathbf{v}$ in Equation (5.2) then gives the fluid velocity field as

$$\mathbf{u} = -\nabla V_f \quad \text{with } V_f = \cos\theta \left[(U_b - u_0)r + \left(\frac{1}{2}U_b + u_0 \right) \frac{r_b^3}{r^2} \right]. \tag{5.6}$$

The vector $\mathbf{u}$ may be expressed alternatively in terms of a Stokes stream function ψ_f, where

$$\psi_f = \frac{r^2 \sin^2\theta}{2} \left[(U_b - u_0) - (U_b + 2u_0)\frac{r_b^3}{r^3} \right]. \tag{5.7}$$

An important feature of the fluid flow field can be seen immediately from (5.7). Provided $U_b > u_0$ there exists a real value of r, given by

$$r = r_c = r_b \left(\frac{U_b + 2u_0}{U_b - u_0} \right)^{1/3}, \tag{5.8}$$

such that ψ_f vanishes when $r = r_c$. Thus the sphere $r = r_c$ is a surface generated entirely by streamlines; no fluid crosses any part of it, in either direction. It is, therefore, a surface of separation between the fluid within and that without. The fluid outside this surface flows downward past the bubble, in the present reference frame, while the fluid inside remains always in the neighbourhood of the bubble and is usually referred to as the "Davidson cloud".

Figure 5.4 shows the streamlines of the flow (5.7) for $U_b/u_0 = 1.30$. We see that the fluid in the cloud circulates as a toroidal vortex concentric with the bubble. Equation (5.8) shows that $r_c/r_b \to \infty$ when $U_b \to u_0$, while $r_c/r_b \to 1$ when $U_b \to \infty$. Thus, for small particles (small u_0) and large bubbles (large U_b), the cloud penetrates only into a thin skin outside the bubble surface. As the particle size is increased and/or the bubble size is decreased the cloud becomes larger in relation to the bubble, finally expanding without bound as the bubble velocity decreases to match the interstitial fluid velocity.

When $U_b < u_0$ the nature of the fluid flow relative to the bubble is quite different. There is now no longer a circulating cloud; instead the presence of the bubble merely serves to concentrate the fluid streamlines in its vicinity, as seen in Figure 5.5. This is not surprising since its particle-free interior offers a low resistance path for the passage of the fluidizing medium. The change in the fluid flow pattern between Figures 5.4 and 5.5 is clearly a result of the fact that

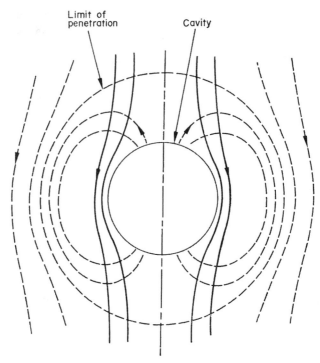

Figure 5.4. Streamlines of fluid flow (broken lines) and particle flow (unbroken lines). $U_b = 25.9$ cm/s; $u_0 = 19.9$ cm/s. (From Davidson, 1961.)

the motion far from the bubble is downward relative to the bubble in Figure 5.4 whereas it is upward in Figure 5.5.

The Davidson analysis can also be applied to a two-dimensional bubble, where the third component of velocity of both phases is constrained to vanish. The resulting stream function for the fluid is then

$$\psi_f = r \sin \theta \left[(U_b - u_0) - (U_b + u_0) \frac{r_b^2}{r^2} \right], \qquad (5.9)$$

and the corresponding flow field differs only quantitatively from that described by (5.7).

Davidson's results, particularly the prediction of a cloud of gas accompanying large bubbles, immediately stimulated experiments on the nature of the gas flow. The definitive early work is that of Rowe, Partridge & Lyall (1964), who introduced the technique of NO_2 injection. A two-dimensional bed of small particles was fluidized with air to a point where it was just on the verge of bubbling; then a bubble of substantial size was introduced by injecting a pulse

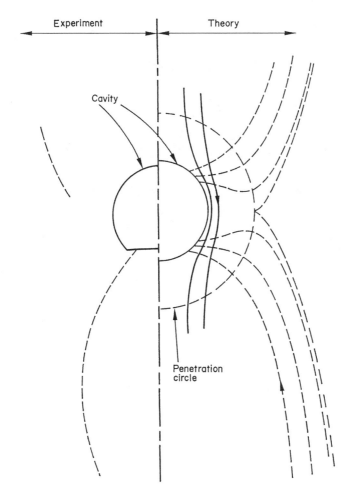

Figure 5.5. Streamlines of fluid flow (broken lines) and particle flow (unbroken lines). $U_b = 39.6$ cm/s; $u_0 = 68.5$ cm/s. (From Davidson, 1961.)

of NO_2 near the bottom of the bed. Because of its contrasting brown colour the motion of the NO_2 could be followed photographically, and it was found that a region coloured brown accompanied the bubble as it rose, with relatively minor dispersion of the brown colour into the rest of the flowing gas. The boundary of the brown region, which should delineate the boundary of the Davidson cloud, was fairly sharp, so the profile of the cloud could be traced quite accurately for comparison with the predictions of Davidson's model. Figure 5.6 shows the typical appearance and location of the cloud in relation to the bubble.

In general terms the measurements support the main predictions of Davidson's analysis. There is, indeed, a cloud of gas that remains confined to

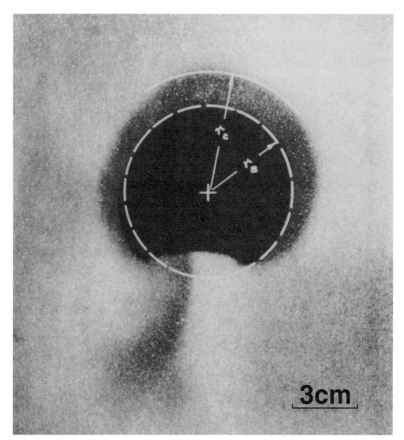

Figure 5.6. Photograph illustrating cloud identification by NO_2 injection. $U_b/u_0 = 2.5$. (From Davidson & Harrison, 1971.)

the neighbourhood of the bubble and, as seen from Figure 5.7, its size does not differ greatly from that predicted theoretically. Also, as predicted, for given particles the ratio of the cloud size to the bubble size decreases as the size of the bubble increases. The main difference between theory and observation is the location of the cloud. Davidson predicted the cloud profile to be a circle, concentric with the circular bubble boundary, but the experiments showed the centre of the cloud to be displaced significantly above the equatorial plane of the bubble. He also predicted a cloud of somewhat larger diameter than was observed in the experiments.

NO_2 injection can also be used in another way to investigate the gas flow field. If it is injected as a constant stream at a fixed point in the bed the trace of colour downstream from the injection point maps a streakline of the gas flow field (not

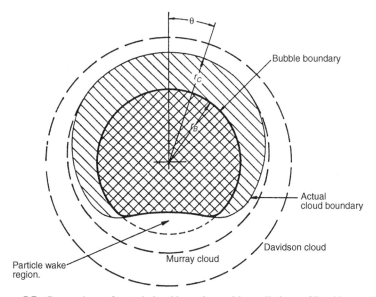

Figure 5.7. Comparison of actual cloud boundary with predictions of Davidson and of Murray. $U_b/u_0 = 2.0$. (From Davidson & Harrison, 1971.)

a streamline, since streamlines and streaklines differ in unsteady flows). Thus, by injecting streams of NO_2 at a set of points spaced along a horizontal line, the field of streaklines can be visualized (Wace & Burnett, 1961; Rowe, 1962).

Though the experiments reveal no serious weakness in Davidson's description of the fluid flow round a bubble his analysis leaves entirely open the question of the particle motion, since the form of this is simply assumed as a starting point for the analysis. Consequently, although Davidson's velocity fields satisfy the continuity equations for both phases and the momentum equation (which reduces to a force balance, since fluid inertia is neglected) for the fluid phase, the question of how they relate to the particle phase momentum balance is not addressed at all. Two attempts to complete the mechanical picture therefore followed quite shortly after Davidson's work. These will now be described in the order of their appearance.

5.3 Other Early Analyses of Bubble Motion

To appreciate the nature of the problem to be faced in attempting to satisfy all the equations of motion, for both fluid and particles, it is useful simply to count variables and equations. There are two scalar continuity equations and two vector momentum equations, for a total of eight differential equations, and to satisfy these we have available the three components of velocity for each of the phases, the pressure of the fluid (assumed incompressible), and the solids volume fraction. If the solids volume fraction is constrained to be constant, as in Davidson's theory, then to satisfy the equations of motion another variable, namely the particle-phase pressure p^p, must be left available. (This is, of course, familiar from the mechanics of incompressible fluids.) However, with Davidson's particle velocity field (5.1), the resulting p^p would neither vanish nor even be constant on the postulated spherical surface of the bubble. Since the interior of the bubble is assumed to be free of particles there would, consequently, be a mechanically unacceptable discontinuity in p^p at its surface. This difficulty can be avoided if the solids volume fraction is not constrained to be constant, but then the equations of motion become much more difficult to solve. Two strategies to overcome, or circumvent, this difficulty, and hence to satisfy a momentum equation for the particle phase, were proposed shortly after Davidson's work became known.

Jackson (1963) left the particle volume fraction ϕ unspecified but, at the same time, omitted all terms representing stresses transmitted by the particle phase. Thus his equations of motion were the two continuity equations, supplemented by a force balance for the fluid (whose inertia was neglected) and a momentum balance for the particles. In the frame of reference of the bubble the motion was assumed to be steady, so these reduced to the following form:

$$\nabla \cdot (\phi \mathbf{v}) = \nabla \cdot [(1 - \phi)\mathbf{u}] = 0, \tag{5.10}$$

$$\nabla p^f + \beta(\phi)(\mathbf{u} - \mathbf{v}) = 0, \tag{5.11}$$

$$\rho_s \phi \mathbf{v} \cdot \nabla \mathbf{v} - \rho_s \phi \mathbf{g} - \beta(\phi)(\mathbf{u} - \mathbf{v}) = 0. \tag{5.12}$$

A simplification was then introduced by noting that, for a dense suspension near the point of random close packing, β varies quite rapidly with ϕ. Thus, if the "emulsion phase" outside the bubble is near minimum fluidization, it may be justified to set $\phi = \phi_0 = $ constant in the equations, everywhere except in the function $\beta(\phi)$. Then, writing $\rho_s \phi_0 = \rho_0$ reduces the equations to

$$\nabla \cdot \mathbf{u} = \nabla \cdot \mathbf{v} = 0, \tag{5.10'}$$

$$\nabla p^f + \beta(\mathbf{u} - \mathbf{v}) = 0, \tag{5.11'}$$

$$\rho_0 \mathbf{v} \cdot \nabla \mathbf{v} - \rho_0 \mathbf{g} - \beta(\mathbf{u} - \mathbf{v}) = 0, \tag{5.12'}$$

where β now replaces ϕ as one of the unknowns. These equations in turn can be rearranged in the following form:

$$\nabla \cdot \mathbf{v} = 0, \tag{5.13}$$

$$\rho_0 \mathbf{v} \cdot \nabla \mathbf{v} + \mathbf{i}\rho_0 g + \nabla p^f = 0, \tag{5.14}$$

$$\nabla \cdot \left(\frac{1}{\beta} \nabla p^f \right) = 0, \qquad (5.15)$$

$$\mathbf{u} = \mathbf{v} - \frac{1}{\beta} \nabla p^f, \qquad (5.16)$$

where $\mathbf{i}$ denotes the unit vector in the upward vertical direction. Equations (5.13) and (5.14) are now formally identical with the equations of motion of an incompressible, inviscid fluid, and they are subject to the boundary conditions that $\mathbf{v} \to -\mathbf{i}U_b$ at points well above the bubble, where the bed is undisturbed by its presence, and $p^f = $ constant within the bubble. Both the rise velocity, U_b, and the shape of the bubble remain to be determined. This is now mathematically identical with the problem of determining the motion of a gas bubble rising through an inviscid liquid with zero surface tension, and the classic solution of Davies & Taylor (1950) can be adopted without change.

Assume initially that the bubble is spherical and take a system of polar coordinates with origin at its centre and polar axis pointing vertically upward. Then the particle velocity field is given by (5.1), as in Davidson's analysis. But now the particle-phase momentum equation can be invoked in the form (5.14) to determine the fluid pressure field. Apart from an integration constant this is given by

$$\frac{p^f}{\rho_0} = -gr\cos\theta - \frac{U_b^2 r_b^3}{2r^3}\left[\left(1 + \frac{r_b^3}{4r^3}\right) - 3\cos^2\theta\left(1 - \frac{r_b^3}{4r^3}\right)\right]. \qquad (5.17)$$

This cannot be made to satisfy the condition of constant pressure on the bubble surface, $r = r_b$, by any choice of value for U_b, which implies that a sphere is not the correct choice for the shape of the bubble. However, from (5.17) we easily find that

$$\frac{\partial p^f}{\partial \theta} = 0, \quad \frac{\partial^2 p^f}{\partial \theta^2} = gr_b - \frac{9}{4}U_b^2 \quad \text{at } \theta = 0,$$

so the first two terms in a Taylor expansion of $p(r_b, \theta)$ in powers of θ, about $\theta = 0$, can be made to vanish by choosing

$$U_b = \frac{2}{3}\sqrt{gr_b}. \qquad (5.18)$$

Then the pressure is approximately constant over a spherical cap surrounding the nose of the rising bubble. This indicates that, whatever the shape of the complete bubble, its velocity of rise is given to a good approximation by (5.18), the part of the surface near the apex is approximately a spherical cap, and the velocity field (5.1) is a good approximation to the particle motion in a region above and adjacent to the bubble.

The above is merely a recapitulation of the arguments of Davies & Taylor; to go further and determine the shape of the bubble and the velocity field completely would require the solution of a much more difficult free boundary problem. However, bearing in mind the limited domain in which it is expected to be reliable, the pressure field (5.17) can be substituted into (5.15), which then becomes an equation for β; indeed, defining $\gamma = \beta(\phi_0)/\beta$ reduces it to the following linear partial differential equation for γ:

$$\nabla\gamma \cdot \nabla p^f + \gamma \nabla^2 p^f = 0$$

or, in the polar coordinates,

$$\frac{\partial p^f}{\partial s}\frac{\partial \gamma}{\partial s} + \frac{\partial p^f}{\partial \theta}\frac{1}{s}\frac{\partial \gamma}{\partial \theta} + \gamma \nabla^2 p^f = 0, \qquad (5.19)$$

where $s = r/r_b$ and ∇^2 denotes the Laplacian with s as the radial coordinate. This cannot be solved for γ in closed form but a numerical solution can easily be generated by the method of characteristics. In terms of an arbitrary parameter h the differential equations for a characteristic curve are

$$\frac{ds}{dh} = \frac{\partial p^f}{\partial s}, \quad \frac{d\theta}{dh} = \frac{1}{s}\frac{\partial p^f}{\partial \theta}, \quad \frac{d\gamma}{dh} = -\gamma \nabla^2 p^f, \qquad (5.20)$$

and these can be integrated downward from points well above the bubble where the bed is undisturbed and $\gamma = 1$. By starting the integration from a number of points $(s_0, \theta_0, 1)$ on a plane some distance above the apex of the bubble the spatial dependence of γ can be determined throughout the region where (5.17) is expected to provide a good representation of the pressure field. The result is shown in Figure 5.8 as a contour map of $1/\gamma = \beta/\beta_0$ in a plane containing the vertical axis through the centre of the bubble. Only the region above the equatorial plane is shown, since the pressure field used is certainly unreliable below this plane. We see that the upper part of the bubble is predicted to be surrounded by a mantle in which the value of β, and hence the particle concentration, is lower than in the bulk of the bed at points distant from the bubble. (Note that this is exactly the opposite of early views on the packing of particles above the bubble.) At the nose of the bubble β is reduced to half its value in the bulk of the bed but, because of our initial assumption that β is a rapidly varying function of the particle concentration, this will be accompanied by a much smaller percentage reduction in the value of ϕ.

Once γ has been found it can be used, together with the known fields **v** and p^f, to calculate **u** using (5.16). The resulting fluid velocity fields are shown in Figures 5.9 and 5.10, for cases where $U_b < u_0$ and $U_b > u_0$, respectively.

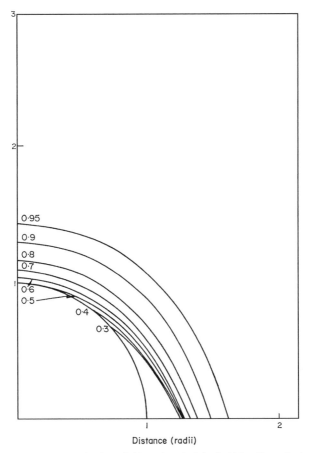

Distance (radii)

Figure 5.8. The ratio β/β_0 in the neighbourhood of the bubble. (From Jackson, 1963.)

Compared with the results of Davidson we see immediately that the streamline patterns are shifted upward relative to the bubble, but a more detailed discussion of these predictions will be deferred until a second approximate theory has been described.

In summary, Jackson's analysis permits the particle concentration to vary slightly with position and uses this freedom to satisfy both the continuity equations and both the momentum equations, to a first approximation, in a region around and above the bubble. Because the approximations break down below the bubble equator, the velocity fields in this region and the shape of the lower part of the bubble surface cannot be found.

An alternative approach to the problem has been proposed by Murray (1965) based on an approximation of the Oseen type. In the absence of a bubble

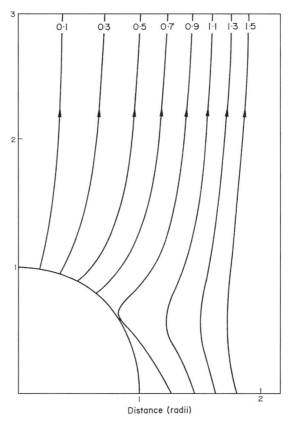

Figure 5.9. Streamlines of fluid flow in the neighbourhood of the bubble. $u_0/U_b = 1.730$. (From Jackson, 1963.)

we have

$$\mathbf{u} = (u_0 - U_b)\mathbf{i}, \quad \mathbf{v} = -U_b\mathbf{i}, \quad \phi = \phi_0, \quad p^f = p_0^f - \beta_0 u_0 x$$

in a frame moving upward with speed U_b, where x is a coordinate measured vertically upward. In the presence of a bubble these values should be approached at large distances from the bubble, so we can write

$$\mathbf{u} = (u_0 - U_b)\mathbf{i} + \mathbf{u}_1, \quad \mathbf{v} = -U_b\mathbf{i} + \mathbf{v}_1,$$

$$\phi = \phi_0 + \phi_1, \quad p^f = p_0^f - \beta_0 u_0 x + p_1^f, \tag{5.21}$$

where $\mathbf{u}_1$, $\mathbf{v}_1$, ϕ_1, and p_1^f may be regarded as small perturbations at points not too near the bubble. Then, introducing (5.21) into Equations (5.10)–(5.12) and

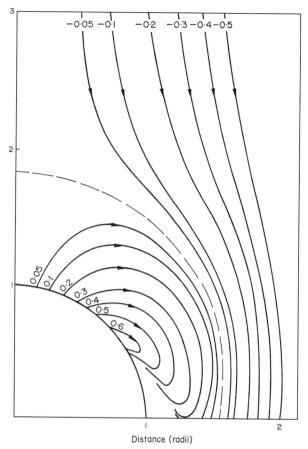

Figure 5.10. Streamlines of fluid flow in the neighbourhood of the bubble. $u_0/U_b = 0.768$. (From Jackson, 1963.)

linearizing with respect to these perturbations, we obtain

$$\phi_0 \nabla \cdot \mathbf{v}_1 - U_b \frac{\partial \phi_1}{\partial x} = 0, \tag{5.22}$$

$$(1 - \phi_0)\nabla \cdot \mathbf{u}_1 - (u_0 - U_b)\frac{\partial \phi_1}{\partial x} = 0, \tag{5.23}$$

$$\nabla p_1^f + \mathbf{i} u_0 \beta_0' \phi_1 + \beta_0(\mathbf{u}_1 - \mathbf{v}_1) = 0, \tag{5.24}$$

$$\rho_p \phi_0 U_b \frac{\partial \mathbf{v}_1}{\partial x} - \mathbf{i}\rho_p g \phi_1 + \mathbf{i} u_0 \beta_0' \phi_1 + \beta_0(\mathbf{u}_1 - \mathbf{v}_1) = 0, \tag{5.25}$$

where β_0' denotes $d\beta/d\phi$, evaluated at $\phi = \phi_0$. These equations should determine the perturbing fields at distances from the bubble sufficiently large that terms of the second order in these perturbations are negligible compared to terms of the first order. They are equivalent to the Oseen approximation in the mechanics of a single-phase fluid, where this approximation is used everywhere outside the body (or bubble) round which the fluid is flowing. The justification for applying the approximation near the body is that terms quadratic in the perturbations, while no longer negligible compared to the linear terms, are here small compared to other terms arising from viscous forces. Since these are linear, the problem is thus describable, to this approximation, by linear equations throughout the domain of interest. This argument cannot be used to justify the use of (5.22) to (5.25) everywhere outside the bubble in the present case, since the original equations of motion, (5.10)–(5.12) contained no viscous terms. Nevertheless, bearing in mind this reservation and following Murray, we explore the consequences of (5.22)–(5.25).

A differential equation to be satisfied by ϕ_1 can be obtained by taking the divergence of (5.25) then using (5.22) and (5.23) to eliminate $\nabla \cdot \mathbf{u}_1$ and $\nabla \cdot \mathbf{v}_1$ from the result, giving

$$\frac{\partial^2 \phi_1}{\partial x^2} = A \frac{\partial \phi_1}{\partial x},$$

where

$$A = \left(1 / \rho_p U_b^2\right)[\rho_p g - u_0 \beta_0' + U_b \beta_0 / \phi_0 (1 - \phi_0) - u_0 \beta_0 / (1 - \phi_0)].$$

This has the general solution

$$\phi_1 = f(y, z) + g(y, z) e^{Ax},$$

where f and g are arbitrary functions of y and z, the Cartesian coordinates in the plane normal to x. But this increases in magnitude without bound, either as $x \to \infty$, if $A > 0$, or as $x \to -\infty$, if $A < 0$, unless $g \equiv 0$. Since ϕ_1 must tend to zero at large distances from the bubble we can conclude that g must vanish. By the same token f must also vanish, so the relevant solution is simply $\phi_1 \equiv 0$. Thus, at distances from the bubble so large that terms quadratic in $\mathbf{u}_1$, $\mathbf{v}_1$, and p_1^f can be neglected, terms *linear* in ϕ_1 can be neglected.

Setting $\phi_1 \equiv 0$ in (5.22)–(5.25) then rewriting these equations in terms of $\mathbf{u}$, $\mathbf{v}$, and p^f, with the help of (5.21), we obtain the following:

$$\nabla \cdot \mathbf{v} = 0, \tag{5.26}$$

$$\nabla \cdot \mathbf{u} = 0, \tag{5.27}$$

$$\nabla p^f + \beta_0 (\mathbf{u} - \mathbf{v}) = 0, \tag{5.28}$$

$$\rho_p \phi_0 U_b \frac{\partial \mathbf{v}}{\partial x} - \rho_p \phi_0 g \mathbf{i} + \beta_0 (\mathbf{u} - \mathbf{v}) = 0, \tag{5.29}$$

where use has also been made of the force balance $\beta_0 u_0 = \rho_p \phi_0 g$ in the undisturbed bed. These provide eight equations and, since there are now only seven unknown functions, they cannot be independent; indeed, if **u** and **v** satisfy (5.26) and (5.29) then (5.27) is necessarily satisfied. The first three equations, (5.26)–(5.28), are identical with Equations (5.2) and (5.3) of Davidson's analysis so, as in that case, it follows that p^f must satisfy Laplace's equation. However, as we shall see, to satisfy the approximate form (5.29) of the particle momentum equation Murray relaxes the requirement that p^f should be constant over the whole surface of the bubble. His solution is initiated by taking the particle velocity field to be that of potential flow round a sphere or, in the case of a two-dimensional bubble, a cylinder. This coincides with the assumptions of both Davidson and Jackson. In the two-dimensional case this means that the particle velocity can be found from the following complex potential:

$$w_p = U_b \left(z + \frac{r_b^2}{z} \right). \tag{5.30}$$

The fluid velocity field then follows from (5.29). It is also irrotational and is determined by the complex potential

$$w_f = -u_0 z + w_p - \frac{u_0 U_b}{g} \frac{d w_p}{dz} + w_{f0}. \tag{5.31}$$

Knowing the particle and fluid velocity fields the pressure is then determined by integrating (5.28), with the result

$$p^f - p_0^f = \beta_0 \mathrm{Re}\,(w_f - w_p). \tag{5.32}$$

The integration constants w_{f0} and p_0^f merely make the pressure arbitrary to the extent of an additive constant. Then, using (5.30) and (5.31) and dropping integration constants reduces (5.32) to

$$\frac{p^f}{u_0 \beta_0} = \mathrm{Re} \left(\frac{U_b^2}{g} \frac{r_b^2}{z^2} - z \right).$$

On the surface of the bubble where $z = r_b e^{i\theta}$ this gives

$$\frac{p^f}{u_0 \beta_0} = \frac{U_b^2}{g} \cos 2\theta - r_b \cos \theta, \tag{5.33}$$

which is clearly not constant. However, as in Jackson's analysis, the first two terms in the Taylor expansion of p^f in powers of θ can be made to vanish by a suitable choice for the value of the bubble velocity U_b. Expanding (5.33) about $\theta = 0$ gives

$$\frac{p^f}{u_0 \beta_0} = \left(\frac{U_b^2}{g} - r_b \right) + \left(\frac{r_b}{2} - \frac{2 U_b^2}{g} \right) \theta^2 + O(\theta^4).$$

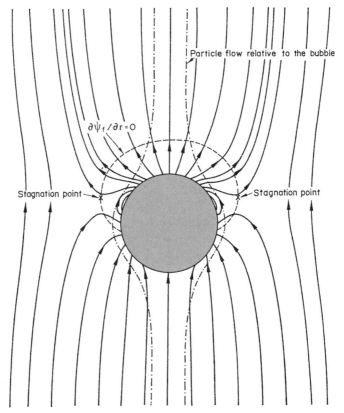

Figure 5.11. Streamlines of fluid flow in the neighbourhood of a two-dimensional bubble. $U_b/u_0 = 0.6$. (From Murray, 1965.)

The term of $O(\theta^2)$ vanishes when

$$U_b = \frac{1}{2}\sqrt{gr_b},$$

which is the appropriate expression for the rise velocity of a Davies–Taylor bubble in two dimensions. Murray's solution is now complete, and the predicted streamlines for fluid flow are shown in Figures 5.11 and 5.12, for $U_b/u_0 = 0.6$ and 2.4, respectively.

To recapitulate briefly, Murray's analysis shows that, if the requirement of constant pressure over the whole bubble surface is relaxed, demanding instead only that p^f should be approximately constant over a limited domain about the nose of the bubble, then an Oseen type of approximation to the particle-phase momentum balance can be satisfied while maintaining a constant value ϕ_0 for the volume fraction everywhere outside the bubble. Furthermore, the solution

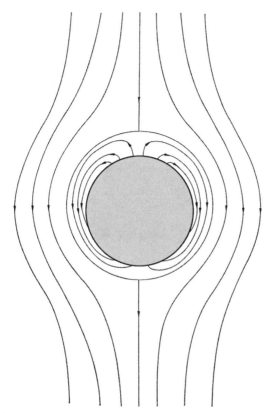

Figure 5.12. Streamlines of fluid flow in the neighbourhood of a two-dimensional bubble. $U_b/u_0 = 2.4$. (From Murray, 1965.)

can be obtained in a closed form and the streamlines of both particle flow and fluid flow can, consequently, be traced throughout the external region, as in Figures 5.11 and 5.12. However, some circumspection must be exercised in interpreting these streamline maps. Since the pressure boundary condition is only approximated well over part of the surface near the apex of the bubble, the predicted pressure and velocity fields should not be taken too seriously much further down. (This is the same situation as in Jackson's analysis where, for this reason, the streamlines were not continued below the equatorial plane of the bubble. See Figures 5.9 and 5.10.) Jackson's analysis, in contrast, leaves the volume fraction free and can therefore, in principle, satisfy the particle-phase momentum balance without invoking the Oseen approximation. Consequently it should yield a better approximation to the complete solution, but this advantage is offset by the fact that the results can no longer be obtained in closed form.

Comparisons of the two solutions and their relation to experimental observations will be deferred to Section 5.4.

Figures 5.1 and 5.2 show that the shape of real bubbles is by no means spherical (or circular in the two-dimensional case). The observed bubbles are strongly indented, with bases that are convex upward, so all the above analyses start from a bubble whose lower part has an unrealistic shape. This is another compelling reason for doubting their predictions below the bubble equator. However, as pointed out by Collins (1965a), in two dimensions it is not difficult to extend Davidson's type of analysis to bubbles of more realistic shape using conformal mapping to generate an irrotational flow field for the particles round a void of selected shape.

The complex potential for the particle flow round a circular, two-dimensional bubble has already been given as Equation (5.30). The corresponding complex potential for the fluid flow, with stream function (5.9), is given by

$$w_f = (U_b - u_0)z + (U_b + u_0)\frac{r_b^2}{z}. \tag{5.34}$$

Collins explored conformal transformations of the form

$$z = \zeta + b - \frac{c^2}{\zeta + b} \tag{5.35}$$

and showed that, with the parameters $b = 5r_b/7$, $c = 2r_b/7$, a circle of radius r_b centred at the origin of the ζ plane is transformed into a closed curve in the z plane that resembles closely the observed shapes of bubbles in two-dimensional beds. Applying the same transformation of variables to the argument z of (5.30) and (5.34) will then give the velocity potentials for the particle and fluid flow fields round a bubble of this transformed shape. Figure 5.13, from Collins (1965a), shows the streamlines of fluid flow, obtained in this way, for the two cases $U_b/u_0 = 3/5$ and $U_b/u_0 = 5/3$. Except in the neighbourhood of the indentation at the base of the bubble they differ only slightly from the streamlines for the circular bubble found by Davidson.

Collins (1965b) has also pointed out an interesting feature of Davidson's analysis, compared with the later treatments of Jackson and Murray. Davidson's flow fields do not satisfy the particle-phase momentum balance, while both the latter treatments, though approximate, make an attempt to include this equation. As a result they are able to determine the bubble rise velocity in terms of its radius, by essentially the same approach as that used by Davies & Taylor for large gas bubbles in a liquid. However, Collins showed that, by suitable choice of the bubble rise velocity, Davidson's solution could be made to satisfy the peripheral component of the particle momentum equation, exactly at the

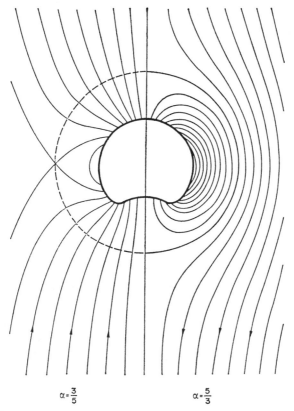

$$\alpha = \frac{3}{5} \qquad\qquad \alpha = \frac{5}{3}$$

Figure 5.13. Streamlines of fluid flow in the neighbourhood of a two-dimensional bubble for $U_b/u_0 = 3/5$ and $U_b/u_0 = 5/3$. Reprinted from Collins (1965a). Copyright 1965, with permission from Elsevier Science.

nose of the bubble and approximately in some neighbourhood of this point. Furthermore, the value of the velocity needed to accomplish this coincides with the Davies–Taylor velocity.

Finally, Collins (1965a) extended the Davidson analysis to the case of a two-dimensional bubble rising in a bed confined between parallel vertical walls, and Stewart & Davidson (1967) used the same ideas to describe large slugs of gas rising in fluidized beds confined within vertical tubes.

All the work described so far has addressed the steady motion of a fully developed bubble. However, in a paper often overlooked, Murray (1967) used his approximation to the equations of motion to investigate the initial motion of a circular or spherical void, imagined somehow to be introduced into the bed. He was then able to trace the evolution of the bubble shape and follow the development of the upward indentation of the base that is so characteristic of observed bubbles. The predictions were found to be in qualitative agreement with experiments of Partridge & Lyall (1967).

5.4 Experimental Tests of Theories of the Motion of a Fully Developed Bubble

Each of the theoretical approaches to bubble motion described above predicts the fluid and particle velocity fields in the reference frame of the bubble, together with the fluid pressure field, though the results would not be expected to be accurate below the equatorial plane of the bubble. The bubble rise velocity is predicted to be related to the radius of curvature of its apex by the Davies–Taylor expression. Jackson's analysis also predicts the variations in volume fraction of solids to be expected above the upper part of the bubble surface. Clearly these results provide fodder for comparisons with experiments that may indicate which approach is best.

Although there have been many measurements relating the rise velocity of bubbles to their size, the results have proved disappointing as a test of the theories. There are two reasons for this. First, it is not easy to measure the volume of a rising bubble accurately; second, the Davies–Taylor formula relates the rise velocity not to the volume, but to the radius of curvature of the bubble surface at its apex. A unique relation between this and the volume exists only if bubbles of all sizes are geometrically similar. Since there are significant variations in the shapes of observed bubbles a definitive, experimentally established relation between rise velocity and surface curvature, for comparison with the theoretical predictions, has not been found.

For bubbles large enough that $U_b/u_0 = \alpha > 1$, which are accompanied by circulating "clouds" of gas, observations of the shape and size of the cloud using the NO_2 injection technique described above can be compared with predictions from the theories. Figure 5.6 shows a photograph of the cloud accompanying a two-dimensional bubble with $\alpha = 2.5$. It is seen to be displaced upward relative to the centre of the bubble, as predicted by Jackson and by Murray. This photograph also serves to establish a definition for a "cloud radius", r_c. A circle is first fitted to the outline of the bubble, ignoring the indentation in its base. The radius of this circle is then identified as r_b and its centre is regarded as the centre of the bubble. The cloud radius, r_c, is then defined as the vertical distance separating the highest point of the cloud from the centre of the bubble. Measurements of cloud size then relate $S (= r_c/r_b)$ to the ratio U_b/u_0, and these measurements can be compared with predictions from the theories. Expressions relating U_b/u_0 to S, from each theory, are given by Stewart (1968) and these are reproduced in Table 5.1.

Table 5.1. *Theoretical expressions relating cloud*
size and U_b/u_0

| Theory | Expressions for $U_b/u_0(=\alpha)$ | |
	Two dimensions	Three dimensions
Davidson	$\dfrac{S^2+1}{S^2-1}$	$\dfrac{S^3+2}{S^3-1}$
Murray	$\dfrac{S^2+\frac{1}{2S}}{S^2-1}$	$\dfrac{S^3+\frac{1}{S}}{S^3-1}$
Jackson	$\dfrac{S^3+\frac{4}{3S}\left(1-\frac{1}{S^3}\right)}{S^3-1}\gamma$	$\dfrac{S^3+\frac{4}{3S}\left(1-\frac{1}{S^3}\right)}{S^3-1}\gamma$

Davidson's and Murray's analyses both give explicit expressions for U_b/u_0, while in the case of Jackson's analysis the result contains $\gamma(=\beta_0/\beta)$ as a factor, where γ is to be evaluated at $r=r_c, \theta=0$. Thus γ is itself a function of S, and this must be found by numerical integration down the axis. In the three-dimensional case the relevant equation is (5.19), which reduces to

$$\frac{d\gamma}{ds}+\gamma\frac{\nabla^2 p^f}{\frac{dp^f}{ds}}=0.$$

With the rise velocity given by the Davies–Taylor expression (5.18) the ratio of $\nabla^2 p^f$ to dp^f/ds is independent of r_b or U_b, so $\gamma(S)$ can be determined by a single numerical integration of the above equation. Similar considerations apply in the two-dimensional case.

Figure 5.14 shows curves relating S to u_0/U_b, computed from each of the theories using the results in Table 5.1, superimposed on some experimental cloud size measurements of Stewart. Notice that Jackson's and Murray's analyses predict almost identical values of the cloud radius, which are significantly smaller than those from Davidson's analysis. The experimental points are very scattered because of the difficulty of determining cloud sizes photographically in three-dimensional beds, but they are mostly contained between the theoretical curves. In two-dimensional beds the cloud size can be found with greater certainty and Figure 5.15 presents experimental results of Rowe et al. (1964), in the form of $((U_b/u_0)-1)S^2$ plotted against U_b/u_0. The results, though still scattered, clearly favour the predictions of Murray or Jackson, which again agree closely but lie well below those of Davidson.

Figures 5.14 and 5.15 show that the cloud radii predicted by Murray and by Jackson agree closely. It should be noted in passing that agreement between the

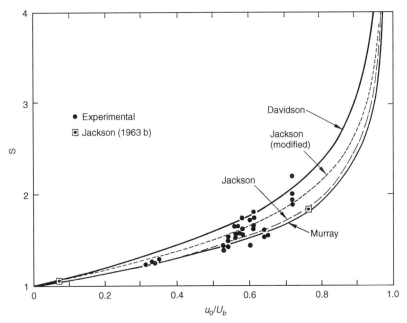

Figure 5.14. Ratio of cloud radius to bubble radius for three-dimensional bubbles. Comparison of measured values and theoretical predictions. (From Stewart, 1968.) ("Jackson (modified)" refers to Jackson analysis, but with bubble velocity $k\sqrt{gr_b}$, with k chosen to give best match with measured velocities.)

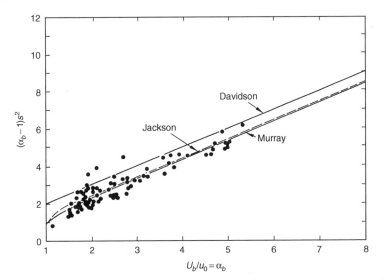

Figure 5.15. Comparison of measured cloud sizes for two-dimensional bubbles with theoretical predictions. (From Stewart, 1968.)

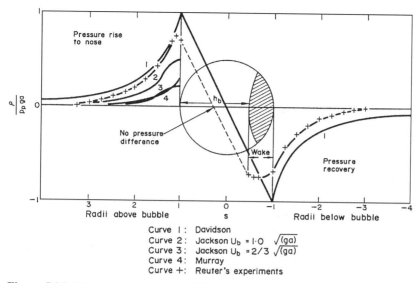

Figure 5.16. Dimensionless pressure difference between points on bubble axis and points far from bubble at same level, for three-dimensional bubbles. (From Stewart, 1968.)

two methods goes further; Rowe et al. (1965) have plotted the complete profiles of the cloud boundary, calculated by both methods for $U_b/u_0 = 1.3$, and found them to be almost identical.

Each of the theories also predicts the gas pressure field and these predictions have been compared with some experimental measurements by Reuter (1963a,b). See Figure 5.16, taken from Stewart (1968). Once again Jackson's and Murray's methods give very similar results, but both predict significantly smaller values of the pressure rise above the bubble than does Davidson's method. Reuter's experimental points lie between the two, but they are closer to Davidson's predictions. However, the experiments indicate some downward turn in the pressure curve just above the apex of the bubble, as predicted by Jackson's theory.

Finally, Jackson's analysis addresses the question of variations in the particle concentration and predicts the existence of a mantle of somewhat reduced concentration immediately above the bubble, so it is important to ask whether this can be confirmed experimentally. Though measurement of instantaneous values of the volume fraction at identified positions near a rising bubble presents considerable difficulties, Lockett & Harrison (1967) reported results from a two-dimensional bed using a capacitance method to determine the void fraction and synchronized cine-photography of the bed to locate the instantaneous position of the capacitance measurement relative to the rising bubble.

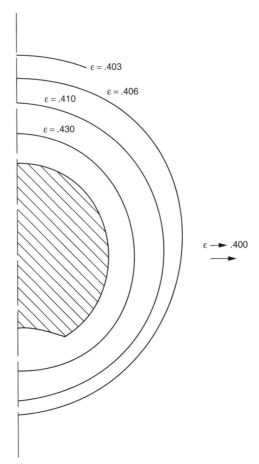

Figure 5.17. Void fraction in the neighbourhood of a rising bubble. Contours fitted to measurements of Lockett & Harrison (1967).

Two types of particles were used; both were glass beads with a mean diameter of approximately 0.5 mm, but in one case the size range was narrow, while in the other it was much broader. Figure 5.17 shows a contour map of the experimentally determined void fraction around a bubble rising in a fluidized bed of the closely sized beads with an undisturbed void fraction $\varepsilon_0 = 0.4$. A corresponding theoretical contour map is presented in Figure 5.18, generated from values of β/β_0 calculated using the two-dimensional form of Jackson's analysis by using the Carman–Kozeny equation to translate β into ε.

The general picture is similar in both figures but the experimentally determined mantle of reduced particle concentration is larger in extent than that predicted theoretically. There are several possible explanations for this. First, the

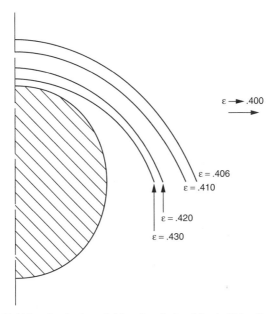

Figure 5.18. Void fraction in the neighbourhood of a rising bubble. Contours predicted by Lockett & Harrison (1967) using a two-dimensional form of Jackson's method.

equations of motion used in the theory are gross approximations, neglecting all terms representing stresses transmitted in the particle phase; second, the equations used to predict β are only a first approximation to these equations of motion; and, third, the Carman–Kozeny formula is probably better replaced by the Richardson–Zaki equation as a means of translating between β and ε. This replacement would spread the contours in Figure 5.18 a little wider, but the effect is small. On the experimental side the electrodes used in the capacitance measurement of ε have a nonvanishing area (they are 1.6 cm × 0.5 cm strips of aluminium foil) and this will smooth out the true variations in volume fraction, thereby extending them over a wider area. Also, as pointed out by Lockett & Harrison, the "two-dimensional" bed actually has a finite thickness (1 cm) in the third dimension, so the bubbles observed lie somewhere between ideal two- and three-dimensional bubbles, and the theory for the three-dimensional case predicts a somewhat larger mantle of reduced density. It is likely that the biggest contributor to the discrepancy between predicted and measured results is the crudely approximate nature of the equations of motion; each of the remaining factors would reduce the discrepancy between the predicted and observed concentration distributions but it is unlikely that they could account for all of it. Thus, the most that can be said is that the experiments establish, beyond reasonable doubt, the existence of the predicted mantle of reduced particle concentration, and they rule out earlier ideas of a "roof" to the bubble, where the particle concentration is higher than that in the bulk of the bed.

Though this, and the previous two sections have been concerned primarily with early theoretical and experimental work, it is worth remarking that some much more recent X-ray measurements on three-dimensional bubbles (Yates & Cheesman, 1992; Yates et al., 1994) confirm the existence of a region of reduced particle concentration surrounding bubbles. Buyevich et al. (1995) have compared these experimental results with theoretical predictions of the particle volume fraction around the bubble in its equatorial plane.

Summarizing the material of Sections 5.2 to 5.4, we see that three early attempts at theoretical descriptions of the motion of particles and gas near rising bubbles all succeed in capturing correctly the salient features of the velocity fields. The fields predicted by Jackson and by Murray appear to be in close agreement with each other, and in better quantitative agreement with observations than the predictions of Davidson. Only Jackson's theory takes account of variations in the particle concentration, which it predicts to be lower near the bubble than in the bulk of the bed. This is also borne out experimentally and the reductions in concentration near the bubble are similar to those predicted by Jackson, though they are spread over a wider region. Thus, this early work proved remarkably successful in elucidating the mechanics of bubble motion. What it could not do, since it focussed attention on fully developed bubbles, was shed any light on the genesis of bubbles or account for the fact that they are commonly observed in gas fluidized beds, but not in most liquid fluidized beds. In the rest of this chapter we shall turn attention to these and related questions.

5.5 Extensions of Stability Analysis and the Genesis of Bubbles and Other Structures

5.5.1 Introduction

In Chapter 4 conventional linear stability analysis was used to examine the stability of the state of uniform fluidization and it was found that most fluidized beds of particles with size and density in the ranges of technical interest are unstable, whether fluidized by a gas or a liquid. Careful observations of water fluidized beds were shown to have confirmed the presence of instabilities of the sort predicted, though bubbles never appeared. In the linearised theoretical treatment gas and liquid fluidized systems were found to be distinguished primarily by the rate of growth of small perturbations, which is typically much greater in gas fluidized beds than in liquid fluidized beds. But this alone does not explain why the gas fluidized beds mostly bubble, while the liquid fluidized beds do not, however great their depth. The cause of this difference in behaviour must, therefore, be sought in something beyond linear stability analysis. Accordingly, over the years, a number of attempts have been made to follow the development of disturbances beyond the point at which they can be described adequately as perturbations so small in amplitude that the linearised equations of motion suffice to describe their propagation. Recall that the dominant (i.e., fastest growing) disturbance in the linear theory was found to be a plane wave propagating vertically upward through the bed. The particle concentration and the local average velocities of fluid and particles are then functions only of time and a single space coordinate, measured vertically upward; in other words, the waves are one dimensional. The generalizations to be described below are of two kinds. In the first the one-dimensional nature of the motion is preserved but the effect of nonlinearities in the equations of motion is taken into account while, in the second, departures from the one-dimensional nature of the disturbances are explored.

5.5.2 Nonlinear Treatments of One-Dimensional Waves

On the basis of their dispersion relation for the velocity of small-amplitude waves Pigford & Baron (1965) speculated about the possibility of shock formation in unstable beds. However, the first detailed investigation of the effect of nonlinearities in the equations of motion on vertically propagating plane waves appears to be that of Fanucci et al. (1979). Recall from Chapter 4 that the fluid-phase momentum equation can be dispensed with when attention is limited to one-dimensional motions. Then, if an equation of the type (2.34) is used for the particle momentum balance, the fluid pressure no longer appears in the problem. Fanucci et al. adopted equations of this sort, omitting terms representing viscous stresses from the particle-phase momentum balance and assuming that the interaction force $n\mathbf{f}_2$ is simply proportional to the velocity difference $\mathbf{u} - \mathbf{v}$. Then the particle momentum equation and the two continuity equations form a hyperbolic set of partial differential equations whose solution can be reduced by the method of characteristics to the integration of ordinary differential equations.

Fanucci et al. performed this integration numerically and Figure 5.19 is typical of their results. This shows one wavelength of the profile of solids volume fraction ϕ (denoted by ε_p on the diagram), at a succession of times, and corresponding positions in the bed. The relation between time and position then gives the velocity of propagation. Two sets of profiles are shown; the broken curves are from solutions of a linearised form of the equations, whereas the unbroken curves come from the full, nonlinear characteristic equations. The initial perturbation is sinusoidal and the solution of the linearised equations retains this form, but with the full equations the ϕ profile becomes progressively

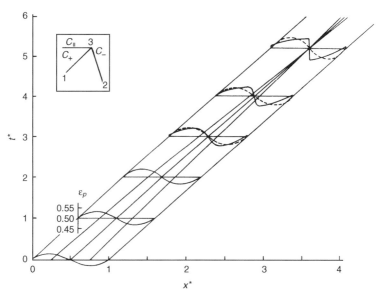

Figure 5.19. Propagation of the one-dimensional ϕ-wave. Broken curves: linearised equations. Unbroken curves: full equations. Inset shows characteristics. (From Fanucci et al., 1979.)

more distorted, eventually forming a shock across which ϕ jumps discontinuously. The particle concentration then increases continuously with height to a maximum value, at which point it decreases discontinuously, thus initiating another identical cycle. Fanucci et al. arbitrarily identify the appearance of a shock with bubble formation, though the shock and the pattern of concentration that accompanies it bear almost no resemblance to a bubble in a fluidized bed. For a given amplitude of initial disturbance the time to shock formation is found to increase with a decrease in particle size, a decrease in solids density, and a decrease in the bed expansion. All these are changes that also decrease the initial rate of growth of small disturbances, as predicted by the linearised equations. An interesting and perhaps unexpected result of the analysis is that shocks can form, in certain circumstances, even when the amplitude of the initial disturbance decreases monotonically with increasing time. Unfortunately since all the reported results are for a fluid density of 1.2 kg/m³, which is typical of a gas fluidized bed, they do not shed any light on the question of why typical liquid fluidized beds do not bubble. In a second paper (Fanucci et al., 1981) the same authors introduce viscous terms into their momentum equations and go on to investigate the structure and stability of the shocks.

Subsequent work on one-dimensional disturbances has followed three lines: first, direct numerical integration of the full equations of motion starting from the

Table 5.2. *Parameter values*

	Glass beads in air	Glass beads in water
ρ_s	2.2 g/cm^3	2.2 g/cm^3
ρ_f	0.0013 g/cm^3	1.0 g/cm^3
a	100 μm	0.5 mm
μ^f	0.0181 cp	1.0 cp
v_t	142 cm/s	14.3 cm/s
n	4.25	3.65
ϕ_m	0.65	0.65
M (Eq. 5.42)	0.571p (0.5 p)	0.571p (0.5 p)
r (Eq. 5.41)	0.3	0.3
P (Eq. 5.41)	1.08 Pa	0.027 Pa
ϕ_c	0.576	0.576
ϕ_0	0.57	0.57
δ	0.00059	0.455
γ	0.00036	0.02
α	0.00024	0.00059
Ω	113.3	3.62
V	22.2 cm/s	3.18 cm/s

small-amplitude sinusoidal wave that satisfies the linearised equations, second, bifurcation theory and computational searches to identify fully developed wave patterns that propagate without change of form, and, third, analyses yielding explicit results obtainable when the waves can be assumed to be only weakly nonlinear. Results have accumulated gradually over the three decades since the first paper of Fanucci et al., with considerable overlap between different workers so, rather than attempting a presentation in historical order, we shall first examine the situation as revealed by the most recent work, then comment on the earlier contributions in relation to this.

At the time of writing the most recent and extensive computational approach is reported in three papers (Anderson et al., 1995; Glasser et al., 1996, 1997), the first of which focusses entirely on transient integration of the full equations of motion, while the last two study branches of fully developed waves and explore their bifurcations. All three papers address the question of two-dimensional structures as well as plane waves, but we shall defer discussion of this aspect of the problem until the next section.

Two model systems are common to both the above groups, namely 200 μm diameter glass beads fluidized by air at ambient conditions and 500 μm diameter glass beads fluidized by water. Their properties are listed in Table 5.2. The former is typical of gas fluidized beds that bubble as soon as they are expanded, while the latter is a typical liquid fluidized bed in which bubbles are not seen.

The same equations of motion were used by both groups, though this is not immediately apparent since they quote the momentum equations as different linear combinations, with Anderson et al. using the form found in Chapter 2 as Equations (2.33) and (2.34), whereas Glasser et al. present the equations in the equivalent form (2.31) and (2.32). For convenience we reproduce (2.33) and (2.34) below:

$$\rho_f \frac{D_f \mathbf{u}}{Dt} = \nabla \cdot \mathbf{S}^f - n\mathbf{f}_2 + \rho_f \mathbf{g}, \qquad (5.36)$$

$$\rho_s \phi \frac{D_p \mathbf{v}}{Dt} = \nabla \cdot \mathbf{S}^p + n\mathbf{f}_2 + (\rho_s - \rho_f)\phi \mathbf{g} + \rho_f \phi \frac{D_f \mathbf{u}}{Dt}. \qquad (5.37)$$

The simplest credible expressions were chosen for the stress tensors $\mathbf{S}^f$ and $\mathbf{S}^p$ and the interaction force $n\mathbf{f}_2$, namely

$$\mathbf{S}^f = -p^f \mathbf{I} + \mu^f \left[\nabla \mathbf{u} + (\nabla \mathbf{u})^T - \frac{2}{3}(\nabla \cdot \mathbf{u})\mathbf{I} \right], \qquad (5.38)$$

$$\mathbf{S}^p = -p^p \mathbf{I} + \mu^p \left[\nabla \mathbf{v} + (\nabla \mathbf{v})^T - \frac{2}{3}(\nabla \cdot \mathbf{v})\mathbf{I} \right], \qquad (5.39)$$

$$n\mathbf{f}_2 = \beta(\phi)(\mathbf{u} - \mathbf{v}), \qquad (5.40)$$

with the following expressions for p^p, μ^p and β:

$$p^p = P\phi^3 \exp\left(\frac{r\phi}{\phi_m - \phi} \right), \qquad (5.41)$$

$$\mu^p = \frac{M\phi}{1 - (\phi/\phi_m)^{1/3}}, \qquad (5.42)$$

$$\beta = \frac{(\rho_s - \rho_f)\phi g}{v_t(1 - \phi)^{n-1}}. \qquad (5.43)$$

These contain the parameters P, r, M, n, ϕ_m, and v_t, where ϕ_m represents the volume fraction of solids at random close packing and v_t is the terminal velocity of fall of an isolated particle. The parameter values are specified in Table 5.2. The forms of (5.41) and (5.42) are such that p^p and μ^p both increase monotonically with increasing ϕ, starting from zero at $\phi = 0$ and diverging as $\phi \to \phi_m$. This divergence of p^p is important since it prevents the particles from packing closer than ϕ_m and hence, as we shall see, is responsible for the existence of fully developed waves in the bed.

Some comments are necessary on the choice of parameter values in Table 5.2. Two values are recorded for M, the first used by Anderson et al. and the second by Glasser et al. The difference is not enough to affect the results significantly, and the value is such that μ^p in the unperturbed bed (with $\phi = \phi_0 = 0.57$) is 6.65 p, which is in the range of typical measured values for the effective viscosities of both liquid and gas fluidized beds. The values quoted for P are determined in the following way. Both systems are of the sort that are observed to become unstable as soon as they are expanded, and expansion is expected to begin at about the same particle concentration in each case. Thus it is reasonable to require ϕ_c, the volume fraction at which the uniform bed becomes unstable, to be the same in both cases. This is achieved by adjusting the value of P, and we see from Table 5.2 that ϕ_c takes the value 0.576 for both beds, while P is much larger for the gas fluidized bed than for the water fluidized bed. This is to be expected, since the particle-phase pressure exerts a stabilizing influence and the gas fluidized bed is expected to be harder to stabilize than the water fluidized bed. In each case the unperturbed bed has $\phi_0 = 0.57$, so each represents a dense fluidized bed expanded to the same extent, just beyond the point of limiting stability, $\phi = \phi_c$.

Glasser et al. render their equations dimensionless using the following scaling quantities:

$$\text{length: } l = (M v_t / \rho_s g)^{1/2}, \quad \text{mass: } m = \rho_s l^3, \quad \text{time: } t = l/v_t. \quad (5.44a)$$

The resulting equations then contain four dimensionless parameters, namely

$$\delta = \frac{\rho_f}{\rho_s}, \quad \gamma = \frac{\mu^f}{M}, \quad \alpha = \frac{P}{\rho_s v_t^2}, \quad \Omega = \left(\frac{\rho_s v_t^3}{Mg}\right)^{1/2}. \quad (5.44b)$$

The physical meaning of the first three is obvious while the fourth may be regarded either as a Reynolds number $Re = \rho_s v_t l / M$ or a Froude number $Fr = v_t^2/gl$. At first sight these scalings do not resemble those used in Chapter 4 when dealing with stability and low-amplitude wave propagation. However, closer examination reveals that the two scalings differ by factors that depend only on the volume fraction ϕ and the density ratio δ, and these factors are such as to make the magnitudes of the scaling quantities used in Chapter 4 closer to physically relevant sizes. This advantage is gained at the expense of extra complexity compared with the simple scalings above. The values of the parameters (5.44b) for the systems studied are quoted in Table 5.2.

The dispersion relations for the growth rates of small-amplitude waves in the two systems are shown in Figures 5.20(a) and (b). These are presented in dimensionless form, with the dimensional growth rate scaled by v_t/l and the dimensional wavenumber by $1/l$. The points A and C identify the critical conditions; the uniformly fluidized bed is stable to disturbances with larger wavenumbers and unstable to those with smaller wavenumbers. The points Z and X identify the disturbances whose initial rate of growth is largest.

The question then arises of whether there are circumstances in which periodic waves can propagate without change of amplitude or form and, if so, for what

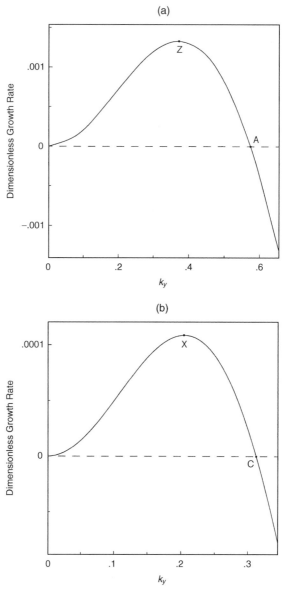

Figure 5.20. Dispersion relations for small perturbations of the uniform beds specified in Table 5.2. (a) Air fluidized bed; (b) water fluidized bed. (From Glasser et al., 1997.)

values of the wavenumber this is possible. In seeking such waves their velocity of propagation, c, is unknown and must be found as part of the solution. However, since c is constant, a solution of this sort is a periodic function of Y only, where $Y = y - ct$, and it can therefore be expanded in a Fourier series of the form

$$\mathbf{A}(Y) = \mathbf{b}_0 + \sum_{n=1}^{\infty} [\mathbf{a}_n \sin(nk_y Y) + \mathbf{b}_n \cos(nk_y Y)], \qquad (5.45)$$

where $\mathbf{A}$ represents the vector (ϕ, p^f, u_y, v_y). An approximation to the desired solution can then be found by truncating the above series at some finite number of terms $n = N$, substituting the result into the equations of motion, and equating to zero the inner product of the results with each of the Fourier basis functions. This generates a set of equations for the unknown velocity c and the coefficients in the truncated expansion (5.45). Details of the procedure and the method used to solve these equations can be found in Glasser et al. (1996) and in Glasser (1996), though computations of this type were first reported by Hernández & Jiménez (1991). A complete branch of solutions is generated by continuation, starting from the critical condition (A or C in Figure 5.20) where they degenerate to zero amplitude.

The amplitude of the periodic variations in the volume fraction of solids can be defined by

$$\|A_\phi\| = \left\{ \sum_{n=1}^{N} [(a_{n\phi})^2 + (b_{n\phi})^2] \right\}^{1/2},$$

and the results of the computations are summarized in Figure 5.21(a) and (b) by plotting this quantity as a function of the dimensionless wavenumber. As in Figure 5.20 A and C denote the critical conditions of marginal stability. To the right of these points, where the uniform bed is stable, $\|A_\phi\|$ is represented by the continuous line $\|A_\phi\| = 0$. The extrapolation of this to the left of the critical points is represented by a broken line to indicate that the uniform bed, while still a solution of the equations of motion, is unstable. In addition a new branch of solutions, with nonvanishing amplitude, is seen to bifurcate from the uniform state at the critical point and to grow in amplitude as k_y decreases. The points Z and X on this branch in Figures 5.21(a) and (b), respectively, are located at values of k_y for which the growth rates of infinitesimal periodic perturbations of the uniform state are largest, and at each of these points an inset shows one period of the corresponding volume fraction profile. The amplitudes of these fully developed waves are similar in both the gas and water fluidized beds but the profile, which is almost symmetric in the water fluidized bed, exhibits a marked asymmetry in the case of the gas fluidized bed.

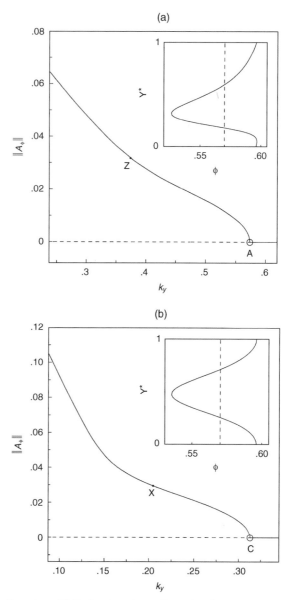

Figure 5.21. Branches of fully developed one-dimensional waves, represented by showing $\|A_\phi\|$ as a function of k_y. (a) Air fluidized bed; (b) water fluidized bed. (From Glasser et al., 1997.)

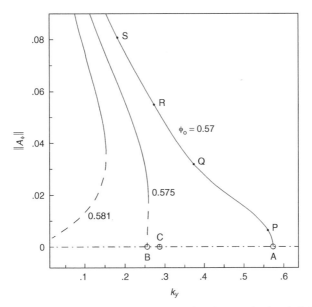

Figure 5.22. Bifurcation diagrams for one-dimensional waves in the air fluidized bed at three values of ϕ_0. (From Glasser et al., 1996.)

In Figure 5.21 the branches of fully developed periodic waves are represented by unbroken lines, indicating that they are stable motions. However, this statement requires some qualification. These solutions are, indeed, stable if they are constrained to retain the periodicity indicated by the corresponding value of k_y; in other words, they are stable to perturbations that share this periodicity. What happens if the perturbations are not constrained in this way will be discussed below.

It is interesting to explore how the bifurcation diagram changes if ϕ_0, the particle concentration in the unperturbed uniform bed, is increased. Figure 5.22 superimposes the bifurcation diagrams in the gas fluidized bed for $\phi_0 = 0.57$, 0.575, and 0.581, the last of which corresponds to a uniform bed that is stable for all values of k_y. For $\phi_0 = 0.57$ the branch of periodic waves bifurcates supercritically from the uniform state at the critical point, but when ϕ_0 is increased to 0.575 the bifurcation becomes subcritical. The initial part of the branch of periodic waves is then unstable, as indicated by the broken line, with stability restored only at the point where the tangent to the curve is vertical. For k_y larger than the critical value (point B), but sufficiently close to it, the uniform bed is stable to sufficiently small perturbations, but a large enough perturbation could displace the system onto the stable part of the upper branch, representing a periodic wave motion. When $\phi_0 = 0.581$ the uniform state is stable for all k_y

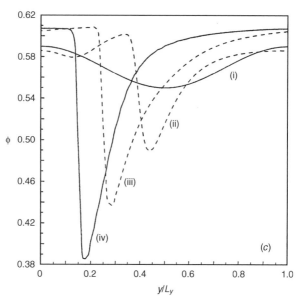

Figure 5.23. Development in time of a one-dimensional wave in the air fluidized bed. $\phi_0 = 0.57$; $k_y = 0.123$. (i) $t = 0$; (ii) $t = 0.97$ (0.67 s); (iii) $t = 2.58$ (1.78 s); (iv) $t = 3.84$ (2.65 s). Initial wave speed $= 0.975V$. Final wave speed $= 0.889V$. (From Anderson et al., 1995.)

and there is no critical point. Nevertheless periodic wave solutions still exist, but now they appear as an isolated branch. They could presumably be accessed from the uniform state by a sufficiently large perturbation of long wavelength.

The amplitudes of the fully developed volume fraction waves at points Z and X in Figure 5.21 are less than 0.07 in both cases, so they represent only modest variations in ϕ. This amplitude increases monotonically on moving along the branch to smaller values of k_y, but as the value of k_y is decreased the number of modes needed to generate an accurate solution increases and eventually becomes impracticably large. An alternative approach is then to approximate the fully developed structure by integrating the equations of motion forward in time, starting from the low-amplitude, exponentially growing sine wave that satisfies their linearised form. This is the method that was used by Anderson et al. (1995), and it succeeds up to wavelengths significantly longer than those for which the Fourier method is practicable. Figure 5.23 shows a succession of volume fraction profiles in the air fluidized bed, generated in this way, for $k_y = 0.123$. (The scaled time t used by Anderson et al. is equal to the dimensional time multiplied by the growth rate of a small-amplitude perturbation of the same wavelength, as determined from the eigenvalues of the linearised equations. No significance should be attached to the relative horizontal displacements of the

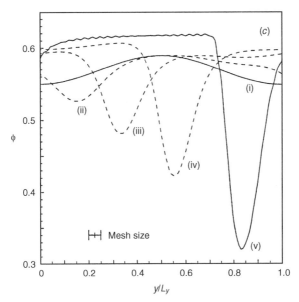

Figure 5.24. Development in time of a one-dimensional wave in the water fluidized bed. $\phi_0 = 0.57$; $k_y = 0.068$. (i) $t = 0$; (ii) $t = 0.64$ (23 s); (iii) $t = 1.20$ (43 s); (iv) $t = 1.52$ (55 s); (v) $t = 2.16$ (78 s). Initial wave speed $= 0.975V$. Final wave speed $= 0.984V$. (From Anderson et al., 1995.)

profiles; they are arbitrary and do not indicate displacement of the profiles in time.) These profiles start from the low-amplitude sinusoidal wave labelled (i) and end with a profile labelled (iv) that is no longer changing significantly and can therefore be assumed to be a good approximation to the fully developed wave. The amplitude of this fully developed wave is 0.22 and there is a very marked asymmetry; indeed, the periodic wave now consists of a sequence of plugs of densely packed particles, with $\phi > 0.6$, separated by relatively narrow bands of significantly lower density. The upper surface of each plug is quite sharp but its lower boundary is diffuse as the density decreases gradually to a minimum value. This type of motion is commonly observed in practice in gas fluidized beds contained in narrow tubes. The lower end of each plug is unstable and erodes, giving rise to a region of gradually falling density as the particles accelerate downward. These falling particles are then decelerated suddenly when they encounter the sharp upper boundary of the next plug below, and their transfer from the lower surface of one plug to the upper surface of the plug below is responsible for the upward movement of the whole density profile.

Figure 5.24 is similar to Figure 5.23 but shows successive volume fraction profiles computed by Anderson et al. for the water fluidized bed at a dimensionless wavenumber $k_y = 0.068$. The amplitude of the final, approximately fully developed profile is approximately 0.3, rather larger than that of the wave in the gas fluidized bed pictured in Figure 5.23.

The captions of Figures 5.23 and 5.24 list the velocities of propagation for both the initial small-amplitude wave and the final fully developed structure; they also indicate the time elapsed, from the start of the integration, for each profile shown. The velocities are quoted in dimensionless terms as multiples of the continuity wave velocity V (see Eq. (4.15)) in the uniform bed. The times are given in dimensionless form as multiples of the characteristic exponential growth time ($1/\mathrm{Re}(s)$ or $1/s_r$) for small disturbances of the uniform state and are also quoted in dimensional terms. Values of the continuity wave velocity are listed in Table 5.2. In each case we see that the velocity of the fully developed wave is close to the continuity wave speed, while the elapsed time to reach the fully developed solution is between two and four times the small-amplitude growth time. Note also that the speed of the fully developed wave is somewhat smaller than that of the initial small-amplitude wave in the air fluidized system. In dimensional terms then, the fully developed waves in the air fluidized bed rise with about seven times the speed of those in the water fluidized bed, and small disturbances grow to the fully developed form about fifty times as quickly. Apart from these quantitative differences, which are not unexpected in view of what we already know from the behaviour of small-amplitude perturbations, there is only one significant qualitative distinction between the fully developed waves in the two systems, namely the marked asymmetry of the profile for the air fluidized bed. The corresponding profile in the water fluidized bed shows some asymmetry of the same sort, but there is no analogue in Figure 5.24 of the sharp, shocklike drop in density at the top of a plug of high density material that we see in Figure 5.23.

Finally, let us return to the question of the stability of the fully developed waves. We noted above that, wherever the branch representing these waves is drawn as a continuous line, the waves are stable against one-dimensional perturbations that retain the same periodicity. This leaves open two questions: first, what happens if we remove the constraint that the perturbations have the same periodicity? Second, what happens if we permit perturbations that are no longer required to be one-dimensional? The first question we shall address now; the second turns out to be intimately related to the mechanism of bubble formation, as will be shown in the next section.

Consider, in particular, the wave shown as the inset in Figure 5.21(a). This has a wavelength that coincides with the wavelength λ_m for maximum growth rate of small perturbations. Of course $2\lambda_m$ is also a periodicity of this wave and we can examine its stability to perturbations with this wavelength. It is then found to be unstable, with a complex conjugate pair of leading eigenvalues in the right half-plane. A small increment of the corresponding eigenfunction can

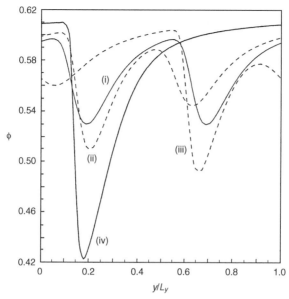

Figure 5.25. Coalescence of adjacent fully developed waves in the air fluidized bed. $k_y = 0.37$ initially. $k_y = 0.185$ finally. (i) $t = 0$; (ii) $t = 3.19$; (iii) $t = 5.81$; (iv) $t = 9.0$. (From Anderson et al., 1995.)

then be added to the original wave and the result can be used as the starting point for a numerical integration forward in time. The result of this is shown in Figure 5.25. After some rather complicated behaviour at intermediate times the profile labelled (iv) is approached, and this appears to be fully developed. Indeed, it coincides with the fully developed profile found by Glasser et al., using the Fourier technique, for wavelength $2\lambda_m$. Thus, the fully developed wave of wavelength λ_m, although stable if constrained to retain this periodicity, becomes unstable if the spatial periodicity constraint is relaxed to $2\lambda_m$. The result is that adjacent periods coalesce, forming a fully developed wave of twice the original wavelength. This process can be repeated up to longer and longer wavelengths, but there are some indications that it may terminate eventually. Very long waves may break up, rather than coalesce, degenerating into a time-dependent behaviour that does not settle into any pattern. This may be related to an upper bound for the wavelength of stable one-dimensional waves predicted by Needham & Merkin (1986).

Parts of the picture outlined above have been anticipated in a number of publications since the early work of Fanucci et al. Unfortunately, in most of this work ρ_f and μ_f have been set equal to zero for simplicity, so the results are applicable only to gas fluidized beds. In a trio of papers Needham & Merkin (1983, 1984, 1986) examined various aspects of the propagation of disturbances in fluidized beds. In addition to the above simplifications they also assumed that $p^p \propto \phi$, which, as we shall see, loses important features of the behaviour that result from the physically necessary divergence of p^p as $\phi \to \phi_m$. In their first paper they developed solutions of the linearised equations of motion starting from initial conditions of the form

$$\phi(x, 0) = \phi_0 + \alpha\phi_1(x), \quad v(x, 0) = 0, \quad u(x, 0) = u_0,$$

where u_0 and ϕ_0 are the gas velocity and solids volume fraction in the unperturbed uniform bed, α is small, and $\phi_1(x)$ takes nonvanishing values only in some bounded interval of x with length h. This choice of initial conditions leads to a more difficult mathematical problem than the propagation of a sinusoidal disturbance of infinite extent, as studied in Chapter 4. In the latter case the form of the solution locks the phases of the variables ϕ, u, and v into an appropriate relationship at all times. With the above initial conditions, in contrast, there is an initial interval of rapid change, during which an analogue of these phase relations is established, followed by a slower growth that plays a role similar to the exponential growth of the unbounded sine wave solution. Accordingly, it is necessary to develop the solution separately in "inner" and "outer" regions. The conclusions add little to the results of earlier stability analyses, but the authors point out the importance of the quantity

$$\frac{p_0^{p'}}{\rho_s v_t^2} - n^2\phi_0^2(1 - \phi_0)^{2(n-1)}.$$

Referring to (4.40) we see that this is related simply to the stability criterion. Indeed, since ρ_f and the virtual mass coefficient C_v are zero in the work of Needham & Merkin, the stability criterion (4.40) reduces to the requirement that the above quantity should be positive. Needham & Merkin note that it plays a role in the equations equivalent to a diffusion coefficient that can have either sign. Instability, which occurs when it is negative, can thus be attributed to a reversal of the usual effect of diffusion, leading to a "focussing" of voidage gradients. As in the work of Fanucci et al. Needham & Merkin found that shocks could form when the disturbance grew beyond the range of validity of linearisation.

   Needham & Merkin's second paper (1984), which was devoted to two-dimensional disturbances, will be discussed in Subsection 5.5.5. Their third paper (1986) addressed the possibility of fully developed periodic wave solutions of the complete equations of motion, of the sort studied by Glasser et al. and described above. In a reference frame moving with the wave profile the volume fraction has to satisfy a second-order ordinary differential equation in the space variable, and this can be replaced by a pair of first-order equations for ϕ and ω, where $\omega = d\phi/dy$. The solutions can then be examined through their phase trajectories in the (ϕ, ω)-plane. The uniform bed is the fixed point

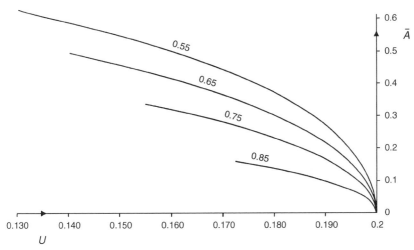

Figure 5.26. Amplitude of fully developed waves as a function of their speed, U. $\varepsilon_0 = 0.85, 0.75, 0.65, 0.55$. (From Needham & Merkin, 1986.)

($\phi_0, 0$) and Needham & Merkin showed that a periodic orbit bifurcates from this fixed point through a Hopf bifurcation at the condition of limiting stability. Periodic orbits of small amplitude were approximated by a series expansion method, while orbits of all sizes were generated by transient numerical integration of the differential equations, and the results were exhibited by plotting the amplitude of the periodic wave (defined by $\bar{A} = (\phi_{max} - \phi_{min})$) against its speed of propagation, U.

Figure 5.26 shows the resulting diagrams for some subcritical cases. Unlike the subcritical cases computed by Glasser et al. the branches representing the periodic waves do not reverse slope on moving along them away from the branch point. Needham & Merkin also exhibit some ϕ profiles at points on a supercritical branch, and these show little evidence of the asymmetry between the leading and trailing edges of a plug that is such a striking feature of Figure 5.23. This lack of asymmetry most likely reflects the linear relation assumed between p^p and ϕ though in some cases it could simply be a result of the small amplitude.

The above work (Needham & Merkin, 1986) can be regarded as a supplement to an earlier study of the possibility of fully developed periodic waves by Liu (1983). Liu's work was limited to weakly nonlinear behaviour, and hence to periodic waves of low amplitude in the close vicinity of the critical point. He concluded that, under different conditions, a branch of periodic waves could bifurcate either supercritically or subcritically from the solution representing the uniform bed. He also showed that solutions close to the bifurcation point on a supercritical branch are approached as $t \to \infty$ from a slightly perturbed uniform bed, whereas those in a similar position on a subcritical branch are approached as $t \to -\infty$. In other words, a branch that bifurcates subcritically consists, near the bifurcation point, of periodic states that are unstable and will collapse back to the uniform state. These findings are consistent with the results of Glasser et al. shown in Figure 5.22, and also with those of Needham & Merkin (1986), who go further to examine the extension of the branches of periodic waves away from the critical point, where the approximation of weak nonlinearity is no longer valid.

A more recent publication of Ganser & Drew (1990) also relates closely to the work of Liu (1983) and of Needham & Merkin (1986), though the authors were apparently unaware of the latter paper. Ganser & Drew took issue with Liu's findings regarding the circumstances in which there is a supercritical bifurcation to an equilibrated periodic wave. They constructed the branch of solutions that bifurcates from the uniform bed at the critical point, both computationally and, in the neighbourhood of the critical point, by an approximation using the method of multiple time scales. Only subcritical bifurcations were found. However, they limited attention to values of ϕ_0 only slightly smaller than the particle concentration at which the uniform bed becomes linearly stable against perturbations of all wavelengths. In these circumstances, as we have already seen (Glasser et al., 1996), the bifurcation is indeed subcritical; supercritical bifurcation occurs for smaller values of the particle concentration. The subcritical branches found by Ganser & Drew and by Needham & Merkin share the property that the sign of their slope does not reverse on moving along them away from the bifurcation point, unlike the subcritical branches computed by Glasser et al. Ganser & Drew speculate that this is a consequence of the form of their expression for the particle-phase pressure, which does not diverge on approaching close packing, and this speculation is confirmed by the results of Glasser et al. shown in Figure 5.22.

In all the examples described so far we have seen a single branch of periodic waves bifurcating from the uniform bed solution at the critical value of the wavenumber k_y, where the uniform solution undergoes a transition from stability to instability. However, when the bed parameters are such that $-c_1^2/4 < d - \frac{1}{2}u_0c_1 < 0$ it was shown in Section 4.3 that there are actually *two* critical values of the wavenumber. The dispersion relation for small-amplitude waves then has the form illustrated in Figure 4.1(iv). In such a case branches of solutions representing fully developed periodic waves can bifurcate from the uniform bed at both the critical values of k_y and Göz (1993) indicates that, in certain conditions, these may join together to form a single branch that emerges from the uniform state at one critical point and rejoins it at the other.

Finally, Harris (1996) recently published an analytical study of the development from small-amplitude to fully developed periodic wave motions and, among other things, was able to confirm the unusual observation of Anderson et al. (1995) that the waves may slow down as their amplitude increases.

In addition to the work described above on periodic wave motions, in the past few years a number of publications have considered the propagation of solitons and solitary waves in fluidized suspensions (Crighton, 1991; Komatsu & Hayakawa, 1993; Harris & Crighton, 1994; Hayakawa et al., 1994). The significance of these results in relation to observations remains to be seen.

5.5.3 Computational Exploration of Two-Dimensional Transients and Fully Developed Structures

Plane wave disturbances of the sort discussed above are easily observed in small diameter liquid fluidized beds of transparent particles, such as glass beads, as seen in Figure 4.8. However, as the lateral dimensions of the bed are increased more complicated motions develop. Didwania & Homsy (1981) observed the behaviour of water fluidized beds of glass beads confined in a "two-dimensional" container whose cross section was rectangular, with dimensions 30 cm × 3.15 cm. When the bed was viewed through the smaller dimension, plane waves of varying void fraction were seen to develop on moving up from the distributor. The wavefronts were horizontal and they grew at about the same rate as the waves observed in narrower beds. However, as the waves moved up the bed they began to develop structure in the transverse direction. The originally horizontal wavefronts became corrugated, with a characteristic wavelength that was comparable with the separation of the wavefronts in the vertical direction, and this suggested an instability of the initially planar waves. Since bubbles are large-amplitude two- or three-dimensional phenomena it was tempting to speculate that an analogous phenomenon in gas fluidized beds might have something to do with the mechanism of bubble formation.

In a second paper Didwania & Homsy (1982) speculated that the observed corrugations might be a result of nonlinearities in the equations of motion inducing a resonant interaction between the transverse structure and the second harmonic of the primary plane wave instability, following a mechanism described by Phillips (1974). Subsequently, interest in this phenomenon has grown and two-dimensional disturbances have now been investigated theoretically by a number of workers. Here we shall follow the plan of the previous section by describing some of the most recent results first. Then, in the next subsection, these will be related to other investigations. The computational explorations of Anderson et al. (1995) and Glasser et al. (1996, 1997), presented in some detail above, were not limited to one-dimensional disturbances. Figures 5.21 and 5.22 show how branches of fully developed, one-dimensional waves can bifurcate from the uniform bed solution at its critical point of limiting stability, and each of these branches contains structures that are stable, in the sense defined there. Specifically, each structure is stable against one-dimensional perturbations that depend only on the vertical coordinate and have the same periodicity. We have already seen how this conclusion is modified if the period of the perturbation is allowed to be longer; we must now go on to investigate the effect of allowing the perturbation to have a periodic structure in the horizontal direction.

Both Anderson et al. and Glasser et al. approach this problem in the same way, achieving essentially the same results. The fully developed one-dimensional wave, which is a steady solution of the equations of motion when viewed from a frame moving with its own velocity, is subjected to a small perturbation that shares its periodicity in the vertical direction but is also periodic in the horizontal direction (attention is limited to two dimensions) with wavenumber k_x. The initial development of this perturbation, described by linearisation of the equations of motion about the one-dimensional wave solution, is exponential in time with an exponent that is an eigenvalue of the linearised problem. Of course the one-dimensional wave, which is the base state for this stability analysis, is available only as a computed approximation and, correspondingly, the eigenvalues belong to the numerical algorithm used to approximate the equations of motion, rather than the equations of motion themselves. Since Anderson et al. used a Galerkin finite element method, while Glasser et al. generated their transients by a quite different, time-dependent generalization of the Fourier decomposition described above, it is encouraging that their results agree closely.

For large values of k_x all the eigenvalues are found to lie in the left half-plane. Hence the one-dimensional waves are stable against perturbations with a periodicity of short wavelength in the transverse direction. However, as k_x is decreased an eigenvalue migrates along the real axis into the right half-plane, indicating instability. As k_x is decreased further this eigenvalue moves out into the right half-plane along the real axis, and it may be followed by other eigenvalues, but eventually the motion reverses with the most dangerous eigenvalue returning to the origin of the complex plane as $k_x \to 0$. This behaviour is illustrated in Figure 5.27, which plots the leading eigenvalue (i.e., the dimensionless growth rate), as a function of k_x, for each of several fully developed one-dimensional waves in the air fluidized bed identified in Table 5.2. The waves in question are identified by capital letters in Figure 5.22 and the corresponding letters are used to label the curves in Figure 5.27. Each curve passes through a maximum value and the corresponding value of k_x determines the transverse wavelength of the most rapidly growing disturbance. For small-amplitude one-dimensional waves, for example those corresponding to point P in Figure 5.22, this wavelength is much longer than the wavelength in the vertical direction. However, for large-amplitude one-dimensional waves (see point S in Figure 5.22) the two

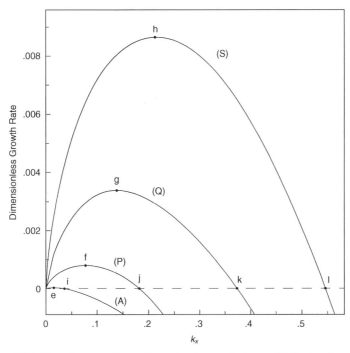

Figure 5.27. Growth rates of two-dimensional perturbations of fully developed one-dimensional waves, as functions of the transverse wavenumber k_x. The curves labelled A, P, Q, and S belong to the fully developed one-dimensional waves associated with the points labelled by the corresponding letters in Figure 5.22. (From Glasser et al., 1996.)

are of comparable size. In the gas fluidized bed the growth rates of the dominant two-dimensional perturbations from the fully developed one-dimensional waves are found to be of the same order of magnitude as the growth rates of small-amplitude one-dimensional perturbations from the uniform fluidized bed. The nature of the eigenfunctions is also of interest, and this is explored in some detail by Glasser et al. (1996, 1997). In Figure 5.28 we illustrate just one of these, namely that belonging to the maximum point h of the curve S in Figure 5.27. This represents the fastest growing perturbation of a large-amplitude one-dimensional wave. A contour map of the perturbation in ϕ and vector plots of the perturbations in $\mathbf{u}$ and $\mathbf{v}$ are shown, together with the ϕ profile of the unperturbed one-dimensional wave.

The above discussion has referred entirely to the air fluidized bed, but the picture is broadly the same for the water fluidized bed identified in Table 5.2. Figure 5.29 shows the growth rate, as a function of k_x, for the dominant two-dimensional perturbation of the fully developed one-dimensional wave in the water fluidized bed represented by point X in Figure 5.21(b).

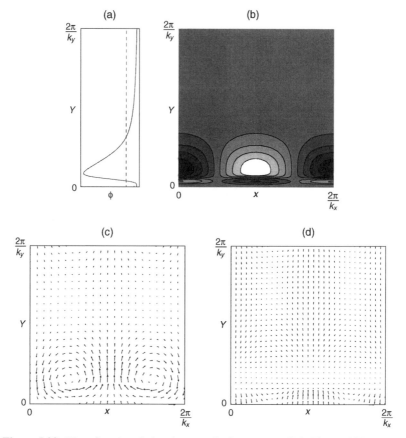

Figure 5.28. Eigenfunctions belonging to point h on curve (S) in Figure 5.27. (a) Unperturbed wave; (b) ϕ; (c) **u**; (d) **v**. (From Glasser et al., 1996.)

The corresponding eigenfunction is exhibited in Figure 5.30. It is more symmetric in the vertical direction than the eigenfunction of Figure 5.28, simply because the profile of the one-dimensional wave itself is more symmetric. Note that the rate of growth of this two-dimensional structure in the water fluidized bed is of the same order of magnitude as that found above in the case of the air fluidized bed, despite the fact that small one-dimensional waves in the water fluidized bed grow much more slowly than those in the air fluidized bed. In the example from the air fluidized bed we noted that the rates of growth were comparable for the initial one-dimensional perturbation of the uniform bed and for the two-dimensional perturbation of the one-dimensional wave. This is clearly no longer true in the water fluidized bed, where the one-dimensional instability of the uniform bed grows an order of magnitude more slowly than the two-dimensional instability of the fully developed wave.

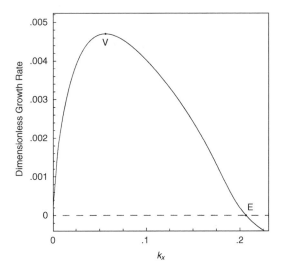

Figure 5.29. Growth rate of a two-dimensional perturbation of the fully developed one-dimensional wave in the water fluidized bed belonging to the point labeled X in Figure 5.21(b). (From Glasser, 1996.)

The eigenfunctions for the two-dimensional perturbations determine the initial stage in their development from a one-dimensional wave towards some two-dimensional structure. Clearly this could be followed further by numerical integration of the full, nonlinear equations of motion, but it would also be valuable to investigate whether there exist fully developed two-dimensional structures that can propagate without change of form.

The Fourier method of Glasser et al., described above, can be generalised to identify fully developed two-dimensional structures in just the same way as it was used for one-dimensional waves. The expansion (5.45) must be replaced by the general expansion of a function periodic in both x and Y, namely

$$
\begin{aligned}
\mathbf{A}(x, Y) = \mathbf{b}_{00} &+ \sum_{n=1}^{\infty} \{\mathbf{a}_{n0} \sin(nk_y Y) + \mathbf{b}_{n0} \cos(nk_y Y)\} \\
&+ \sum_{m=1}^{\infty} \left[\mathbf{b}_{0m}^c + \sum_{n=1}^{\infty} \{\mathbf{a}_{nm}^c \sin(nk_y Y) + \mathbf{b}_{nm}^c \cos(nk_y Y)\} \right] \cos(mk_x x) \\
&+ \sum_{m=1}^{\infty} \left[\mathbf{b}_{0m}^s + \sum_{n=1}^{\infty} \{\mathbf{a}_{nm}^s \sin(nk_y Y) + \mathbf{b}_{nm}^s \cos(nk_y Y)\} \right] \sin(mk_x x),
\end{aligned}
$$

$$(5.46)$$

where $Y = y - ct$, as before. The procedure for determining the coefficients is then the same as for the one-dimensional case, though considerably more

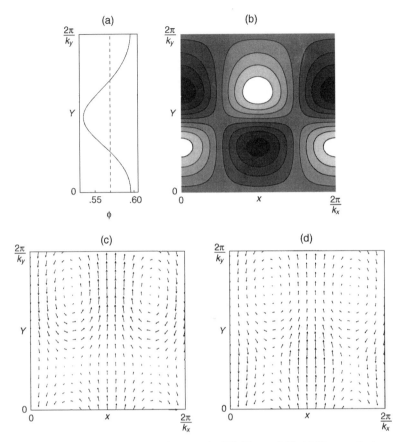

Figure 5.30. Eigenfunctions belonging to point V in Figure 5.29. (a) Unperturbed wave; (b) ϕ; (c) **u**; (d) **v**. (From Glasser et al., 1997.)

demanding computationally. Convergence would be very difficult to achieve starting from arbitrary initial values of the coefficients so, as in the one-dimensional case, the procedure adopted is to generate a whole branch of two-dimensional waves by continuation, starting at the point where it bifurcates from the branch of fully developed one-dimensional waves.

Because there are now three parameters to consider – ϕ_0, k_y, and k_x – to present the results as a graph, two of these, or two relations between them, must be specified. In Figure 5.31, which refers to the air fluidized bed, we have chosen to specify that $\phi_0 = 0.57$ and $k_x = k_y$. The diagram then shows the amplitude of the variations in ϕ, for each type of fully developed wave, as a function of k_y. As before, a continuous line indicates a stable structure and a broken line an unstable structure, where stability and instability are defined relative to perturbations that retain the same periodicity as the structure itself, in both x

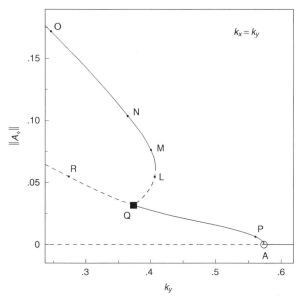

Figure 5.31. Bifurcation diagram showing the uniform bed and fully developed one-
and two-dimensional wave solutions for the air fluidized bed. $\phi_0 = 0.57$; $k_x = k_y$. Points
labelled P, Q, and R correspond to points identified by the same letters in Figure 5.22.
(From Glasser et al., 1996.)

and y. The branches representing the uniform bed and one-dimensional waves
are identical to those in Figure 5.22 for the same value of ϕ_0. However, the branch
of one-dimensional waves, which was stable for all values of k_y in Figure 5.22,
now suffers a transition to instability at Q, where $k_y \approx 0.37$, and for smaller
values of k_y, it is unstable against two-dimensional perturbations. At the critical
point Q a branch QLMNO of waves with two-dimensional structure bifurcates
from the one-dimensional branch. (This is actually a pitchfork bifurcation but
the two branches differ from each other only by a phase shift of half a wavelength
in the x direction. They are therefore equivalent physically and, since they
have the same values of $\|A\|$, they appear superimposed in the diagram.) The
bifurcation at Q is subcritical and the first part of the new branch represents
unstable, small-amplitude, two-dimensional waves. However, at L the waves
become stable and remain so for as far as the branch was followed. This is
not as far as one would wish since, as k_y decreases, the structures on the two-
dimensional branch exhibit ever sharper spatial gradients, so the number of
Fourier modes needed to retain acceptable accuracy grows rapidly.

Figures 5.32 and 5.33 illustrate the nature of the propagating structures at
points M and O, respectively, in Figure 5.31. The figures show only one period
of the solution, of course; the complete solution is constructed by duplicating

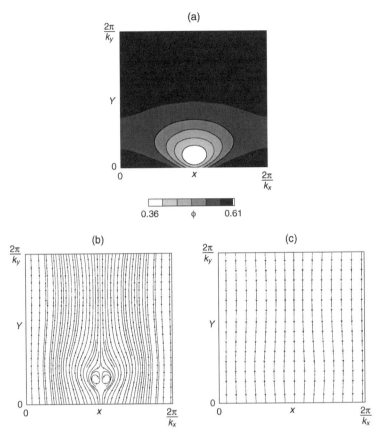

Figure 5.32. Two-dimensional travelling wave corresponding to point M in Figure 5.31. (a) ϕ; (b) gas streamlines; (c) particle streamlines. (From Glasser et al., 1996.)

the figures in both the horizontal and vertical directions to give a doubly infinite array. In both cases the contour plots of ϕ reveal a compact "hole" of low particle concentration, in the neighbourhood of which the gas streamlines show a pair of vortices within which the gas circulates. At point M the lowest value of ϕ is about 0.4, but at O this is reduced to about 0.2, and the linear dimensions of the structure are larger by a factor of 1.6. In general, on moving out along this branch the size of the structure increases, the "hole" of low particle concentration becomes deeper, and there is some change in its shape, but the general features of the structure remain the same.

One cannot fail to notice that the structure of Figure 5.33 has much in common with a bubble in a gas fluidized bed, though the minimum particle concentration is still far from zero. In particular, it is accompanied by a circulating vortex of gas, analogous to the Davidson cloud, and its upper surface

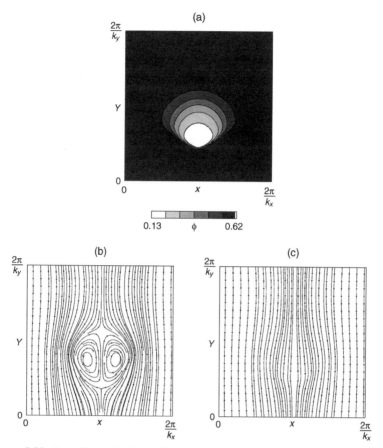

Figure 5.33. Two-dimensional travelling wave corresponding to point O in Figure 5.31. (a) ϕ; (b) gas streamlines; (c) particle streamlines. (From Glasser et al., 1996.)

is surrounded by a mantle of reduced particle density. Its lower surface does not have the characteristically indented shape typical of a bubble but this discrepancy can be removed by a minor change in the equations of motion. If the expression (5.42) for the particle phase viscosity is simply replaced by an assumption of constant viscosity, setting $\mu^p = 6.65p$, with no other change in the equations, the bifurcation diagram is changed to that shown in Figure 5.34. This differs very little from Figure 5.31, as one might anticipate in view of the minor alteration to the equations of motion. However, the structure at point F of Figure 5.34 is now that shown in Figure 5.35. It retains all the bubble-like features commented on above and, in addition, it now has an appropriately indented base.

It is now natural to ask whether similar, fully developed bubble-like structures also exist in the water fluidized bed. The bifurcation diagram for this case, which

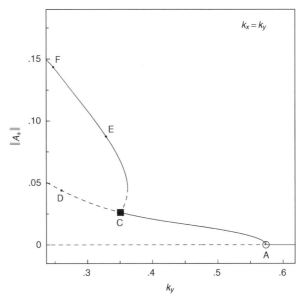

Figure 5.34. Bifurcation diagram for the same conditions as in Figure 5.31, but with constant viscosity $\mu^p = 6.65p$, $\phi_0 = 0.57$, and $k_x = k_y$. (From Glasser et al., 1996.)

is shown in Figure 5.36, differs very little in appearance from the corresponding diagram for the air fluidized bed (Figure 5.31). Furthermore, the bed structure at the point labelled S, which is shown in Figure 5.37, is a "hole" just as deep as any seen in the air fluidized bed and, like them, it is accompanied by a circulating vortex of fluid. It differs from the structures seen in the air fluidized bed only by a minor change in the shape, and we have seen how easily details of the shape can be changed by small adjustments to the equations of motion. Thus, even the complete branching structures deduced from the full equations of motion provide no insight into the question of why bubbles are seen in the air fluidized bed, but not in the water fluidized bed. To investigate this further we must explore the growth of initially small, two-dimensional disturbances beyond the range of validity of linearisation.

The progress from the fully developed, one-dimensional waves towards the above two-dimensional structures can be followed by numerical integration of the equations of motion forward in time. The one-dimensional wave is perturbed initially by adding to it a small multiple of the eigenfunction belonging to a two-dimensional perturbation of selected wavenumber k_x. Then, with this as an initial condition, the equations of motion are integrated to generate the developing transient motion. Calculations of this sort are reported by both Anderson et al. (1995) and Glasser et al. (1997) using their respective, and quite different,

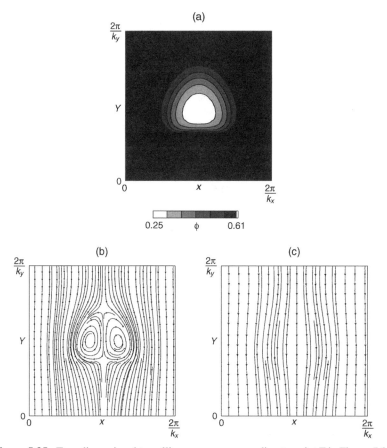

Figure 5.35. Two-dimensional travelling wave corresponding to point F in Figure 5.34. (a) ϕ; (b) gas streamlines; (c) particle streamlines. (From Glasser et al., 1996.)

integration algorithms. The Galerkin finite element method of Anderson et al. is slower than Glasser et al.'s Fourier method, but it can handle larger structures, for which steep gradients develop in the dependent variables. For smaller structures, where both methods can be used, their results are in good agreement.

The development in the air fluidized bed from a one-dimensional wave towards the two-dimensional structure corresponding to point O in Figure 5.31 (as shown in Figure 5.33) is traced in Figure 5.38. This shows the distribution of particle concentration at four successive times, as obtained by integration of the full equations of motion using the method of Glasser et al. The time elapsed from the beginning of the integration is given for each panel, both in the dimensionless form defined in (5.44a) and in dimensional terms. Panel (a) represents the initial condition, a slightly perturbed one-dimensional

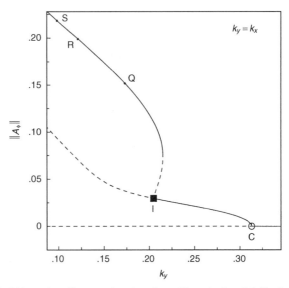

Figure 5.36. Bifurcation diagram showing the uniform bed and fully developed one- and two-dimensional wave solutions for the water fluidized bed. $\phi_0 = 0.57$; $k_x = k_y$. (From Glasser et al., 1997.)

wave, while the concentration distribution shown in panel (d) is not easily distinguishable from the fully developed structure of Figure 5.33. The total elapsed time is 0.95 s, which is comparable to the time of growth of the one-dimensional wave itself from the uniform bed. A similar sequence for a larger structure in the air fluidized bed, obtained by Anderson et al., is shown in Figure 5.39. In this case $k_y = 0.123$, so the one-dimensional wave from which the process is initiated is that represented by the profile labelled (iv) in Figure 5.23. The value specified for k_x is $3k_y/2$ or 0.185. As before, panel (a) shows the slightly perturbed one-dimensional wave that serves as the initial condition, while the structure shown in panel (d) does not change significantly if the integration is extended to longer times. It can therefore be assumed to be a good approximation to a fully developed two-dimensional wave. The dimensionless time t now represents a multiple of the growth time for a small-amplitude perturbation of the uniform bed with wavenumber k_y, as determined from the linearised problem. In dimensional terms the total elapsed time is 0.52 s.

  The structure in Figure 5.39(d) is the largest fully developed two-dimensional wave computed in the air fluidized bed, so it might be expected to be the best representation of a real bubble. The fluid and particle velocity fields are therefore presented in detail in Figure 5.40, together with a contour map of the solids volume fraction. The lowest value of the volume fraction is approximately

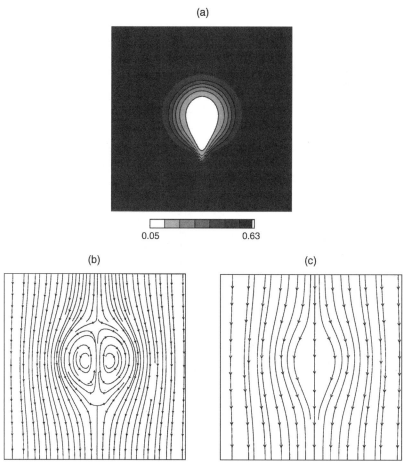

Figure 5.37. Two-dimensional travelling wave corresponding to point S in Figure 5.36. (a) ϕ; (b) liquid streamlines; (c) particle streamlines. (From Glasser et al., 1997.)

0.15. The vortex that forms the Davidson cloud is very well defined, and one can even see that it is displaced somewhat above the centre of the bubble, defined as the point of lowest particle concentration. The mantle of reduced particle concentration is also very evident. The process by which the bubble forms can be traced in Figure 5.39. Initially the upper surface of the band of low concentration bulges slightly upward. Subsequently, the whole band begins to buckle upward, but the upward bulge of the lower boundary is less than that of the upper boundary, and a compact region of even lower concentration develops between them. As time passes this trend continues; the originally one-dimensional wave becomes very distorted, buckling sharply upward in the

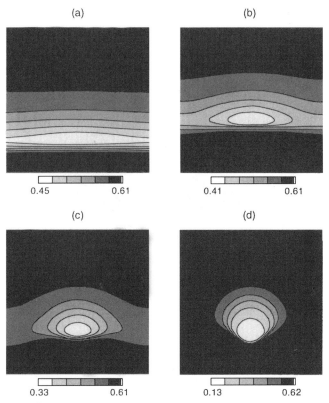

Figure 5.38. Growth of the two-dimensional travelling wave of Figure 5.33 from a slightly perturbed one-dimensional wave. (a) $t = 0$; (b) $t = 234\,(0.30\text{ s})$; (c) $t = 391(0.50$ s); (d) $t = 742\,(0.95\text{ s})$. (From Glasser et al., 1997.)

centre of the computational cell, and a deepening "hole" of low concentration grows in the apex of this wave, eventually becoming the bubble.

The above computations describe the formation of a bubble as a two-stage process: first a one-dimensional instability grows into a fully developed one-dimensional wave; then this wave develops an instability in the horizontal direction that initiates a further progression to a bubble. One should also enquire whether a bubble can grow directly from a uniform bed as a consequence of a small initial perturbation with two-dimensional structure, so that the transient is two dimensional from the beginning. Anderson et al. (1995) examined this question by integrating the full equations of motion forward from an initial condition for which the particle volume fraction is given by

$$\phi(x, y, 0) = \phi_0 + \phi' \cos(k_y y)[(1 - \varepsilon) + \varepsilon \cos(k_x x)]. \qquad (5.47)$$

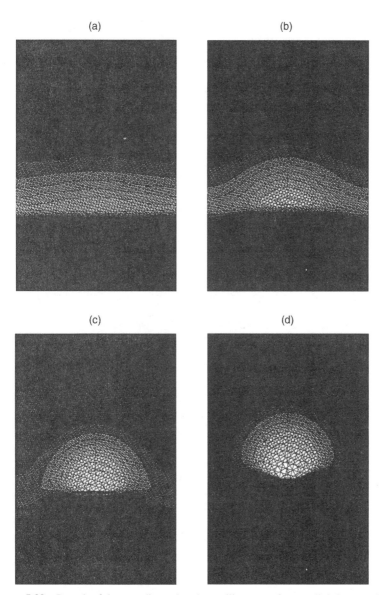

Figure 5.39. Growth of the two-dimensional travelling wave from a slightly perturbed one-dimensional wave of Figure 5.23. $\phi_0 = 0.57; k_y = 0.123; k_x = 0.185$. (a) $t = 0$; (b) $t = 0.36$ (0.25 s); (c) $t = 0.58$ (0.40 s); (d) $t = 0.75$ (0.52 s). (From Anderson et al., 1995.)

(a)

(b)

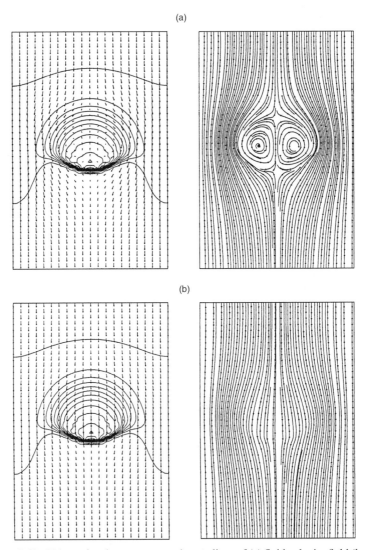

Figure 5.40. Volume fraction contours and steamlines of (a) fluid velocity field (largest magnitude) $= 0.183v_t$) and (b) particle velocity field (largest magnitude $= 0.291v_t$) for solution of Figure 5.39(d). (From Anderson et al., 1995.)

Then ϕ' and ε are measures of the amplitude and the two-dimensional character of the perturbation respectively. Figure 5.41 illustrates the sequence of volume fraction distributions found in this way, with $\phi' = 0.02$ and $\varepsilon = 0.1$, and with values of k_y and k_x the same as for Figure 5.39. Panel (d) of Figure 5.41 is seen to resemble closely panel (c) of Figure 5.39, from which point on the

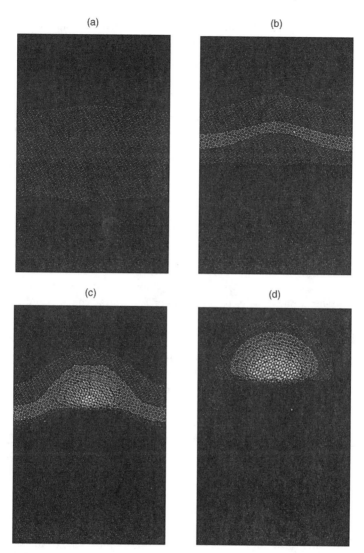

Figure 5.41. Growth of an initially two-dimensional small perturbation of the uniform fluidized bed. $\phi' = 0.02; \varepsilon = 0.1$ (Eq. 5.47). $\phi_0 = 0.57, k_y = 0.123$, and $k_x = 0.185$, as for Figure 5.39. (a) $t = 0$; (b) $t = 0.44$ (0.30 s); (c) $t = 0.91$ (0.63 s); (d) $t = 1.35$ (0.93 s). (From Anderson et al., 1995.)

structures remain similar in both cases, establishing that the two-stage process is not essential for the growth of the fully developed, bubble-like structure.

The mechanism by which the concentration of particles within the bubble decreases can be clarified by examining the flux of solid material, as a function of position, across one of the closed contours of constant concentration (Anderson et al., 1995). It is found that the particles depart from the interior of the contour by moving away down the low density band, clearly visible in Figure 5.41(c), more quickly than they enter across the upper part of the contour, with the net result that the inventory of particles within the contour decreases.

The above results establish the existence of a mechanism by which bubbles can form in the air fluidized bed. No mechanical assumptions need be invoked other than those implicit in the form of the equations of motion (5.36) to (5.43). Furthermore, Glasser et al. (1996) have examined the robustness of the results against changes in the closures (5.41) and (5.42) for the pressure and viscosity of the particle phase. Both the magnitudes of these properties, determined by the constants P and M, and the algebraic forms of the closures were varied. Both P and M could be changed by an order of magnitude with no serious disruption of the picture, and we have already seen that although replacing the expression (5.42) by a constant changes the shape of the computed bubble, it does not affect its ability to grow. We can therefore conjecture that the rapid growth of bubbles in a typical gas fluidized bed is a robust feature of the equations of motion.

It remains a puzzle why bubbles do not form in a typical liquid fluidized bed, since the existence in the water fluidized bed of fully developed motions clearly resembling bubbles has already been demonstrated (see Figures 5.36 and 5.37). Let us therefore examine whether these structures can be grown from the corresponding fully developed one-dimensional waves, just as we did in the case of the air fluidized bed. First consider the two-dimensional structure corresponding to point Q in Figure 5.36, with $k_y = k_x = 0.173$. Glasser et al. integrated the equations of motion from an initial condition representing the one-dimensional wave with $k_y = 0.173$, perturbed by adding a small multiple of the eigenfunction for a horizontal instability with $k_x = k_y$, and the sequence of concentration distributions they found is shown in Figure 5.42. This is similar to what we have seen for the air fluidized bed, with panel (d) very close to the fully developed two-dimensional wave, but the elapsed time is much longer. However, something quite new happens if we attempt to grow larger structures in this way.

As we have seen already there is a limitation on the size of the structures that can be handled by the method of Glasser et al. with a practicable number of Fourier modes, but the numerical method used by Anderson et al. is less limited in this respect, and Figure 5.43 shows the course of the solution in the water flu-idized bed, starting from a slightly perturbed form of the one-dimensional wave with $k_y = 0.068$ (Figure 5.24). This is a structure roughly two and a half times the size of the one whose growth is followed in Figure 5.42. The initial stages of the motion bear some resemblance to what was seen in the air fluidized bed. The initial band of low particle concentration begins to buckle, and a compact region of low concentration forms at its apex, as seen in panel (b). However, instead of

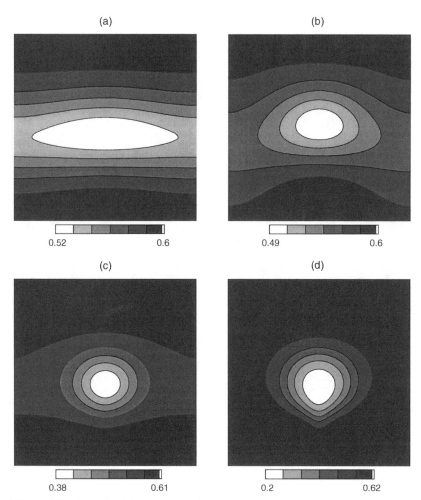

Figure 5.42. Growth of the two-dimensional travelling wave from a slightly perturbed one-dimensional wave in the water fluidized bed. $\phi_0 = 0.57; k_x = k_y = 0.173$. (a) $t = 0$; (b) $t = 1,300$ (5.24 s); (c) $t = 3,980$ (16.04 s); (d) $t = 5,370$ (21.64 s). (From Glasser et al., 1997.)

remaining compact and continuing to grow, as was the case in the air fluidized bed, it soon starts to break up. The beginning of this can be seen in panel (c), where a small region of high density has intruded into the larger region of low density, while complicated low-amplitude structures are beginning to appear in other parts of the field. The velocities of rise of the different features of the concentration distribution then diverge, so the pattern rapidly loses its cohesion and, in panel (d), only some low-amplitude concentration variations remain. For

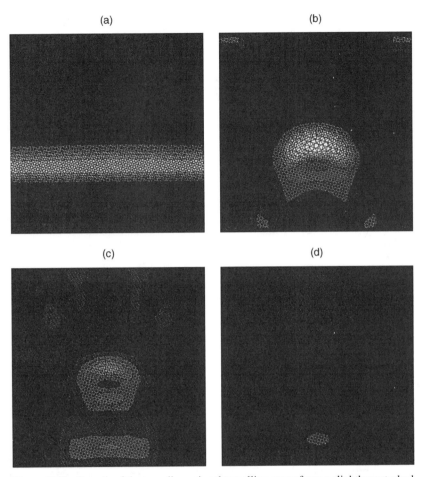

Figure 5.43. Growth of the two-dimensional travelling wave from a slightly perturbed one-dimensional wave of Figure 5.24. $\phi_0 = 0.57; k_y = 0.068; k_x = 0.068$. (a) $t = 0$; (b) $t = 0.033$ (1.2 s); (c) $t = 0.075$ (2.7 s); (d) $t = 0.18$ (6.6 s). (From Anderson et al., 1995.)

as long as the integration was continued there was no sign that the concentration distribution was settling into any invariant form. Thus, in this case the system does not develop to a bubble-like structure but appears to be headed to some low-amplitude attractor whose nature is unknown. Note that the dimensionless time for development to the bubble-like structure seen in panel (b) is $t = 0.033$, compared with a value $t = 0.75$ for the bubble in the air fluidized bed found in panel (d) of Figure 5.38. In dimensional terms these times are 1.2 s for the water fluidized bed and 0.52 s for the air fluidized bed, so the bubble-like structures grow from the one-dimensional waves with comparable speed in both cases.

Anderson et al. also started an integration from initial conditions representing a small and slightly two-dimensional perturbation of the uniform bed in the form (5.47). In contrast to the result for the air fluidized bed, presented in Figure 5.41, no structure with significant variations of concentration formed at any time, nor did the solution settle to any invariant propagating waveform.

The results of these transient integrations in the air and water fluidized beds appear to have identified the difference in behaviour between the two systems that accounts for the presence of bubbles in the former and their absence from the latter. It is not surprising that this difference had proved elusive, since it appears only in the last stages of the transients, where the nonlinear nature of the equations of motion has the greatest effect. In summary, the equations indicate that bubble-like structures will grow in a typical gas fluidized bed, either directly from a two-dimensional perturbation of the uniform bed or indirectly, via an intermediate one-dimensional wave, in a time of the order of a fraction of a second. For a typical liquid fluidized bed, in contrast, there may be transient growth of a bubble-like structure from a one-dimensional wave but this structure subsequently disperses, and nothing resembling a bubble can be grown directly from a two-dimensional perturbation of the uniform bed. In this connection it is worth noting that Didwania & Homsy (1981) have reported experimental observations of the transient appearance and subsequent dispersal of bubble-like voids in water fluidized beds of glass beads.

In view of the very different course of development of the large-scale, two-dimensional structures generated by instability of the one-dimensional waves in the gas and liquid fluidized beds, it is important to identify any physical features of the systems that might have a bearing on this difference. As pointed out by Anderson et al. (1995) one of these might be the relative lengths of the growth times for the primary, one-dimensional instability and the secondary, two-dimensional instability. In the case of the air fluidized bed the total elapsed time for growth of the one-dimensional wave of Figure 5.23(iv) is 2.65 s, while the elapsed time to grow the bubble of Figure 5.39(d) from a perturbed form of this wave is 0.52 s. In the water fluidized bed, however, the elapsed time for the growth of the fully developed wave of Figure 5.24(v) is 78 s, while that for the development of the bubble-like structure of Figure 5.43(b) from this one-dimensional wave is only 1.2 s. Unlike the case of the air fluidized bed these times are no longer comparable. If one speculates that a significant growth of one-dimensional structure is a necessary prelude to the birth of a sustainable bubble these figures are important. In the case of the water fluidized bed the one-dimensional wave is likely be dispersed by the much faster growing two-dimensional instability before it has a chance for much growth.

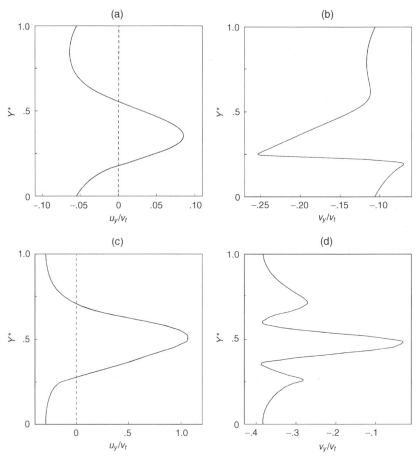

Figure 5.44. Fluid and particle velocities, u_y/v_t and v_y/v_t, along the centreline of bubble-like structures in the air and water fluidized beds. (a), (b) Air fluidized bed (Figure 5.33); (c), (d) water fluidized bed (Figure 5.37). (Adapted from Glasser, 1996.)

As pointed out by Glasser et al. (1997) another difference between the two cases is reflected in the way the velocity of a particle varies during its descent through the fully developed bubble. Figure 5.44(b) shows the vertical component of the solids velocity, as a function of dimensionless height, along the centreline of the bubble depicted in Figure 5.33, while Figure 5.44(d) similarly shows this velocity for the bubble of Figure 5.37. The former belongs to the air fluidized bed whereas the latter is one of the fully developed bubbles that we know can also exist in the water fluidized bed. In both cases the velocities are measured relative to the rest frame of the bubble. The profile of

Figure 5.44(b) shows that the particles accelerate smoothly downward until they decelerate sharply when they rejoin the dense bed below the bubble. In contrast, the profile of Figure 5.44(d) reveals that the downward acceleration is interrupted by an interval of deceleration in which the velocity of fall relative to the bubble is reduced to about $0.03v_t$ before the material reaccelerates on its way to the base of the bubble. The reason for this difference in behaviour is made clear by the fluid velocity profiles, shown as Figures 5.44(a) and (c), respectively. In the air fluidized bed (Figure 5.44(a)) the fluid velocity reaches a maximum upward value of less than $0.1v_t$, while in the water fluidized bed (Figure 5.44(c)) its maximum value actually exceeds v_t. This immediately calls to mind an early criterion of Davidson & Harrison (1963) for the stabilty of a bubble. These authors argued that, if the upward velocity of fluid within the bubble exceeds v_t, the bubble will be destroyed by particles entrained from the dense bed below. However, Glasser et al. were unable to find any fully developed structure in which the upward fluid velocity actually caused a reversal of the downward motion of the particles, as suggested by Davidson & Harrison.

5.5.4 Structures Other Than Bubbles

The results described so far all refer to dense fluidized beds with $\phi_0 = 0.57$, and the fully developed structures that have been identified bear credible comparison with slugs in narrow fluidized beds and bubbles in wider beds. But other types of structure have been observed in beds at higher expansions; for example, compact clusters and streamers with elevated particle concentration are widely reported to be present in dilute beds. The question therefore arises whether these structures might emerge from the equations of motion if we extended the methods used above to beds of smaller ϕ_0. This has recently been addressed by Glasser et al. (1998).

Recall that the search for strucures that propagate without change of form involved three parameters, ϕ_0, k_x, and k_y, and in exhibiting the branching structure of these solutions we fixed two of these and plotted the amplitude of the spatial variations in concentration against the third, which hitherto was always selected to be k_y. This was a reasonable procedure when studying beds with a single, specified value of ϕ_0. However, since we are now interested in the effect of changing the value of ϕ_0, it is useful to replace k_y with this variable in exploring the branching structure. This is the approach adopted by Glasser et al. (1998) who study an air fluidized bed of 200 μm diameter glass beads, with the properties specified in Table 5.2, except that the particle-phase viscosity is

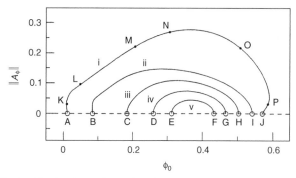

Figure 5.45. Amplitude of one-dimensional structures ($k_x = 0$) as a function of ϕ_0 for various values of k_y. (i) $k_y = 0.33$; (ii) $k_y = 0.66$; (iii) $k_y = 4.64$; (iv) $k_y = 5.30$; (v) $k_y = 5.64$. (From Glasser et al., 1998.) Copyright 1998 by the American Physical Society.

assigned the fixed value 6.65 p and the particle-phase pressure is represented by an expression of the form

$$p^p = \frac{P\phi}{(\phi_m - \phi)^2}$$

with $P = 0.0388$ Pa.

First limit attention to one-dimensional structures, setting $k_x = 0$. Then the amplitude of the spatial variations in fully developed one-dimensional waves can be plotted against ϕ_0, for various fixed values of k_y, as shown in Figure 5.45. For any given value of k_y there exist two critical values of ϕ_0. The uniform bed is unstable to small perturbations with wavenumber k_y when ϕ_0 lies between these critical values and stable for other values of ϕ_0. A single branch of one-dimensional structures bifurcates from the uniform bed solution through Hopf bifurcations at each of the critical points. For the smallest value of k_y both bifurcations are subcritical but, as k_y is increased, first the bifurcation at the higher value of ϕ_0 becomes supercritical and then that at the lower value of ϕ_0.

The volume fraction profiles for the solutions corresponding to the points labeled KLMNOP in Figure 5.45 are shown in Figure 5.46. In beds with high expansion, represented by points such as L and M in Figure 5.45, we see that the wave structure consists of narrow bands with high particle concentration separated by relatively wide bands where the concentration of particles is low. In dense beds, in contrast, represented by points such as O, the structure consists of narrow bands with low particle concentration separated by relatively wide bands of dense material. The latter type of structure we have already seen and have identified with slugs in dense fluidized beds confined within narrow tubes. The former could be regarded as a one-dimensional analogue of the dense clusters

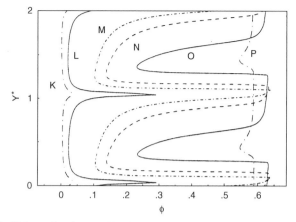

Figure 5.46. Volume fraction profiles for the one-dimensional waves corresponding to points K–P in Figure 5.45. (From Glasser et al., 1998.) Copyright 1998 by the American Physical Society.

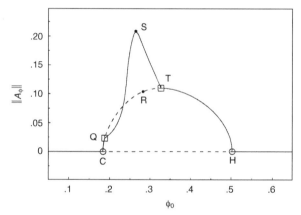

Figure 5.47. Amplitude of one-dimensional structures with $k_x = 0$, $k_y = 4.64$, and two-dimensional structures with $k_x = 3.61$, $k_y = 4.64$, as functions of ϕ_0. (From Glasser et al., 1998.) Copyright 1998 by the American Physical Society.

observed in dilute fluidized beds. There is a continuous transition between the two types of structure on moving along the branch.

The stability of each of the one-dimensional branches of solutions against two-dimensional perturbations can also be tested. Figure 5.47 reproduces the branch CH from Figure 5.45, corresponding to $k_y = 4.64$, and identifies two critical points, labelled Q and T, for stability with respect to two-dimensional perturbations with $k_y = 4.64$ and $k_x = 3.61$. The arc QRT is shown as a broken line since it represents one-dimensional waves that are unstable with respect to two-dimensional perturbations with this periodicity. The unbroken curve QST joining the critical points represents a branch of fully developed two-dimensional structures with these values of k_y and k_x. Very near the point Q these are just slightly distorted forms of the one-dimensional waves with narrow dense bands separated by wider bands of low density. In the same way, very near the point T they are slightly distorted forms of the one-dimensional waves consisting of narrow bands of low density separated by wider bands of high density. However, on moving further away from Q the structure develops into compact clusters of high density material dispersed in a backgound of low density bed, while on moving further away from T one finds compact "bubbles" of low density dispersed in a background of high density bed. Thus, structures resembling bubbles can propagate in dense beds and structures resembling clusters in dilute beds. At intermediate values of ϕ_0 structures are found that could be interpreted as "streamers". From this exploration, of admittedly limited scope, it therefore appears that the equations of motion may be capable of predicting the existence of each of the main types of observed structure, and of indicating under what conditions they will be found. Much more exploratory work of this sort is needed to settle this issue.

5.5.5 Other Work on Two-Dimensional Disturbances and Bubbles

In Section 5.5.3 above we mentioned the early work of Didwania & Homsy (1982) on instabilities of one-dimensional waves that can give rise to transverse structure. This was followed in 1984 by an investigation of the stability of a uniform fluidized bed against small, two-dimensional disturbances (Needham & Merkin, 1984), which showed, not unexpectedly, that the one-dimensional wave is the fastest growing disturbance and is the only disturbance that can grow when the bed is very narrow. However, when the bed is unstable, increasing its width allows an increasing number of modes with two-dimensional structure to grow. Using a Fourier method Needham & Merkin were able to compute and exhibit the evolution of the particle concentration for a mode of this type.

The transient computations described in the last subsection have shown that the fully developed one-dimensional waves are unstable to a motion that initially causes them to "buckle" in the horizontal direction. The buckling is accompanied by a further depletion of particles from high points along a buckled band of low concentration, and this in turn initiates the formation of the compact regions of low concentration that eventually develop into bubbles in gas fluidized beds but fail to do so in liquid fluidized beds. The early stages of this behaviour, which are common to both types of bed, could be anticipated from an analysis of the stability of a stratified fluid due to Batchelor & Nitsche (1991). These authors consider a fluid whose density, in the base state, varies sinusoidally with height:

$$\rho = \rho_0(1 + A \sin(\kappa y)), \qquad (5.48)$$

where the density variation might be thought of as due to variations in concentration of suspended, nonsedimenting particles. Then if $\mathbf{u}_v$ is the volume average velocity for the suspension as a whole and ϕ represents the volume fraction of the particles, $\mathbf{u}_v = \phi \mathbf{v} + (1 - \phi)\mathbf{u}$, and a continuity equation for the mixture can be written in the form

$$\frac{\partial \phi}{\partial t} + \mathbf{u}_v \cdot \nabla \phi = \nabla \cdot (D\nabla \phi) = D\nabla^2 \phi, \qquad (5.49)$$

where D is a diffusion coefficient (assumed constant) for the particles in the fluid. The suspension is also regarded as having an effective kinematic viscosity ν.

The stability of the state (5.48) against perturbations of the form $e^{\gamma t} \cos(\alpha x)$ is then investigated, where x denotes a horizontal coordinate. A dimensionless growth rate and transverse wavenumber can be defined by

$$s = \frac{\gamma}{(\kappa g A)^{1/2}}, \quad a = \frac{\alpha}{\kappa},$$

which are found to be related as follows:

$$\left(s R'^{1/2} + a^2 \right) \left(s R'^{1/2} + a^2 + 1 \right) \left(s \frac{\nu}{D} R'^{1/2} + a^2 \right)$$

$$\times \left(s \frac{\nu}{D} R'^{1/2} + a^2 + 1 \right) = \frac{1}{2} \left(\frac{\nu}{D} \right)^2 \frac{R'^2 a^2}{1 + a^2}. \qquad (5.50)$$

This expression contains two dimensionless groups, namely

$$R' = \frac{gA}{\nu^2 \kappa^3} \quad \text{and} \quad \frac{\nu}{D}. \qquad (5.51)$$

Using (5.50) it is possible to find the dimensionless wavenumber a_m for which s is largest, as well as the corresponding value of s_m, which represents the fastest growth rate, as functions of these dimensionless groups.

The dominant wavenumber a_m and corresponding growth rate s_m, found as above, can be compared with the corresponding quantities computed for two-dimensional perturbations of the fully developed one-dimensional waves in a fluidized bed. Then A is identified with $\Delta\rho/2\rho_0$, where $\Delta\rho$ is the difference between the maximum and the minimum density for the fully developed one-dimensional wave, while α is the wavenumber of the fastest growing two-dimensional perturbation of that wave. The kinematic viscosity ν can be identified with $\mu^p/\rho_s \phi_0$ and it remains only to find a value for D in order to

Table 5.3. *Wave numbers and growth rates for the dominant
two-dimensional instability*

	R'	v/D	s_m (B&N)	a_m (B&N)	s_m (A et al.)	a_m (A et al.)
Air fluidized bed	35.05	5.5	0.567	0.97	0.387	≈ 1.5
Water fluidized bed	7.10	542	0.491	0.88	0.244	≈ 1.0

determine the values of the groups (5.51). A description of the motion of the particles relative to the fluid in terms of a dispersion coefficient is appropriate when inertial effects are negligible and the relative motion of the two phases is driven by a gradient in the particle concentration. But then, from (5.37), (5.39), and (5.40) it follows that

$$\mathbf{u} - \mathbf{v} = \mathbf{w} = -\frac{\nabla p^p}{\beta} = -\frac{p^{p'}}{\beta}\nabla\phi. \tag{5.52}$$

Also, the continuity equation for the particles can be written in terms of $\mathbf{u}_v$ and $\mathbf{w}$, when it takes the form

$$\frac{\partial \phi}{\partial t} + \mathbf{u}_v \cdot \nabla \phi = -\nabla \cdot [\phi(1-\phi)\mathbf{w}]$$

or, substituting for $\mathbf{w}$ from (5.52),

$$\frac{\partial \phi}{\partial t} + \mathbf{u}_v \cdot \nabla \phi = \nabla \cdot \left[\frac{\phi(1-\phi)p^{p'}}{\beta}\nabla\phi \right],$$

which has the same form as (5.49). The two equations are identical if we take

$$D = \frac{\phi(1-\phi)p^{p'}(\phi)}{\beta(\phi)}. \tag{5.53}$$

The parameters R' and v/D can now be found for the one-dimensional waves that formed the starting points for the two-dimensional solutions illustrated in Figures 5.39 and 5.43, and the rates of initial growth and the wavenumbers for the dominant two-dimensional perturbations of these one-dimensional waves can be compared with the rates of growth and wavenumbers determined, as described above, from (5.50). The results are presented in Table 5.3. They are of the same order of magnitude whether obtained from the computations of Anderson et al. or from the simple stratified fluid theory of Batchelor & Nitsche. This suggests that the initial buckling of the one-dimensional waves in a fluidized bed occurs simply as a result of the density stratification, by a mechanism analogous to that proposed by Batchelor & Nitsche.

In a later publication, Batchelor (1993) extended the theory to take into account the motion of the particles relative to the fluid when gravity is present, as in a fluidized bed. This introduces one more dimensionless group, namely $J = V/\kappa(Dv)^{1/2}$, but the analysis resembles closely that of Batchelor & Nitsche. In the work of Anderson et al. and of Glasser et al. only a small number of examples have been explored numerically. (In addition to the work described here Glasser et al. (1997) studied a bed of lead shot fluidized by water.) More general questions regarding the types of instability and the bifurcation scenarios associated with equations of motion of the form (5.36)–(5.43) have been addressed in a number of papers by Göz (1992, 1993, 1995). In addition to bifurcations of the sort studied by Glasser et al., which have their origin in a single bifurcation of a branch of one-dimensional waves from the uniform bed solution, Göz (1993) pointed out that there may be two such bifurcations, corresponding to the existence of two critical wavenumbers, for certain ranges of parameter values. The branches originating at these bifurcations may either remain separate or they may join together to form a single branch joining the bifurcation points. Göz also indicated that other branches may exist that correspond to homoclinic orbits in a phase plane, which are physically descriptive of pulses, and to heteroclinic orbits, which describe propagating wavefronts. In later work Göz (1995) examined the question of the stability of one-dimensional waves to two-dimensional perturbations. Since explicit results were sought, he was forced to limit attention to one-dimensional waves in their initial stages of growth, when they are small in amplitude and sinusoidal in form. Then he was able to show that they become unstable when their amplitude has grown to a value proportional to the square of the wavenumber of the transverse perturbation. Thus, they become unstable to very long wave transverse perturbations while their amplitude is small, but they must grow to larger amplitudes if this instability is to extend to smaller transverse wavelengths. As a consequence of this it is to be expected that growing one-dimensional waves will retain their stability to larger amplitudes in narrow beds than in wide beds, in agreement with the conclusion of Needham & Merkin (1984) mentioned above.

Harris & Crighton (1994) and Harris (1996) have obtained analytical results on the growth of fully developed one-dimensional waves from small perturbations of a uniform fluidized bed. These are consistent with the numerical computations carried out for specific cases by Anderson et al. (1995) and Glasser et al. (1996, 1997, 1998) and lend some confidence that the features observed in these computations are not limited to these cases but have more general validity. In parallel with this work are the papers by Hayakawa and coworkers mentioned earlier (Komatsu & Hayakawa, 1993; Hayakawa et al., 1994) starting from the same differential equation for the particle concentration, valid only in the weakly nonlinear range.

The computations of Anderson et al. (1995) grew bubbles from small perturbations of fluidized beds. To be more precise, because of the periodicity conditions imposed, they grew a regular two-dimensional array of identical bubbles in an infinite fluidized bed. A similar situation has also been addressed by Hernández & Jiménez (1991). In practice, however, bubbles usually form at the distributor of a bed of finite depth. Having broken away from the distributor they then move up through the bed and burst through its upper surface. A number of authors have simulated this situation by integrating equations of motion similar to those used here. The earliest work of this sort may be that of Pritchett et al. (1978), though with the numerical methods employed at that time it is unlikely that their results represented actual solutions with much precision. The modelling of fluidized bed behaviour by direct numerical integration of continuum equations of motion was pioneered by Gidaspow and coworkers. Gidaspow et al. (1986) describe results specific to bubbles, and an extensive survey of the work of this group can be found in Gidaspow (1994).

Figure 5.48, from Gidaspow et al. (1986), shows a comparison between photographs of a bubble in a two-dimensional air fluidized bed of 800 μm diameter glass beads and a bubble generated computationally in a simulation of the same conditions. The equations of motion included a term representing a particle-phase pressure dependent on the particle concentration, but no viscous terms were present. The bed was two dimensional and expanded just to minimum fluidization, and the bubble was generated by injecting a larger gas flow locally at the centre of the distributor, starting at time zero. The times indicated in the figure are elapsed times in seconds since the gas injection began. The bright points in the photographs of the experimental bubble are, of course, individual particles, but those in the computationally generated bubbles are not. Instead they are simply points distributed locally at random, with a surface concentration proportional to the computed density of the bed. The centre of the computed bubble appears almost black, but this is an illusion resulting from the factor of proportionality used in determining the concentration of the dots; in fact, there is still a significant concentration of particles within the computed bubble. Nevertheless, the shape and size of the bubble are well represented and the diffuse nature of the roof compared to the floor of the bubble is evident.

A rather more extensive series of experiments and computational simulations of bubbles generated in air fluidized beds of glass beads has been reported more recently by Kuipers et al. (1991). The system investigated was the same as that of Gidaspow et al., namely a two-dimensional bed with a jet of gas injected through the plane of the distributor, starting at time zero. However, the authors appear to have been unaware of the earlier work by Gidaspow's group.

In a separate, more ambitious work Gidaspow et al. (1992) have attempted to simulate the motion of a fluidized suspension within a complete circulation loop, consisting of a riser and a fluidized standpipe with a quite complex geometry.

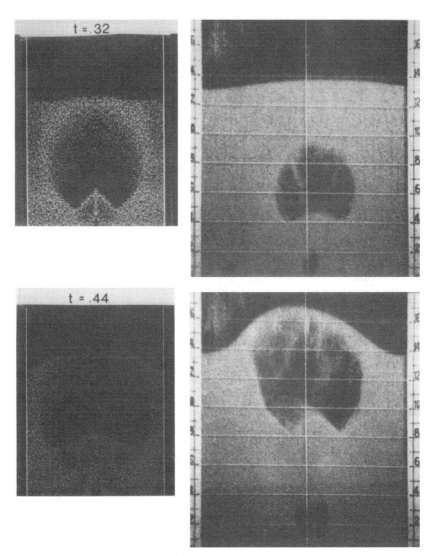

Figure 5.48. Comparison of computer simulated (left) and experimentally observed (right) bubbles induced by injecting a gas jet, starting at $t = 0$, into a bed of 800 μm diameter glass beads. Elapsed times in seconds are indicated.

The size of the computational grid was probably too large to yield quantitatively reliable predictions, but the simulated motion painted a picture of the system that was not at all unreasonable on physical grounds. Regardless of its quantitative accuracy this work points a way that is likely to be followed increasingly in the future. If successful it could revolutionize our approach to the design and analysis of practical devices based on fluidized suspensions.

References

Anderson, K., Sundaresan, S., & Jackson, R. 1995. Instabilities and the formation of bubbles in fluidized beds. *J. Fluid Mech.* **303**, 327–366.

Batchelor, G. K. 1993. Secondary instability of a gas fluidized bed. *J. Fluid Mech.* **257**, 359–371.

Batchelor, G. K. & Nitsche, J. M. 1991. Instability of stationary unbounded stratified fluid. *J. Fluid Mech.* **227**, 357–391.

Buyevich, Y. A., Yates, J. G., Cheesman, D. J., & Wu, K-T. 1995. A model for the distribution of voidage around bubbles in a fluidized bed. *Chem. Eng. Sci.* **50**, 3155–3162.

Collins, R. 1965a. An extension of Davidson's theory of bubbles in fluidized beds. *Chem. Eng. Sci.* **20**, 747–755.

Collins, R. 1965b. The rise velocity of Davidson's fluidization bubble. *Chem. Eng. Sci.* **20**, 788–789.

Crighton, D. G. 1991. Nonlinear waves in fluidized beds. In *Nonlinear Waves in Real Fluids*, ed. A. Kluwick, pp. 83–90. Springer-Verlag.

Davidson, J. F. 1961. Symposium on fluidization–discussion. *Trans. Inst. Chem. Eng.* **39**, 230–232.

Davidson, J. F. & Harrison, D. 1963. *Fluidized Particles*. Cambridge University Press.

Davidson, J. F. & Harrison, D. 1971. *Fluidization*. Academic Press.

Davies, R. M. & Taylor, Sir Geoffrey. 1950. The mechanics of large bubbles rising through extended liquids and through liquids in tubes. *Proc. Roy. Soc. London*, A **200**, 375–390.

Didwania, A. K. & Homsy, G. M. 1981. Flow regimes and flow transitions in liquid fluidized beds. *Int. J. Multiphase Flow.* **7**, 563–580.

Didwania, A. K. & Homsy, G. M. 1982. Resonant side-band instabilities in wave propagation in fluidized beds. *J. Fluid Mech.* **122**, 433–438.

Fanucci, J. B., Ness, N., & Yen, R-H. 1979. On the formation of bubbles in gas–particulate fluidized beds. *J. Fluid Mech.* **94**, 353–367.

Fanucci, J. B., Ness, N., & Yen, R-H. 1981. Structure of shock waves in gas–particulate fluidized beds. *Phys. Fluids* **24**, 1944–1954.

Ganser, G. H. & Drew, D. A. 1990. Nonlinear stability analysis of a uniform fluidized bed. *Int. J. Multiphase Flow* **16**, 447–460.

Geldart, D. 1973. Types of gas fluidization. *Powder Technol.* **7**, 285–292.

Glasser, B. J. 1996. One- and two-dimensional traveling wave solutions in fluidized beds. Ph.D. dissertation, Princeton University.

Glasser, B. J., Kevrekidis, I. G., & Sundaresan, S. 1996. One- and two-dimensional travelling wave solutions in gas-fluidized beds. *J. Fluid Mech.* **306**, 183–221.

Glasser, B. J., Kevrekidis, I. G., & Sundaresan, S. 1997. Fully developed travelling wave solutions and bubble formation in fluidized beds. *J. Fluid Mech.* **306**, 183–221.

Glasser, B. J., Sundaresan, S., & Kevrekidis, I. G. 1998. From bubbles to clusters in fluidized beds. *Phys. Rev. Lett.* **81**, 1849–1852.

Gidaspow, D. 1994. *Multiphase Flow and Fluidization*. Academic Press.

Gidaspow, D., Syamlal, M., & Seo, Y. C. 1986. Hydrodynamics of fluidization: supercomputer generated vs. experimental bubbles. *J. Powder & Bulk Solids Tech.* **10**(3), 19–23.

Gidaspow, D., Bezbaruah, R., & Ding, J. 1992. Hydrodynamics of circulating fluidized beds: kinetic theory approach. In *Fluidization VII*, ed. O. E. Potter & D. J. Nicklin, pp. 75–82. Engineering Foundation.

Göz, M. F. 1992. On the origin of wave patterns in fluidized beds. *J. Fluid Mech.* **240**, 379–404.

Göz, M. F. 1993. Bifurcation of plane voidage waves in fluidized beds. *Physica* **D65**, 319–351.

Göz, M. F. 1995. Transverse instability of plane wavetrains in gas-fluidized beds. *J. Fluid Mech.* **303**, 55–81.

Harris, S. E. 1996. The growth of periodic waves in gas-fluidized beds. *J. Fluid Mech.* **325**, 261–282.

Harris, S. E. & Crighton, D. G. 1994. Solitons, solitary waves and voidage disturbances in gas-fluidized beds. *J. Fluid Mech.* **266**, 243–276.

Hayakawa, H., Komatsu, T. S., & Tsuzuki, T. 1994. Pseudo-solitons in fluidized beds. *Physica* **A204**, 277–289.

Hernández, J. A. & Jiménez, J. 1991. Bubble formation in dense fluidized beds. In *Proceedings of NATO Advanced Research Workshop on the Global Geometry of Turbulence*, ed. J. Jiménez, pp. 132–142. Plenum Press.

Jackson, R. 1963. The mechanics of fluidized beds. Part II: the motion of fully developed bubbles. *Trans. Inst. Chem. Eng.* **41**, 22–28.

Jacob, K. V. & Weimer, A. W. 1987. High-pressure particulate expansion and minimum bubbling of fine carbon powders. *AIChE J.* **33**, 1698–1706.

Komatsu, T. S. & Hayakawa, H. 1993. Nonlinear waves in fluidized beds. *Physics Lett.* **A183**, 56–62.

Kuipers, J. A. M., Prins, W., & Van Swaaij, W. P. M. 1991. Theoretical and experimental bubble formation at a single orifice in a two-dimensional gas-fluidized bed. *Chem. Eng. Sci.* **46**, 2881–2894.

Liu, J. T. C. 1983. Nonlinear unstable wave disturbances in fluidized beds. *Proc. Roy. Soc. London*, **A389**, 331–347.

Lockett, M. J. & Harrison, D. 1967. The distribution of voidage fraction near bubbles rising in gas-fluidized beds. In *Proceedings of the International Symposium on Fluidization*, ed. A. A. H. Drinkenburg, pp. 257–267. Netherlands Univ. Press.

Murray, J. D. 1965. On the mathematics of fluidization, Part 2: steady motion of fully developed bubbles. *J. Fluid Mech.* **22**, 57–80.

Murray, J. D. 1967. Initial motion of a bubble in a fluidized bed Part 1. Theory. *J. Fluid Mech.* **28**, 417–428.

Needham, D. J. & Merkin, J. H. 1983. The propagation of a voidage disturbance in a uniform fluidized bed. *J. Fluid Mech.* **131**, 427–454.

Needham, D. J. & Merkin, J. H. 1984. The evolution of a two-dimensional small-amplitude voidage disturbance in a uniformly fluidized bed. *J. Eng. Math.* **18**, 119–132.

Needham, D. J. & Merkin, J. H. 1986. The existence and stability of quasi-steady periodic voidage waves in a fluidized bed. *J. Appl. Math. Phys. (ZAMP)* **37**, 322–339.

Partridge, B. A. & Lyall, E. 1967. Initial motion of a bubble in a fluidized bed Part 2. Experiment. *J. Fluid Mech.* **28**, 429–431.

Phillips, O. M. 1974. Wave interactions. In *Non-Linear Waves*, ed. S. Leibovich & A. R. Seebas, pp. 186–211. Cornell Univ. Press.

Pigford, R. L. & Baron, T. 1965. Hydrodynamic stability of a fluidized bed. *Ind. Eng. Chem. Fundam.* **4**, 81–87.

Pritchett, J. W., Blake, T. R., & Garg, S. K. 1978. A numerical model of gas fluidized beds. *AIChE Symp. Ser. No. 176* **74**, 134–148.

Reuter, H. 1963a. Druckverteilung um Blasen im Gas-Feststoff-Fliessbett. *Chem. Ing. Tech.* **35**, 98–103.

Reuter, H. 1963b. Mechanismus der Blasen im Gas-Feststoff-Fliessbett. *Chem. Ing. Tech.* **35**, 219–228.

Rowe, P. N. 1962. The effect of bubbles on gas–solids contacting in fluidized beds. *Chem. Eng. Progr. Symp. Ser.* **58**, 42–56.

Rowe, P. N., Partridge, B. A., & Lyall, E. 1964. Cloud formation around bubbles in gas fluidized beds. *Chem. Eng. Sci.* **19**, 973–985.

Rowe, P. N., Partridge, B. A., & Lyall, E. 1965. Cloud formation around bubbles in gas fluidized beds. *Chem. Eng. Sci.* **20**, 1151–1153.

Salatino, P., Poletto, M., & Massimilla, L. 1991. Stability of uniform gas fluidized beds opereated with CO_2 in ranges of pressure and temperature between ambient and nearly critical conditions. Paper presented at *IUTAM Symposium on Mechanics of Fluidized Beds*, Stanford University.

Stewart, P. S. B. 1968. Isolated bubbles in fluidized beds–theory and experiment. *Trans. Inst. Chem. Eng.* **46**, 60–66.

Stewart, P. S. B. & Davidson, J. F. 1967. Slug flow in fluidized beds. *Powder Technol.* **1**, 61–80.

Wace, P. F. & Burnett, S. J. 1961. Flow patterns in gas fluidized beds. *Trans. Inst. Chem. Eng.* **39**, 168.

Yates, J. G. & Cheesman, D. J. 1992. Voidage variations in the regions surrounding a rising bubble in a fluidized bed. *AIChE Symp. Ser.* **88** (289), 34–39.

Yates, J. G., Cheesman, D. J., & Sergeev, Y. A. 1994. Experimental observations of voidage distribution around bubbles in a fluidized bed. *Chem. Eng. Sci.* **49**, 1885–1895.

6

Riser flow

6.1 Introduction

In a number of industrially important applications a mixture of particles and fluid flows along a pipe under the influence of gravity, a pressure gradient in the fluid, or a combination of the two. For example, in slurry pipelines the fluid is a liquid and the object is to transport a particulate material over considerable horizontal distances. With a gas in place of the liquid, pneumatic transport lines are used to convey particles both horizontally and vertically, usually within the bounds of a factory. Upward vertical transport of particles by gas occurs in the very important example of a riser reactor; the reactors used for catalytic cracking in the manufacture of gasoline are of this sort, with the transported particulate material being the catalyst for the cracking reactions. Catalytic crackers also afford an example of another important device of this type. To complete the loop through which the catalyst circulates it is necessary to transport the particles from a region of lower gas pressure, at the top of the riser, to a region of higher gas pressure at the bottom. This can be accomplished with the aid of gravity, as we shall see in the next chapter, and the descending pipe used for this purpose is called a standpipe.

If the axis of the pipe is inclined to the vertical gravity will cause the particles to migrate towards the lower part of each cross section, so the distribution of particles over the cross section will clearly be nonuniform. For a vertical pipe, however, it would appear at first glance that the particles should be distributed uniformly over the section, and consequently an analysis of the motion should be straightforward. The earliest attempts to predict the behaviour of gas–particle flow in vertical pipes on the basis of equations of motion took just this point of view; indeed, they can be regarded as quantitative embodiments of the sort of ideas presented in Chapter 1. Almost twenty years ago Arastoopour &

233

Gidaspow (1979) formulated dynamical equations to describe one-dimensional motion in a vertical pipe. These consisted of continuity equations for gas and particulate phases, a mixture momentum equation, and a slip relation to predict the gravitationally induced relative velocity of the two phases. In order to treat developing flow spatial variations were permitted, but only in the axial direction. Then, taking into account the effect of the pipe wall by the traditional device of a "friction factor" it proved possible to predict "Zenz diagrams" with credible shapes, that is, plots of pressure drop versus gas flow for fixed values of the solids flow. As could be anticipated from the simple arguments presented in Chapter 1 the pressure drop was found to decrease, pass through a minimum value, then increase as the gas flow was increased. However, to secure quantitative agreement with measurements it was found necessary to replace the actual particles by "clumps", increasing the true diameter more than sevenfold so as to make the slip velocity between the phases much larger than it would be for a uniform distribution of the particles themselves with the same solids concentration. This is a crude way of acknowledging that the particles are not uniformly distributed, without attempting to probe the mechanism behind the maldistribution. In a subsequent publication (Arastoopour et al., 1982) this type of analysis was extended to encompass particles with a distribution of sizes, and again it was found necessary to introduce effective diameters, obtained by dividing the actual diameters by a coefficient of restitution for particle–particle collisions. Artificially low values of this coefficient had to be used to bring the predictions into agreement with experiments.

Observations of the distribution of particles over the cross section reveal clearly the physical phenomena responsible for the discrepancies that were resolved by the devices just described. When particles are propelled rapidly upward by gas, as in a vertical riser, they are seen on average to congregate preferentially close to the pipe wall (Bartholomew & Casagrande, 1957; Saxton & Worley, 1970; Yerushalmi et al., 1978; Youchou & Kwauk, 1980; Weinstein et al., 1984; Bader et al., 1988; Nieuwland et al., 1996). The core of the pipe is occupied by a dilute suspension transported at high speed by the gas, while particles in the relatively concentrated annulus near the wall move up more slowly or migrate downward against the general direction of flow. Clearly then, neither the relation between flow rates and pressure gradient, nor the performance of the system as a chemical reactor, will be the same as if the particles were uniformly distributed. It is, therefore, important to understand the physical mechanism responsible for the maldistribution of the particles and to be able to predict its quantitative consequences. In addition to this variation in time-average particle concentration, on the scale of the pipe diameter, other inhomogeneities in concentration are observed on smaller scales. The dilute region away from the pipe wall is found to contain clusters and vertical streamers with a higher concentration of particles (Grace & Tuot, 1979; Matsen, 1982; Basu & Nag, 1987; Gidaspow et al., 1989; Tsuo & Gidaspow, 1990). These are transient phenomena, disappearing and reforming continually. Their presence is not surprising in view of the work of Glasser et al. (1998) described in Chapter 5, but they clearly complicate the analysis further.

This chapter will be devoted to an account of some attempts to explain the above observations and even predict their salient consequences, such as the cross-sectional distribution of average particle concentration and the velocity profiles of the two phases, together with the relation between the overall flow rate of each phase and the gas pressure gradient. This work has followed two paths. The first, which will be examined in Sections 6.2 to 6.4, consists of attempts to predict time-averaged quantities such as the radial concentration profile, the flow rates of the two phases, and the axial pressure gradient, without following the complicated, time-dependent motion in detail. The second, which is developing rapidly and attracting much attention, relies on direct, detailed solution of the equations describing time-dependent motions of the gas and the particles.

Before embarking on a description of this work it is important to distinguish a number of different length scales associated with the problem. First there is the particle diameter, which is comparable to the mean spacing between particles at the concentrations of interest in these systems. All physical phenomena on this scale are removed from our explicit attention by the local volume averaging that leads to continuum equations of motion of the sort used throughout this book, and their effects are intended to be subsumed in the closures postulated for the undetermined terms in these equations. The effective diameter of the averaging region sets a second length scale, considerably larger than the first, since it is assumed to contain many particles. Yet a third length scale is associated with the phenomena leading to fluctuations in the local average particle concentration and the velocities of the two phases. These phenomena may be bubbles, clusters, streamers, or pseudoturbulent motions, and, if they are to be describable by the continuum equations of motion, their length scale must, in turn, be considerably larger than the diameter of the local averaging region used in establishing these equations. Finally, there is the macroscopic length scale set by the diameter of the pipe. In the approach of Sections 6.2 to 6.4 details of any fluctuations on the scale of clusters, streamers, turbulence, etc. are erased by time smoothing, leaving only the spatial variations on the scale of the pipe diameter. For this approach to succeed it is essential that there should be a separation between the above length scales, but in the first treatment of this problem, described in Section 6.2, all phenomena at the third length scale are neglected, thus yielding an analogue of the theory of laminar flow for a single-phase fluid in a pipe.

Before proceeding any further we should note that there exists a large and important literature dealing with the effect of a small concentration of suspended particles on the turbulent flow of a fluid through a pipe. Though this is important in its own right it has essentially no bearing on the systems of interest here, since attention is limited to cases where the *mass* fraction of the particles in the mixture is of order one, *at most*. Then the relaxation time between the motions of fluid and particles is of the same order as that for an isolated particle in an infinite fluid, and the presence of the particles can be regarded as exerting a perturbing influence on the structure of the fluid turbulence that would exist in their absence. However, in the systems of present interest the ratio of the mass of particles to the mass of gas is typically between ten and several hundred. In this case, with reference to the discussion in Chapter 2, Section 2.5, the relaxation time is very much shorter than that for an isolated particle and it is the less massive phase, namely the gas, whose local average motion relaxes towards that of the particles forming the more massive phase. The inertia of the solid material then dominates completely, to the extent that it makes little sense to think in terms of particles responding to turbulent fluctuations in the gas velocity. In these dense systems the fluctuations that are observed on the third scale are the result of inertial instabilities, of various types, associated with the collective behaviour of the particle assembly. We have already discussed, in some detail, those instabilities that are precursors of bubbles, clusters, and streamers. However, it has recently been realized that there are also other instabilities exhibited by a sheared assembly of particles, even in the absence of any interstitial fluid (Hopkins & Louge, 1991; Savage, 1992; Goldhirsch et al., 1993; Schmid & Kytomaa, 1994; Wang et al., 1996, 1997; Alam & Nott, 1998). The gas phase merely responds to fluctuations in particle velocity resulting from any or all the above phenomena. Since the fluctuations in concentration and velocity that concern us here are not driven by gas-phase turbulence we shall, with no intended reflection on its value, disregard the whole of the literature devoted to very dilute suspensions.

6.2 Fully Developed "Laminar" Flow of a Gas–Particle Mixture in a Vertical Pipe

A first attempt to elucidate the mechanism responsible for segregation of the particles towards the pipe walls has been reported by Sinclair & Jackson (1989), who neglected all phenomena such as clustering, streamers, and pseudoturbulent fluctuations. Consequently, their treatment yields what might be called a "laminar" solution of the local-averaged equations of motion for a steady, fully developed flow in a vertical pipe. The momentum balances for the fluid and particle phases take the forms (2.33) and (2.34). Since the flow is fully developed the inertial terms on the left-hand sides of these equations vanish. Also, since the fluid in question is a gas, it is permissible to omit the gravity and buoyancy terms that are proportional to ρ_f. Thus the momentum equations reduce to the following force balances:

$$0 = \nabla \cdot \mathbf{S}^f - n\mathbf{f}_2 \tag{6.1}$$

and

$$0 = \nabla \cdot \mathbf{S}^p + n\mathbf{f}_2 + \rho_s \phi \mathbf{g}. \tag{6.2}$$

For the interphase drag force $n\mathbf{f}_2$ the Richardson–Zaki form (2.65) is adopted, namely

$$n\mathbf{f}_2 = \beta(\phi)(\mathbf{u} - \mathbf{v}) = \frac{(\rho_s - \rho_f)\phi g}{v_t \varepsilon^{n-1}}(\mathbf{u} - \mathbf{v}), \tag{6.3}$$

and the stress tensors are assumed to be expressible in the Newtonian form:

$$\mathbf{S}^f = -p^f \mathbf{I} + \mu^f \left[\nabla \mathbf{u} + (\nabla \mathbf{u})^T - \frac{2}{3}(\nabla \cdot \mathbf{u})\mathbf{I} \right], \tag{6.4}$$

$$\mathbf{S}^p = -p^p \mathbf{I} + \mu^p \left[\nabla \mathbf{v} + (\nabla \mathbf{v})^T - \frac{2}{3}(\nabla \cdot \mathbf{v})\mathbf{I} \right]. \tag{6.5}$$

In the gas–particle suspension of interest the Stokes number associated with encounters between particles is expected to be large so, as discussed in Section 2.2, interactions between particles will occur primarily through solid–solid contact forces exerted during collisions. In these circumstances a closure for $\mathbf{S}^p$ of the sort proposed by Koch (1990) or by Buyevich (1994) is appropriate. Each of these adopts expressions for p^p and μ^p that depend on the granular temperature $T(= \frac{1}{3}\langle \mathbf{u}' \cdot \mathbf{u}' \rangle^p)$ as well as the volume fraction ϕ, so this temperature must be found using the equation of balance (2.49) for pseudothermal energy. For steady, fully developed flow this reduces to

$$0 = \mathbf{S}^p : \nabla \mathbf{v} - \nabla \cdot \mathbf{q} + Q_+ - Q_- - Q_c. \tag{6.6}$$

Here the first term represents generation of pseudothermal energy by the deformation of the particle assembly, the second represents accumulation of energy as a result of conduction, the next two describe production and dissipation of the energy of particle motion due to interactions with the gas, and the last gives the dissipation resulting from the inelasticity of collisions between particles.

Since neither Koch's nor Buyevich's closure was available at the time of Sinclair & Jackson's work (Koch's closure is, in any case, limited to particle concentrations smaller than those of interest for riser flow) the equations were closed by adopting a closure for $\mathbf{S}^p$ of the form proposed by Lun et al. (1984) for granular materials (see also Johnson & Jackson (1987)). In the particle momentum equation the pressure and viscosity are then given by

$$p^p = \rho_s \phi T (1 + 4\eta\phi g_0) - \frac{256\eta\phi^2 g_0}{5\pi}\mu\nabla \cdot \mathbf{v}, \qquad (6.7)$$

$$\mu^p = \mu \left\{ \frac{\left(1 + \frac{8}{5}\eta\phi g_0\right)\left(1 + \frac{8}{5}\eta(3\eta - 2)\phi g_0\right)}{\eta(2 - \eta)g_0} + \frac{768\eta\phi^2 g_0}{25\pi} \right\}, \qquad (6.8)$$

where

$$\eta = (1 + e)/2, \quad g_0 = \frac{1}{1 - (\phi/\phi_m)^{1/3}}, \quad \mu = \frac{5m(T/\pi)^{1/2}}{16d^2}, \qquad (6.9)$$

and where e is the coefficient of restitution for collisions between particles, ϕ_m is the solids volume fraction at random close packing, and m and d are the mass and diameter of the particles. In the energy balance (6.6) both stimulation and damping of particle velocity fluctuations by interaction with the gas are neglected, so $Q_+ = Q_- = 0$, while losses due to inelastic collisions are represented by

$$Q_c = \frac{48}{\pi^{1/2}}\eta(1 - \eta)\frac{\rho_s\phi^2}{d}g_0 T^{3/2}. \qquad (6.10)$$

Finally, the pseudothermal energy flux $\mathbf{q}$ is given by

$$\mathbf{q} = -\frac{30\mu}{\eta(41 - 33\eta)g_0}\left[\left(1 + \frac{12}{5}\eta\phi g_0\right)\left(1 + \frac{12}{5}\eta^2(4\eta - 3)\phi g_0\right)\right.$$
$$\left. + \frac{64}{25\pi}(41 - 33\eta)(\eta\phi g_0)^2\right]\nabla T$$
$$+ \left(1 + \frac{12}{5}\eta\phi g_0\right)\frac{12}{5}\eta(\eta - 1)(2\eta - 1)\frac{d}{d\phi}(\phi^2 g_0)\frac{T}{\phi}\nabla\phi. \qquad (6.11)$$

This completes the closure of the particle-phase momentum balance. In the case of the gas, for which compressibility is neglected, p^f is regarded as an independent variable while the effective viscosity μ^f presumably depends on the concentration of the particles. Sinclair & Jackson took the following form:

$$\mu^f = \mu_g(1 + 2.5\phi + 7.6\phi^2)\left(1 - \frac{\phi}{\phi_m}\right), \qquad (6.12)$$

where μ_g is the viscosity of the gas alone. The effective viscosity μ^f increases with particle concentration at low values of the concentration, as might be expected, then decreases as the concentration approaches a value corresponding to random close packing, when the influence of relative motion between two spatially separated parts of the gas might be expected to be screened by the intervening particles.

In cylindrical coordinates the axial component of the gas-phase momentum equation (6.1) becomes

$$-\frac{dp^f}{dz} + \frac{1}{r}\frac{d}{dr}\left(r\mu^f\frac{du_z}{dr}\right) - \beta(\phi)(u_z - v_z) = 0 \tag{6.13}$$

while the radial component simply demands that $\partial p^f / \partial r = 0$. The axial component of the particle-phase momentum equation (6.2) is

$$\frac{dS_{rz}^p}{dr} + \frac{S_{rz}^p}{r} + \beta(\phi)(u_z - v_z) - \rho_s\phi g = 0, \tag{6.14}$$

with S^p given by (6.5), (6.7), (6.8), and (6.9). The radial component of this equation is

$$\frac{dS_{rr}^p}{dr} + \frac{S_{rr}^p}{r} - \frac{S_{\theta\theta}^p}{r} = 0. \tag{6.15}$$

Finally, the pseudothermal energy balance (6.6) reduces to

$$\frac{dq_r}{dr} + \frac{qr}{r} - S_{rz}^p\frac{dv}{dr} + Q_c = 0, \tag{6.16}$$

with Q_c and $\mathbf{q}$ given by (6.10) and (6.11)

Up to this point in our account of the mechanics of fluidized suspensions we have not had to be concerned with boundary conditions to be imposed at a solid wall, but this can no longer be avoided since interaction with the pipe wall has a profound effect on the present problem. Boundary conditions are clearly needed for the tangential components of $\mathbf{u}$ and $\mathbf{v}$ and, since (6.6) is invoked to determine the temperature, a third boundary condition is needed for this equation.

Johnson & Jackson (1987) showed that the form of the boundary conditions on $\mathbf{v}$ and T at the wall can be found by simple force and energy balances. First consider the force balance, which equates the limit of the traction force $-S_{rz}^p$, exerted by the particle phase on the wall, with the rate of transfer of momentum to the wall by particles colliding with it. If c is the average speed of the pseudothermal motion and s the average distance between the wall and the particles nearest to it, then the average frequency of collision of a given particle with the wall is $c/s = (3T)^{1/2}/s$. Thus, if n_a is the average number of particles adjacent to unit area of the wall, the average frequency of collision between particles and unit area of the wall is $n_a(3T)^{1/2}/s$. On average each particle arriving at the wall carries momentum mv_z in the axial direction, and we suppose that the nature of the wall is such that a fraction ϕ' of this is transferred on impact. Then the force balance takes the form

$$-S^p_{rz} = \frac{n_a(3T)^{1/2}}{s}\phi'\rho_s\frac{\pi d^3}{6}v_z.$$

Approximate forms for n_a and s can be found by replacing the random distribution of particles by a simple array with one particle per cell of a cubic lattice with lattice spacing l. Then

$$\phi = \frac{\pi d^3}{6l^3}.$$

The closest packing with this geometry corresponds to $l = d$, so $\phi_m = \pi/6$ and hence $(\phi/\phi_m) = (d/l)^3$. The area of each layer, per particle, is l^2 so, using the last result we get

$$n_a = \frac{1}{d^2}\left(\frac{\phi}{\phi_m}\right)^{2/3}.$$

The spacing between the surfaces of particles in adjacent layers is $l - d$ and this provides a reasonable estimate for s, so

$$s = d\left[\left(\frac{\phi_m}{\phi}\right)^{1/3} - 1\right].$$

Introducing these results into the force balance above gives the required boundary condition in the form

$$S^p_{rz} + \left(\frac{\phi'\sqrt{3}\pi\rho_s(\phi/\phi_m)T^{1/2}}{6\left[1 - (\phi/\phi_m)^{1/3}\right]}\right)v_z = 0. \qquad (6.17)$$

Now we turn our attention to the condition of balance of pseudothermal energy at the wall. There is a flux $-\mathbf{n}\cdot\mathbf{q}$ of energy from the bulk of the particulate material to the interface with the wall, where $\mathbf{n}$ is the unit normal from the wall into the flowing material. Pseudothermal energy is also generated at a rate $-v_z S^p_{rz}$ due to slip at the wall. S^p_{rz}, which represents the force per unit area exerted in the z direction on the particulate material at the wall, has the opposite sign to v_z so the above term is positive. The sum of these terms must be balanced by the rate of dissipation D due to the inelasticity of particle–wall collisions, so the energy balance at the wall takes the form

$$q_r - v_z S^p_{rz} = D.$$

The loss of pseudothermal energy in a single collision between a particle and the wall is $\frac{1}{4}\pi\rho_p d^3 T(1 - e_w^2)$, where e_w is the coefficient of restitution for a

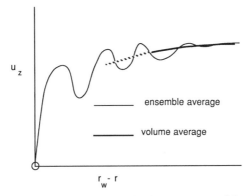

Figure 6.1. Behaviour of ensemble and local volume averages of the axial gas velocity near the wall.

particle–wall collision, and the total rate of dissipation per unit area is found by multiplying this by the number of collisions per unit area, per unit time:

$$D = \left[\frac{1}{4} \pi \rho_s d^3 T \left(1 - e_w^2\right) \right] n_a \frac{(3T)^{1/2}}{s}.$$

Introducing the expressions found above for n_a and s the boundary condition becomes

$$q_r - v_z S_{rz}^p = \frac{\sqrt{3}\pi (\phi/\phi_m)\rho_s T^{3/2} \left(1 - e_w^2\right)}{4 \left[1 - (\phi/\phi_m)^{1/3}\right]}. \tag{6.18}$$

Equations (6.17) and (6.18) provide the boundary conditions that go with the balances of particle-phase momentum and pseudothermal energy, respectively. A corresponding boundary condition for the gas phase is much harder to find. Since the point velocity of the gas satisfies a no-slip condition everywhere on the wall it follows that the ensemble average gas velocity must also vanish at the wall. But our **u** is not an ensemble average; at points sufficiently far away from the wall it is defined by local volume averaging, and at distances from the wall comparable with, or smaller than, the size of the averaging region this definition breaks down. The ensemble average velocity, in contrast, is well defined at any distance from the wall, but when the distance is no more than a few particle diameters this average might be expected to fluctuate significantly as a function of position, reflecting small-scale fluctuations in structure of the suspension induced by the proximity of the wall. This is sketched qualitatively in Figure 6.1.

Our objective is to devise a boundary condition that will determine an extrapolation of the local volume average velocity up to the wall, in such a way that the physical influence of the wall on the flow is represented well, thus subsuming details of the complicated structure very near to the wall. In particular, we shall require the boundary condition to be consistent with known results for the following three situations:

(i) when $\phi \to 0$ the condition should reduce to $u_z = 0$, the no-slip condition for the gas alone,
(ii) when $\phi \to \phi_m$ the particle assembly becomes a packed bed and the fluid is essentially unaware of the presence of the wall, so u_z is determined by a force balance that is almost the same at the wall as within the bed, and
(iii) for all values of ϕ the boundary condition must permit the existence of a solution representing a uniform fluidized bed, with $v_z = 0$ and both u_z and ϕ taking constant values, right up to the wall.

To meet these conditions Sinclair & Jackson proposed the following boundary condition:

$$\frac{\delta_m \phi}{\phi_m} \left[\beta(\phi)(u_z - v_z) + \frac{dp}{dz} \right] + \mu^f(\phi) \left[\frac{du_z}{dr} + \frac{2T\phi_m}{v_t^2 \delta_m \phi} u_z \right] = 0. \qquad (6.19)$$

This has a crude motivation as a force balance on the gas in a thin layer with thickness $\delta_m \phi / \phi_m$, adjacent to the wall. The first two terms represent the forces due to the pressure gradient and to drag exerted on the gas by the particles. The third term is the viscous force on the outer surface of the layer, while the fourth represents the viscous drag force between the gas and the wall. The length δ_m is of the same order of magnitude as the diameter of the local averaging region.

Let us check that (6.19) satisfies the conditions (i) to (iii) above. When $\phi \to 0$ the contribution of the first pair of terms also tends to zero and, since μ^f remains bounded in this limit, the condition reduces to

$$\frac{du_z}{dr} + \frac{2T\phi_m}{v_t^2 \delta_m \phi} u_z = 0.$$

Therefore

$$u_z \approx -\frac{v_t^2 \delta_m \phi}{2T\phi_m} \frac{du_z}{dr}$$

as $\phi \to 0$. Provided du_z/dr remains bounded it follows that $u_z \to 0$ in the limit, so we recover the no-slip condition and satisfy condition (i). When $\phi \to \phi_m$, $\mu^f \to 0$ according to (6.12), so the boundary condition reduces to

$$\beta(\phi)(u_z - v_z) + \frac{dp}{dz} = 0,$$

which is the same as the force balance on the gas in the bulk of the material. Thus the system does not recognize the presence of the wall in this case, as required by condition (ii). Finally, consider a uniform fluidized bed within the pipe. Then $v_z = 0$ everywhere and u_z is also independent of position, satisfying $\beta u_z + dp/dz = 0$ everywhere. Then (6.19) reduces to

$$\frac{2T\phi_m}{v_t^2 \delta_m \phi} u_z = 0$$

at the wall. The velocity u_z does not vanish at the wall in this case but, since the terms Q_+ and Q_- were dropped from the energy equation (6.6), this equation predicts that $T = 0$ when $\mathbf{v} = 0$, as is the case here. Thus condition (iii) above is satisfied.

These arguments justifying the use of a boundary condition in the form (6.19) are clearly less than satisfactory; much remains to be said about this difficult question.

Sinclair & Jackson generated solutions of Equations (6.13)–(6.16) numerically, subject to the boundary conditions (6.17)–(6.19) and the symmetry requirements $du/dr = dv/dr = dT/dr = 0$ at $r = 0$. With a specified value for dp^f/dz, these yield profiles of the gas- and particle-phase velocities, the particle volume fraction, and the particle temperature, and from these the volumetric flow rates of the two phases follow as

$$Q_g = 2\pi \int_0^a \varepsilon u_z r \, dr, \quad Q_s = 2\pi \int_0^a \phi v_z r \, dr. \quad (6.20)$$

The following dimensionless variables were defined:

$$(r^*, z^*) = (r, z)/a, \quad (\mathbf{u}^*, \mathbf{v}^*) = (\mathbf{u}, \mathbf{v})/v_t, \quad T^* = T/v_t^2,$$

$$p^* = p/\rho_p a g, \quad (Q_g^*, Q_s^*) = (Q_g, Q_s)/a^2 v_t,$$

and their results are presented below in terms of these.

An overall view of the possible regimes of flow is obtained by plotting contours of equal pressure gradient in the plane of the flow rates, Q_s^* and Q_g^*, of the two phases. These are shown in Figure 6.2 for a system whose physical properties are listed in Table 6.1. The value of dp^*/dz^* is indicated for each contour. Points lying in the first quadrant represent cocurrent upflow of the two phases, points in the third quadrant represent cocurrent downflow, and points in the second quadrant represent countercurrent flow. The region representing countercurrent flow is actually a small subset of the second quadrant adjacent to the origin, bounded by the envelope of the contours of constant pressure gradient, which defines the condition known as "flooding". This represents the physical limitation that, for each value of the flow rate of particles down the pipe, there is an upper bound on the upflow of gas. Similarly, for each flow of gas up the pipe there is an upper bound on the downflow of particles.

With positive values of dp^*/dz^* (i.e., pressure increasing on moving up the pipe) the contours are confined entirely to the third quadrant, as is to be

Table 6.1. *Physical properties of the system*
studied by Sinclair and Jackson

Particle diameter	$d = 0.00015$ m
Solid material density	$\rho_s = 2,500$ kg/m^3
Pipe radius	$a = 0.015$ m
Gas viscosity	$\mu_g = 0.0365$ cp
Terminal velocity	$v_t = 1.29$ m/s
Coeff. of restitution	
between particles	$e = 1.0$
Coeff. of restitution	
between particle & wall	$e_w = 0.9$
Specularity factor	$\phi' = 0.5$
Maximum solids fraction	$\phi_m = 0.65$

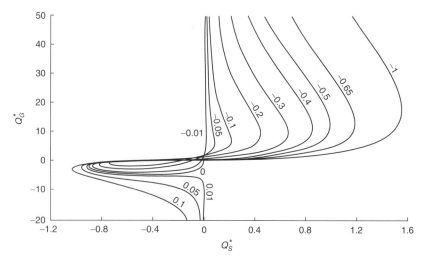

Figure 6.2. Contours of dp^*/dz^* in the (Q_s^*, Q_g^*)-plane for the system defined in Table 6.1. (From Sinclair & Jackson, 1989. Reproduced with permission of the American Institute of Chemical Engineers. Copyright 1989 AIChE. All rights reserved.)

expected, since only cocurrent downflow is possible. However, contours with negative values of dp^*/dz^* also penetrate into the third quadrant if $|dp^*/dz^*|$ is small enough, showing that particles and gas can both move down the pipe despite the fact that the gas pressure is increasing in this direction. As we shall see in the next chapter this behaviour is important in standpipes. On the scale of Figure 6.2 it is clear that some of these contours form closed loops in the third quadrant, but their true form is complex and can be seen only when this region is enlarged, as in Figure 6.3. It is then revealed that the contours with

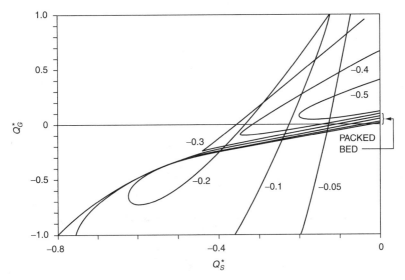

Figure 6.3. Enlargement of the third quadrant of Figure 6.2. (From Sinclair & Jackson, 1989. Reproduced with permission of the American Institute of Chemical Engineers. Copyright 1989 AIChE. All rights reserved.)

$|dp^*/dz^*| < 0.3$ loop round and intersect themselves before ending on the axis $Q_s^* = 0$ where, as we shall see, they represent stationary packed beds traversed by an upflow of gas. For $|dp^*/dz^*| \geq 0.3$, in contrast, the contours turn round to approach the axis $Q_s^* = 0$ without forming loops.

Another complication not visible on the scale of Figure 6.2 becomes apparent when the part of the diagram representing countercurrent flow is enlarged, as in Figure 6.4. Then it is observed that contours may cross from the first into the second quadrant, proceed some distance into this quadrant, then reverse and return to the axis $Q_s^* = 0$. At this point they again reverse and move down through the second quadrant. This zig-zag is referred to by Sinclair & Jackson as a "crossback" region. The resulting two intersections of a given contour with the axis $Q_s^* = 0$, though close together, represent quite different physical configurations, as we shall see.

Perhaps the best way to illustrate the great variety of behaviour predicted by this model is to follow along a single contour from Figure 6.2, showing the radial profiles of gas velocity, particle velocity, and particle volume fraction at selected points. For this purpose Sinclair & Jackson selected the contour $dp^*/dz^* = -0.2$. The part of this lying in the first quadrant is shown in the third panel of Figure 6.5, and the first two panels represent the velocity profiles and the volume fraction profile, respectively, at the point on the contour identified

6. Riser flow

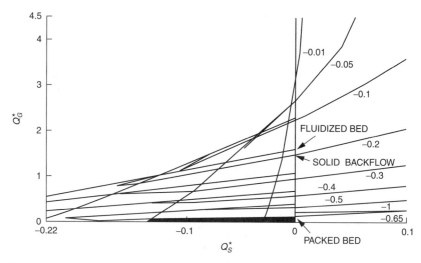

Figure 6.4. Enlargement of the second quadrant of Figure 6.2. (From Sinclair & Jackson, 1989. Reproduced with permission of the American Institute of Chemical Engineers. Copyright 1989 AIChE. All rights reserved.)

by a star. We see that the particle concentration increases on moving towards the wall, reaching a value at the wall that is almost four times that on the axis of the pipe. The velocity profiles show the particles moving up more slowly than the gas, as they must. The reversal of the curvature of the velocity profiles near the wall reflects the effect of gravity on the increased concentration of particles there.

Figure 6.6 again shows velocity and concentration profiles, but now at a point quite close to the intersection of the contour with the ordinate axis. The concentration of the particles near the pipe wall is now so large that they move downward under gravity near the wall, though they still move up in the central part of the cross section and the net flow of solids, though small in magnitude, remains upward. It is clear, from this picture, that the vanishing total solids flux at the point where the contour crosses into the second quadrant is achieved by a balance between upflow in the central part of the pipe and an equal downflow in the peripheral part.

On moving into the second quadrant Figure 6.7 shows that the gas- and particle-phase velocities are directed upward and downward, respectively, and the profile of solids volume fraction has flattened out significantly compared with those presented in Figures 6.5 and 6.6. Figure 6.8, which also belongs to the second quadrant, shows the situation when the contour curve has crossed back to meet the ordinate axis again at the point marked by a star. Here the

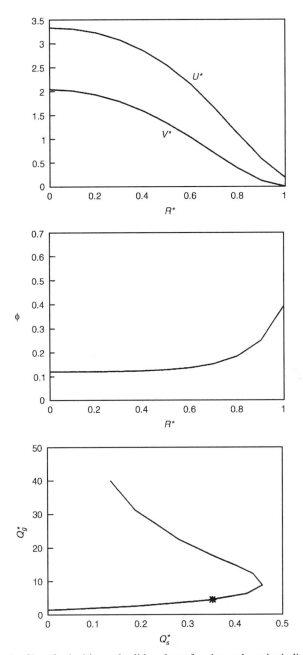

Figure 6.5. Profiles of velocities and solids volume fraction at the point indicated on the contour $dp^*/dz^* = -0.2$. (From Sinclair & Jackson, 1989. Reproduced with permission of the American Institute of Chemical Engineers. Copyright 1989 AIChE. All rights reserved.)

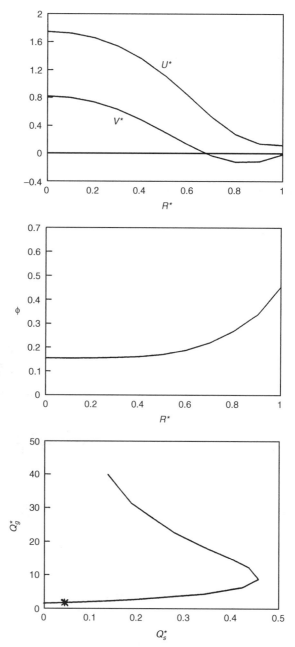

Figure 6.6. Profiles of velocities and solids volume fraction at the point indicated on the contour $dp^*/dz^* = -0.2$. (From Sinclair & Jackson, 1989. Reproduced with permission of the American Institute of Chemical Engineers. Copyright 1989 AIChE. All rights reserved.)

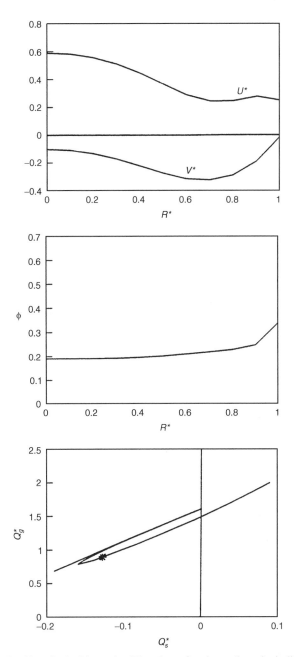

Figure 6.7. Profiles of velocities and solids volume fraction at the point indicated on the contour $dp^*/dz^* = -0.2$. (From Sinclair & Jackson, 1989. Reproduced with permission of the American Institute of Chemical Engineers. Copyright 1989 AIChE. All rights reserved.)

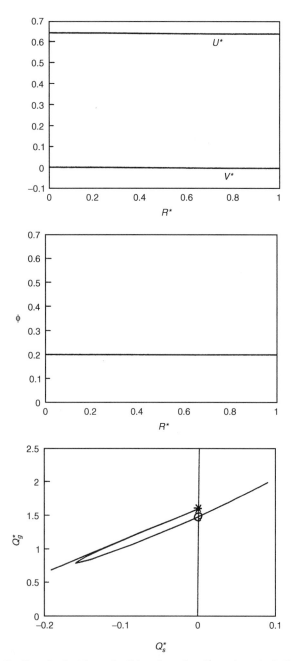

Figure 6.8. Profiles of velocities and solids volume fraction at the point indicated on the contour $dp^*/dz^* = -0.2$. (From Sinclair & Jackson, 1989. Reproduced with permission of the American Institute of Chemical Engineers. Copyright 1989 AIChE. All rights reserved.)

particle velocity vanishes everywhere, while the solids volume fraction and the gas velocity are uniform over the cross section, so the solution corresponds to a uniform fluidized bed at rest in the pipe. Contrast this with the situation at the point circled in the third panel of the figure, where zero net flux of particles was achieved, as we have seen, by a balance between downflow near the wall and upflow near the axis.

Figure 6.9 corresponds to a point lying on the loop formed by the contour in the third quadrant. Though there is an overall flux downward for both phases the gas velocity is directed upward in a narrow annulus adjacent to the pipe wall. The contour of dp^*/dz^* ends where it once more meets the ordinate axis, at the point identified by the star in Figure 6.10. Here again, as in Figure 6.8, the particles are at rest but now $\phi = \phi_m$ everywhere in the cross section, so they form a stationary packed bed traversed by a uniform upward flow of gas.

Figures 6.5 to 6.10 summarize the main types of behaviour to be found in vertical pipe flow though, for different values of dp^*/dz^*, the geometry of the pressure gradient contours may differ somewhat from that illustrated by these figures. For further details the reader should consult the paper of Sinclair & Jackson.

As pointed out in Chapter 1, for the practical purpose of designing or analyzing a vertical transport device it is convenient to use an alternative presentation of the relation among Q_s, Q_g, and dp^f/dz (sometimes known as a Zenz diagram) in which the pressure gradient is plotted against Q_g, for a sequence of fixed values of Q_s. Figure 6.11 shows the first quadrant of Figure 6.2 replotted in this way, and the form of the curves of constant Q_s is familiar from the qualitative arguments given in Chapter 1. Each curve passes through a minimum and the locus of these minima, indicated by the broken line in the diagram, is the demarcation between "dense" and "lean" flow regimes, in the sense of Zenz & Othmer (1960). Points to the left of this line correspond to dense flows, while those to the right represent lean flows. As shown in Chapter 1 the nature of the constraints placed on the operating conditions has a profound effect on qualitative features of the system's behaviour. If Q_s and Q_g are both specified the value of dp^f/dz is determined uniquely, so there is only one possible operating condition. However, if Q_s and dp^f/dz are specified there are two possible values for Q_g, with the smaller corresponding to a "dense" flow and the larger to a "lean" flow. In practice the gas may be supplied by a compressor with a known characteristic curve, and then the possible operating conditions can be found by superimposing this curve on Figure 6.11, as described in Chapter 1.

Just as Figure 6.11 was constructed by replotting the first quadrant of Figure 6.2, contours of constant Q_s in the plane of Q_g and dp^f/dz can be found by replotting the second and third quadrants of Figure 6.2. The result, which is

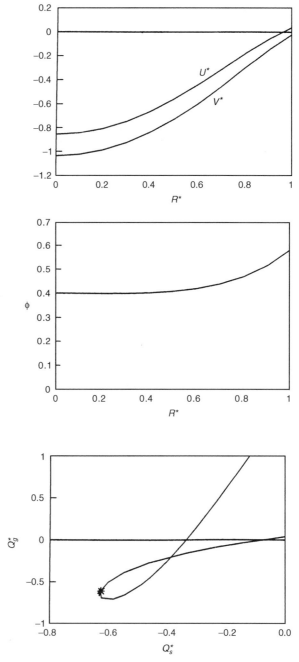

Figure 6.9. Profiles of velocities and solids volume fraction at the point indicated on the contour $dp^*/dz^* = -0.2$. (From Sinclair & Jackson, 1989. Reproduced with permission of the American Institute of Chemical Engineers. Copyright 1989 AIChE. All rights reserved.)

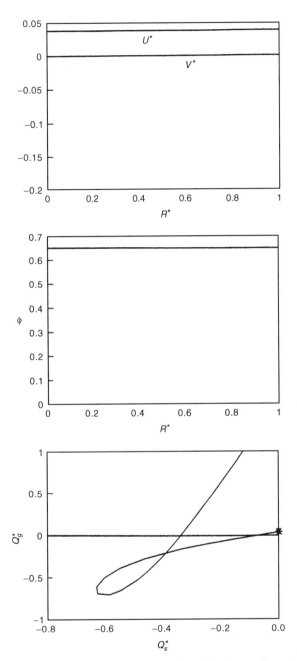

Figure 6.10. Profiles of velocities and solids volume fraction at the point indicated on the contour $dp^*/dz^* = -0.2$. (From Sinclair & Jackson, 1989. Reproduced with permission of the American Institute of Chemical Engineers. Copyright 1989 AIChE. All rights reserved.)

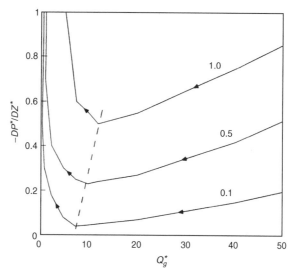

Figure 6.11. Plots of dp^*/dz^* against Q_g^* for the values of Q_s^* indicated. (From Sinclair & Jackson, 1989. Reproduced with permission of the American Institute of Chemical Engineers.)

complicated and offers many opportunities for multiple solutions, can be found in the original paper of Sinclair & Jackson (1989).

The neglect of any effects due to fluctuations associated with streamers, clustering, or pseudoturbulent motion in the above treatment is clearly unjustified, as was recognized by the authors who remarked: "the object of the present work is simply to determine whether collective effects due to interactions between particles are *alone* capable of accounting qualitatively for the rather complex behaviour observed in these systems". To this limited extent the work was successful since it succeeded in accounting for large-scale aspects of the observed behaviour, such as choking and flooding, and also it demonstrated that segregation of particles towards the pipe wall is simply a consequence of demanding that the radial component of the particle-phase momentum balance should be satisfied. This requires that the concentration of the particles should be small in regions where the granular temperature (and hence their mean square fluctuation velocity) is large, and vice versa. However, the prediction of the temperature profile (and hence the concentration profile) is very sensitive to the value of the particle–particle coefficient of restitution, which appears in the dissipation term of the pseudothermal energy balance. Indeed a very small reduction in this, from $e = 1$ to $e = 0.99$, changes the distribution of solids volume fraction completely, as can be seen from Figure 6.12. This shows a concentration profile that takes its largest value on the axis of the pipe, instead of near the wall.

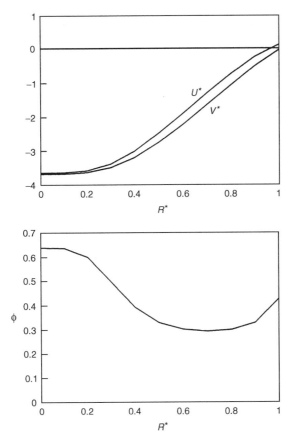

Figure 6.12. Velocity and solids volume fraction profiles for $dp^*/dz^* = -0.2$ with $Q_s^* = -2.10$ and $Q_g^* = -2.83$. $e = 0.99$. (From Sinclair & Jackson, 1989. Reproduced with permission of the American Institute of Chemical Engineers. Copyright 1989 AIChE. All rights reserved.)

The extraordinary sensitivity of the predictions of this model to the value of the coefficient of restitution has also been demonstrated by Pita & Sundaresan (1991). However, there is no *experimental* evidence that the gross performance of systems of this sort, or the tendency of particles to concentrate near to the pipe wall, is critically dependent on the value of the coefficient of restitution. Clearly, then, the particle velocity fluctuations that are mainly responsible in practice for this type of concentration profile must be those associated with phenomena on the scale of clustering or turbulence, rather than those resulting from the pseudothermal motion associated with collisions. We shall refer to this type of motion as pseudoturbulent, reserving the term "turbulent" to describe motion of fluid alone at sufficiently high Reynolds number, as is customary.

6.3 Approximate Treatment of Pseudoturbulent Flow of a
Gas–Particle Mixture in a Vertical Pipe

The easiest way to eliminate undesirable sensitivity to the value of the coefficient of restitution is to revert to a set of equations of motion simpler than those invoked in Section 6.2. In particular the closure of the particle-phase stress tensor provided by Equations (6.7) and (6.8), which involves the granular temperature and therefore introduces the pseudothermal energy balance equation (6.6), may be replaced by simple specifications of the particle-phase pressure and viscosity as prescribed functions of ϕ. The resulting equations then contain no reference to the coefficient of restitution and therefore the question of sensitivity does not arise.

This approach is adopted by Dasgupta et al. (1994), who start from momentum equations in the form of (2.30) and (2.31), which can also be written as

$$\rho_f \left[\frac{\partial}{\partial t}(\varepsilon \mathbf{u}) + \nabla \cdot (\varepsilon \mathbf{u}\mathbf{u}) \right] = \varepsilon \nabla \cdot \mathbf{S}^f - n\mathbf{f}_1 + \rho_f \varepsilon \mathbf{g} \qquad (6.21)$$

and

$$\rho_s \left[\frac{\partial}{\partial t}(\phi \mathbf{v}) + \nabla \cdot (\phi \mathbf{v}\mathbf{v}) \right] = \nabla \cdot \mathbf{S}^p + \phi \nabla \cdot \mathbf{S}^f + n\mathbf{f}_1 + \rho_s \phi \mathbf{g}. \qquad (6.22)$$

The particles are assumed to be small enough that their motion relative to the fluid is Stokesian, so that $n\mathbf{f}_1 = \beta_1(\mathbf{u} - \mathbf{v})$. Other approximations are also introduced from the start. First, gravitational and inertial terms are neglected in the fluid-phase momentum Equation (6.21) because of the small density of the gas relative to that of the solid particles. Second, the relaxation time between gas and particle velocities is assumed to be short compared to all other relevant time scales so that the difference between $\mathbf{u}$ and $\mathbf{v}$ can be neglected everywhere except in the drag force. The drag force itself cannot be neglected, since it is the term responsible for the rapid relaxation; indeed, it remains significant because the small vector $\mathbf{u} - \mathbf{v}$ is multiplied by a large factor β_1. Thus (6.21) and (6.22) become

$$0 = \varepsilon \nabla \cdot \mathbf{S}^f - \beta_1 \mathbf{w} \qquad (6.23)$$

and

$$\rho_s \left[\frac{\partial}{\partial t}(\phi \mathbf{v}) + \nabla \cdot (\phi \mathbf{v}\mathbf{v}) \right] = \nabla \cdot \mathbf{S}^p + \phi \nabla \cdot \mathbf{S}^f + \beta_1 \mathbf{w} + \rho_s \phi \mathbf{g}, \qquad (6.24)$$

where $\mathbf{w} = \mathbf{u} - \mathbf{v}$. Adding (6.23) and (6.24) then gives an overall momentum equation for the suspension in the form

$$\rho_s \left[\frac{\partial}{\partial t}(\phi \mathbf{v}) + \nabla \cdot (\phi \mathbf{v}\mathbf{v}) \right] = \nabla \cdot \mathbf{S} + \rho_s \phi \mathbf{g}, \qquad (6.25)$$

where $\mathbf{S} = \mathbf{S}^p + \mathbf{S}^f$.

The continuity equations for the two phases are

$$\frac{\partial \varepsilon}{\partial t} + \nabla \cdot (\varepsilon \mathbf{u}) = 0, \quad \frac{\partial \phi}{\partial t} + \nabla \cdot (\phi \mathbf{v}) = 0.$$

Adding these and setting $\mathbf{u} = \mathbf{v}$ then gives $\nabla \cdot \mathbf{v} = 0$.

We now average each of these equations of motion with respect to time, denoting time averages by overbars and fluctuations about these averages by primes:

$$\varepsilon = \bar{\varepsilon} + \varepsilon', \quad \phi = \bar{\phi} + \phi', \quad \mathbf{u} = \bar{\mathbf{u}} + \mathbf{u}', \quad \mathbf{v} = \bar{\mathbf{v}} + \mathbf{v}'.$$

Then the continuity equations become

$$\frac{\partial \bar{\varepsilon}}{\partial t} + \nabla \cdot [\bar{\varepsilon}\,\bar{\mathbf{u}} + \overline{(\varepsilon' \mathbf{u}')}] = 0, \quad \frac{\partial \bar{\phi}}{\partial t} + \nabla \cdot [\bar{\phi}\bar{\mathbf{v}} + \overline{(\phi' \mathbf{v}')}] = 0, \tag{6.26}$$

and, if $\mathbf{u}$ is not distinguished from $\mathbf{v}$, $\nabla \cdot \bar{\mathbf{v}} = 0$.

Dividing (6.23) by ε, then time averaging gives

$$\nabla \cdot \overline{\mathbf{S}^f} = \bar{\beta}\bar{\mathbf{w}} + \left(\frac{d\beta}{d\phi}\right)_{\bar{\phi}} \overline{(\phi' \mathbf{w}')}, \tag{6.27}$$

where $\beta = \beta_1/\varepsilon$ and its derivative is evaluated at $\phi = \bar{\phi}$. Finally, the time average of the overall momentum equation (6.25) can be written as

$$\rho_s \left\{ \frac{\partial(\bar{\phi}\bar{\mathbf{v}})}{\partial t} + \nabla \cdot (\bar{\phi}\overline{\mathbf{v}\mathbf{v}}) + \frac{\partial\overline{(\phi' \mathbf{v}')}}{\partial t} \right\}$$

$$= \nabla \cdot [\bar{\mathbf{S}} - \rho_s \{ \bar{\phi}\overline{(\mathbf{v}' \mathbf{v}')} + \overline{(\phi' \mathbf{v}')}\bar{\mathbf{v}} + \bar{\mathbf{v}}\overline{(\phi' \mathbf{v}')} \}] + \rho_s \bar{\phi}\mathbf{g}, \tag{6.28}$$

where a third-order correlation $\overline{(\phi' \mathbf{v}' \mathbf{v}')}$ has been neglected.

We now restrict our attention to a vertical tube of circular section and to flow that is fully developed, in the sense that all time-averaged quantities except the gas pressure depend only on the radial coordinate. Then the time-averaged gas continuity equation reduces to

$$\frac{1}{r}\frac{d}{dr}\left\{ r[\bar{\varepsilon}\,\bar{u}_r + \overline{(\varepsilon' u_r')}] \right\} = 0.$$

Integrating and noting that the expression in square brackets must remain bounded on the axis of the pipe we obtain

$$\bar{\varepsilon}\,\bar{u}_r + \overline{(\varepsilon' u_r')} = 0, \tag{6.29}$$

and in the same way the particle-phase continuity equation gives

$$\bar{\phi}\bar{v}_r + \overline{(\phi' v_r')} = 0. \tag{6.30}$$

The axial and radial components of the gas-phase momentum equation (6.27) are

$$\frac{1}{r}\frac{\partial}{\partial r}\left(r\bar{S}_{rz}^f\right) + \frac{\partial \bar{S}_{zz}^f}{\partial z} = \bar{\beta}\bar{w}_z + \left(\frac{d\beta}{d\phi}\right)_{\bar{\phi}}\overline{(\phi' w_z')} \tag{6.31}$$

and

$$\frac{1}{r}\frac{\partial}{\partial r}\left(r\bar{S}_{rr}^f\right) + \frac{\bar{S}_{\theta\theta}^f}{r} = \bar{\beta}\bar{w}_r + \left(\frac{d\beta}{d\phi}\right)_{\bar{\phi}}\overline{(\phi' w_r')}. \tag{6.32}$$

The axial component of the overall momentum Equation (6.28) for flow that is fully developed, in the present sense, is

$$\frac{\rho_s}{r}\frac{\partial}{\partial r}\{r[\bar{\phi}\bar{v}_r\bar{v}_z + \bar{\phi}\,\overline{(v_r'v_z')} + \overline{(\phi'v_r')}\bar{v}_z + \bar{v}_r\overline{(\phi'v_z')}]\}$$

$$= \frac{1}{r}\frac{\partial}{\partial r}(r\bar{S}_{rz}) + \frac{\partial \bar{S}_{zz}}{\partial z} - \rho_s\bar{\phi}g,$$

and the corresponding radial component is

$$\frac{\rho_s}{r}\frac{\partial}{\partial r}\{r[\bar{\phi}\bar{v}_r\bar{v}_r + \bar{\phi}\,\overline{(v_r'v_r')} + 2\bar{v}_r\overline{(\phi'v_r')}]\} - \frac{\rho_s}{r}\bar{\phi}\overline{(v_\theta'v_\theta')} = \frac{1}{r}\frac{\partial}{\partial r}(r\bar{S}_{rr}) - \frac{\bar{S}_{\theta\theta}}{r}.$$

Now if the difference between $\bar{\mathbf{u}}$ and $\bar{\mathbf{v}}$ is neglected, (6.29) and (6.30) together imply that $\bar{v}_r = 0$. Then using this and (6.30) simplifies the above pair of equations to

$$\frac{\rho_s}{r}\frac{\partial}{\partial r}[r\bar{\phi}\,\overline{(v_r'v_z')}] = \frac{1}{r}\frac{\partial}{\partial r}(r\bar{S}_{rz}) + \frac{\partial \bar{S}_{zz}}{\partial z} - \rho_s\bar{\phi}g \tag{6.33}$$

and

$$\frac{\rho_s}{r}\frac{\partial}{\partial r}[r\bar{\phi}\,\overline{(v_r'v_r')}] - \frac{\rho_s}{r}\bar{\phi}\overline{(v_\theta'v_\theta')} = \frac{1}{r}\frac{\partial}{\partial r}(r\bar{S}_{rr}) - \frac{\bar{S}_{\theta\theta}}{r}. \tag{6.34}$$

To close these equations expressions are needed for $\bar{\mathbf{S}}^f$ and $\bar{\mathbf{S}}^p$ to determine $\bar{\mathbf{S}} = \bar{\mathbf{S}}^f + \bar{\mathbf{S}}^p$. For example the Newtonian closures (2.75) and (2.76) would be a reasonable choice. However, since the effective viscosities in these equations depend on ϕ, some correlations involving fluctuations in ϕ are generated on time averaging to find $\bar{\mathbf{S}}$. To avoid this complication Dasgupta et al. adopted the simpler expedient of assuming a Newtonian expression for $\bar{\mathbf{S}}$ itself in terms of time-averaged quantities. Since $\nabla \cdot \bar{\mathbf{v}} = 0$ with the approximations invoked here, this takes the form

$$\bar{\mathbf{S}} = -[\bar{p}^p(\bar{\phi}) + \bar{p}^f]\mathbf{I} + \mu^e(\bar{\phi})[\nabla\bar{\mathbf{v}} + (\nabla\bar{\mathbf{v}})^T]. \tag{6.35}$$

Thus for the present flow field

$$\bar{S}_{rr} = \bar{S}_{\theta\theta} = \bar{S}_{zz} = -\bar{p}^p(\bar{\phi}) - \bar{p}^f, \quad \bar{S}_{rz} = \mu^e(\bar{\phi})\frac{d\bar{v}_z}{dr},$$

and (6.33) and (6.34) become

$$\frac{\rho_s}{r}\frac{\partial}{\partial r}[r\bar{\phi}\,\overline{(v_r'v_z')}] = \frac{1}{r}\frac{\partial}{\partial r}\left[r\mu^e\frac{\partial\bar{v}_z}{\partial r}\right] - \frac{\partial\bar{p}^f}{\partial z} - \rho_s\bar{\phi}g \qquad (6.36)$$

and

$$\frac{\rho_s}{r}\frac{\partial}{\partial r}[r\bar{\phi}\,\overline{(v_r'v_r')}] - \frac{\rho_s}{r}\bar{\phi}\overline{(v_\theta'v_\theta')} = -\frac{\partial\bar{p}^p}{\partial r} - \frac{\partial\bar{p}^f}{\partial r}. \qquad (6.37)$$

The radial derivative of the gas-phase pressure can be found from (6.32), which becomes

$$-\frac{\partial\bar{p}^f}{\partial r} = \beta(\bar{\phi})\,[\bar{u}_r - \bar{v}_r] + \left(\frac{d\beta}{d\phi}\right)_{\bar{\phi}}[\overline{(\phi'u_r')} - \overline{(\phi'v_r')}]. \qquad (6.38)$$

It is tempting to set $\bar{u}_r = \bar{v}_r$ and $u_r' = v_r'$ and hence conclude from (6.38) that $\partial\bar{p}^f/\partial r = 0$. However, even though $\bar{u}_r - \bar{v}_r$ may be small, $\beta(\bar{\phi})$ is large enough that the first term on the right-hand side of (6.38) cannot be dismissed. But Equations (6.29) and (6.30) provide expressions for $\bar{u}_r$ and $\bar{v}_r$, and, substituting these into (6.38) and setting $u_r' = v_r'$, we find that

$$-\frac{\partial\bar{p}^f}{\partial r} = \frac{\bar{\beta}\,\overline{(\phi'v_r')}}{\bar{\phi}(1-\bar{\phi})}.$$

Substituting this expression for $\partial\bar{p}^f/\partial r$ into (6.37) then gives

$$\frac{\rho_s}{r}\frac{\partial}{\partial r}[r\bar{\phi}\,\overline{(v_r'v_r')}] - \frac{\rho_s}{r}\bar{\phi}\overline{(v_\theta'v_\theta')} + \frac{\partial\bar{p}^p}{\partial r} = \frac{\bar{\beta}\,\overline{(\phi'v_r')}}{\bar{\phi}(1-\bar{\phi})}. \qquad (6.39)$$

Equations (6.36) and (6.39) are the two components of the time-smoothed momentum balance for the suspension.

So far, apart from some approximations associated with the small density of the gas relative to the solid material, the smallness of $|\mathbf{u} - \mathbf{v}|$, and the neglect of a third-order correlation, the derivation has been entirely formal. Thus, the validity of the resulting equations does not depend on any particular physical mechanism that may be responsible for the fluctuations in velocities, particle concentration, and pressures. We have mentioned the occurrence of phenomena such as bubbles, clusters, and streamers, associated with instabilities that result from interaction between the particles and the interstitial fluid, but an assembly of particles flowing in a tube can exhibit inertially driven instabilities even in the absence of any fluid, as shown recently by Wang et al. (1997). For the latter type of instability the dominant modes are travelling waves propagating along the tube, which could be considered to be the analogue, for this granular "fluid", of the small-amplitude waves believed to be the instigators of turbulence in the flow of a molecular fluid.

We might, therefore, refer to the complicated, time-dependent motions observed in the gas–particle flow as "pseudoturbulent", bearing in mind that this is a catchall term for what is, most likely, the result of a variety of contributions from different physical mechanisms. To make further progress closures must be provided for the correlations appearing in the equations of motion, so Dasgupta et al. suggested it would be worth exploring the consequences of simply adopting the well-known K–ε closure that has had some success in closing the time-smoothed equations for turbulent flow of a true molecular fluid. Specifically, they neglected anisotropy of the velocity fluctuations, writing

$$\overline{(v_r' v_r')} = \overline{(v_\theta' v_\theta')} = \frac{2}{3} K, \tag{6.40}$$

and expressed the remaining significant correlations in terms of a turbulent kinematic viscosity v_T, writing

$$\overline{(v_r' v_z')} = -v_T \frac{d\bar{v}_z}{dr}, \qquad \overline{(\phi' v_r')} = -v_T \frac{d\bar{\phi}}{dr}. \tag{6.41a,b}$$

The first of these gives the "turbulent" shear stress and the second represents a "turbulent" dispersion flux in the radial direction.

Using (6.40) and (6.41) transforms Equations (6.36) and (6.39) into

$$\frac{1}{r} \frac{\partial}{\partial r} \left[r(\mu^e + \rho_s \bar{\phi} v_T) \frac{\partial \bar{v}_z}{\partial r} \right] = \frac{\partial \bar{p}^f}{\partial z} + \rho_s \bar{\phi} g \tag{6.42}$$

and

$$\frac{\partial}{\partial r} \left[\frac{2}{3} \rho_s \bar{\phi} K + \bar{p}^p \right] + \frac{\bar{\beta} v_T}{\bar{\phi}(1 - \bar{\phi})} \frac{\partial \bar{\phi}}{\partial r} = 0. \tag{6.43}$$

The "turbulent" kinematic viscosity is expressed as

$$v_T = \frac{C_\mu K^2}{\varepsilon} \tag{6.44}$$

and the quantities K and ε are solutions of the following pair of equations:

$$\frac{1}{r} \frac{\partial}{\partial r} \left[r \left(\frac{\rho_s \bar{\phi} v_T}{\sigma_K} + \mu^e \right) \frac{\partial K}{\partial r} \right] + \rho_s \bar{\phi} v_T \left(\frac{\partial \bar{v}_z}{\partial r} \right)^2 - \rho_s \bar{\phi} \varepsilon = 0, \tag{6.45}$$

$$\frac{1}{r} \frac{\partial}{\partial r} \left[r \left(\frac{\rho_s \bar{\phi} v_T}{\sigma_\varepsilon} + \mu^e \right) \frac{\partial \varepsilon}{\partial r} \right] + C_{\varepsilon 1} \frac{\varepsilon}{K} \rho_s \bar{\phi} v_T \left(\frac{\partial \bar{v}_z}{\partial r} \right)^2 - C_{\varepsilon 2} \frac{\varepsilon^2}{K} \rho_s \bar{\phi} = 0,$$

$$\tag{6.46}$$

where σ_K, σ_ε, C_μ, and $C_{\varepsilon 1}$ are constants whose values are to be assigned, while

$$C_{\varepsilon 2} = C_2 \left[1 - \exp\left(-\frac{K^4 \rho_s^2 \bar{\phi}^2}{\varepsilon^2 (\mu^e)^2} \right) \right] \tag{6.47}$$

and C_2 is another assignable constant.

To close the set of equations it remains only to specify $\bar{p}^p$ and μ^e, and Dasgupta et al. take $\mu^e = \mu^{ep} + \mu^{ef}$, together with

$$\bar{p}^p = A\bar{\phi}(1 + 4\bar{\phi}g_0), \quad \mu^{ep} = B\left[\frac{(1 + 1.6\bar{\phi}g_0)^2}{g_0} + 9.779\bar{\phi}^2 g_0 \right],$$

$$\tag{6.48a,b}$$

and

$$\mu^{ef} = \mu_g (1 + 2.5\bar{\phi} + 7.6\bar{\phi}^2)\left(1 - \frac{\bar{\phi}}{\phi_m} \right). \tag{6.49}$$

The expressions in (6.48) are consistent with the results of kinetic theory applied to granular materials, except that A and B are taken to be constants, rather than functions of the granular temperature. Equation (6.49) is analogous to the expression (6.12) used by Sinclair & Jackson (1989).

With a given value for the axial gas pressure gradient, $\partial \bar{p}^f / \partial z$, (6.42), (6.43), (6.45), and (6.46) comprise four differential equations in four unknown functions of r, namely $\bar{v}_z$, $\bar{\phi}$, K, and ε. To determine a solution boundary conditions for each of these variables must be specified at the pipe wall. A condition to be satisfied at the wall by the instantaneous value of v_z has already been derived in the form of (6.17). For particles as small as those contemplated here Pita & Sundaresan (1991) pointed out that, to a good approximation, this reduces to a no-slip condition. Then time averaging leads immediately to the condition $\bar{v}_z = 0$ at the wall. Since the instantaneous value of the particle velocity vanishes at the wall, this also implies that $K = 0$ at the wall. An appropriate boundary condition for ε has been suggested by Patel et al. (1985) in the form

$$\varepsilon = \frac{2\mu^e}{\rho_s \bar{\phi}} \left(\frac{\partial K^{1/2}}{\partial r} \right)^2.$$

This leaves the value of $\bar{\phi}$ at the wall free to be specified and this is appropriate, since the value chosen then determines the cross-sectional average of $\bar{\phi}$, in other words the particle loading. Thus, the quantities available as independent variables to be specified are the axial pressure gradient and the particle loading. Solution of the equations then yields the radial profiles of $\bar{v}_z$, $\bar{\phi}$, K, and ε, and hence the total flow rates of the two phases, expressible as the superficial velocities (or volume flow rates per unit area of the pipe cross section) $\bar{V}_g$ and $\bar{V}_s$

Table 6.2. *Parameter values for computation of Dasgupta et al.*

Particle diameter	$d = 100\,\mu\text{m}$
Solid material density	$\rho_s = 1{,}500\,\text{kg/m}^3$
Pipe radius	$a = 0.15\,\text{m}$
Gas viscosity	$\mu_g = 4 \times 10^{-5}\,\text{kg/(m}\cdot\text{s)}$
Terminal velocity	$v_t = 0.2\,\text{m/s}$
Richardson–Zaki exponent	$n = 3.5$
Maximum solids volume fraction	$\phi_m = 0.65$
Gas superficial velocity	$\bar{V}_g = 5$ m/s
Solids superficial velocity	$\bar{V}_s = 0.067$ m/s
Parameters in the equations	$A = 20\,\text{N/m}^2,\ B = 0.09\,\text{kg/(m}\cdot\text{s)},$
	$C_\mu = 0.09,\ C_{\varepsilon1} = 1.45,\ C_2 = 2.0,$
	$\sigma_K = 1.0, \sigma_\varepsilon = 1.3.$

for the gas and the particles, respectively. The problem can also be formulated from an alternative point of view, with $\bar{V}_g$ and $\bar{V}_s$ regarded as the quantities specified. Then the radial profiles, together with the axial gas pressure gradient and the solids loading, can be determined from this specification.

Dasgupta et al. (1994) report the results of only one computation of this sort, for the conditions and parameter values listed in Table 6.2. The solids mass flux is then $100\,\text{kg/(m}^2\cdot\text{s)}$ and the solution yields an axial pressure gradient of $598\,\text{N/m}^3$. The physical properties of the particles and the gas and the flow rates are chosen to be in ranges of commercial interest, while the values of the parameters for the K–ε model are typical of those found most suitable for modelling turbulent flow of a single-phase fluid.

Computed radial profiles of $\bar{v}_z$ and $\bar{\phi}$ are shown in Figures 6.13(a) and (b), while corresponding profiles of K, and particle-phase pressure $\bar{p}^p$ and Reynolds stress $\frac{2}{3}\rho_s\bar{\phi}K$, can be found in Figures 6.14(a) and (b), respectively. From Figure 6.13 we see that the solids volume fraction remains fairly constant from the axis out to almost 0.9a, but then there is a large increase in concentration on approaching the pipe wall, as is observed in practice. Despite this, the axial component of the averaged velocity remains positive everywhere, approaching zero at the pipe wall. Figure 6.14(a) shows that the fluctuations in velocity are most intense at about $r = 0.75a$ and they fall to zero at the wall, as they must. The physical reason for the accretion of particles near the wall is made clear by Figure 6.14(b), which should be viewed in relation to Equation (6.43). The last term on the left-hand side of this equation is proportional to the radial component of the turbulent dispersion flux and would therefore vanish if we neglected this flux. Then the equation could be integrated immediately, with the result

$$\frac{2}{3}\rho_s\bar{\phi}K + \bar{p}^p = \text{constant}.$$

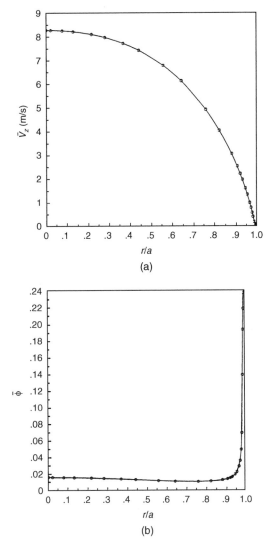

Figure 6.13. Radial profiles of $\bar{v}_z$ and $\bar{\phi}$. (a) Time-averaged axial velocity. (b) Time-averaged volume fraction. (From Dasgupta et al., 1994. Reproduced with permission of the American Institute of Chemical Engineers. Copyright 1994 AIChE. All rights reserved.)

In other words, the sum of the particle-phase pressure and a "Reynolds stress" associated with the fluctuations in particle velocity is required to be the same at all points of the cross section. Referring to Figure 6.14(b) we see that the sharp decrease in the Reynolds stress near the wall, associated with the decrease in K seen in Figure 6.14(a), is compensated by a corresponding increase in $\bar{p}^p$.

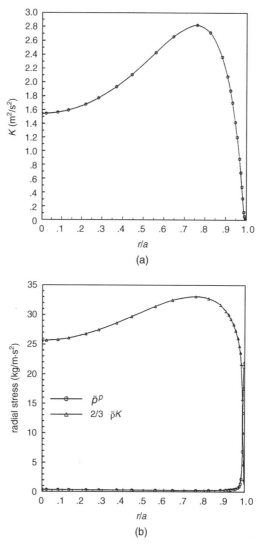

Figure 6.14. Radial profiles of K and contributions to the radial stress. (a) K (b) $\bar{p}^p$ and $2/3\bar{\rho}K$ (From Dasgupta et al., 1994. Reproduced with permission of the American Institute of Chemical Engineers. Copyright 1994 AIChE. All rights reserved.)

The modest remaining variation in the sum of the Reynolds stress and the particle-phase pressure results from the fact that the last term in (6.43) is not completely negligible. This same mechanism is responsible for the increase in particle concentration near the wall found in the earlier work of Sinclair & Jackson (1989), described above. In their case the Reynolds stress is zero, but the particle-phase pressure depends on the granular temperature as well as the concentration of the particles, and the concentration increases near the wall to compensate for the drop in the granular temperature there.

Dasgupta et al. repeated their computations after varying the values of each of the parameters C_μ, $C_{\varepsilon 1}$, C_2, σ_K, and σ_ε about the values quoted in Table 6.2 and found no undue sensitivity of the results to these values. They also checked that the assumptions of separation of scales, discussed in Section 6.1, are satisfied for the solution obtained.

The equations proposed by Dasgupta et al. (1994) have been derived above in some detail, since their simplifying assumptions make the derivation easier to follow than would be the case for some of the improved versions published more recently. However, neglecting the difference between **u** and **v** has one serious consequence, which is not apparent when attention is confined to fully developed flow. In a practical situation, such as a riser reactor, the particles and gas are introduced at the bottom of the pipe and the cross-sectional distribution of the particles at that level is determined entirely by the way in which this is done. On moving up the pipe there must, therefore, be an entry length over which the particles redistribute themselves into the characteristic pattern with an annulus of high concentration adjacent to the wall. But this necessarily involves relative motion of the particles and the gas, so it is precluded by the initial assumptions of Dasgupta et al., and consequently their equations cannot be used to describe the developing flow. This is a serious drawback, since observations suggest that typical entry lengths in commercial risers occupy a significant fraction of the total length. Furthermore, because the computations of Dasgupta et al. were confined to a single set of operating conditions, they failed to elucidate the overall pattern of behaviour in a manner comparable with the results of Sinclair & Jackson (1989), presented in Section 6.2.

The assumption that **u** = **v** is relaxed in later work of Dasgupta et al. (1997, 1998), and computations for an extensive set of operating conditions then yield a map of the overall behaviour of the system similar to that shown in Figure 6.2, based on the theory of Sinclair & Jackson. A complete formulation of the equations can be found in Dasgupta et al. (1998), to which the reader is referred for details. The derivation is closely analogous to that described above, with two notable exceptions. First, momentum equations are formulated for the gas and the particles separately, rather than the gas and the suspension as a whole, and the distinction between **u** and **v** is retained throughout. Second, the expression (6.44) for the turbulent kinematic viscosity is replaced by

$$v_T = C_\mu f_\mu \frac{K^2}{\varepsilon},$$
(6.50)

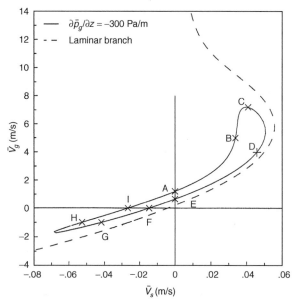

Figure 6.15. Contour of constant pressure gradient, $d\bar{p}^f/dz = -300\,\text{Pa/m}$, in the $(\bar{V}_s, \bar{V}_g)$-plane. (Reprinted from Dasgupta et al., 1998. Copyright 1998, with permission from Elsevier Science.)

with f_μ taking a form proposed by Lam & Bremhorst (1981), namely

$$f_\mu = [1 - \exp(-0.0165\,Re_y)]^2 \left(1 + \frac{20.5}{Re_T}\right), \tag{6.51}$$

where

$$Re_T = \frac{\bar{\rho}K^2}{\varepsilon\mu_e} \quad \text{and} \quad Re_y = \frac{\bar{\rho}K^{1/2}(a-r)}{\mu_e}, \tag{6.52}$$

while $\bar{\rho} = \rho_s\bar{\phi} + \rho_f(1 - \bar{\phi})$ and $\mu^e = \mu^{ep} + \mu^{ef}$. Replacement of (6.44) by (6.50) is found to be necessary if the theory is to predict downflow of particles near the pipe wall in certain circumstances, as observed in practice.

The continuous curve in Figure 6.15 shows a computed contour of constant pressure gradient in the $(\bar{V}_s, \bar{V}_g)$-plane, for $d\bar{p}^f/dz = -300\,\text{Pa/m}$ and in a system with the physical properties listed in Table 6.2. The broken curve represents the result obtained, for the same conditions, after setting $K = \varepsilon = 0$ and thereby eliminating the pseudoturbulent fluctuations. We shall refer to this as the "laminar flow" solution. The mean axial velocity of the mixture, defined by

$$\bar{v}_{mz} = \overline{\left(\frac{\rho_s\phi v_z + \rho_f(1-\phi)u_z}{\rho_s\phi + \rho_f(1-\phi)}\right)}$$

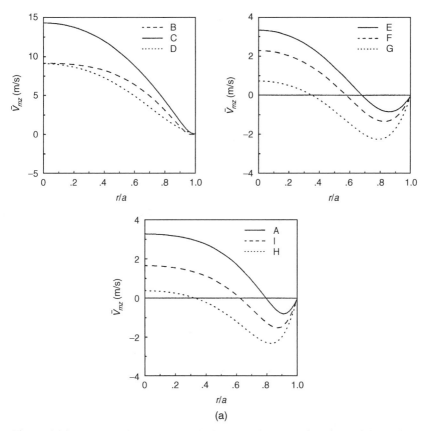

Figure 6.16. Radial profiles at points labelled A to I in Figure 6.15. (a) Axial velocity; (b) solids volume fraction. (Reprinted from Dasgupta et al., 1998. Copyright 1998, with permission from Elsevier Science.)

and the time-averaged volume fraction of solids, $\bar{\phi}$, are shown as functions of radial position in Figures 6.16(a) and (b), respectively, for each of the points identified by the letters A to I in Figure 6.15.

The contour in Figure 6.15 forms a closed loop extending into each of the first three quadrants, so there are two possible values of $\bar{V}_g$ for each solids flux $\bar{V}_s$. This shape is clearly deficient in one respect. For a given value of the axial pressure gradient there must be a solution representing the turbulent flow of gas alone, which will be represented by a point with a large value of $\bar{V}_g$, located on the axis $\bar{V}_s = 0$, and the complete contour should be connected continuously to this. The predicted contour does not include such a point, though it appears that the corresponding "laminar flow" contour may, indeed, intersect the ordinate axis at a large value of $\bar{V}_g$. However, the derivation of the equations depends

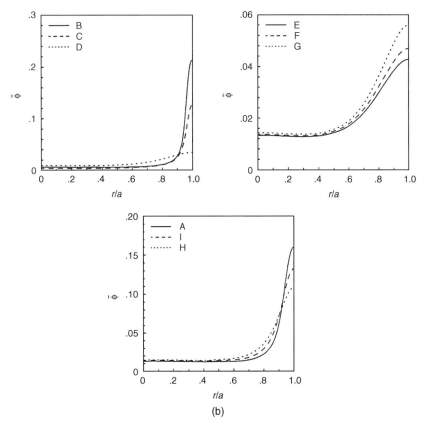

Figure 6.16. (*Cont.*)

on the assumption that the solids loading is large, so one should not, perhaps, expect to be able to extrapolate the results to vanishingly small solids content.

The solids fraction profiles in Figure 6.16(b) show that the radial segregation of the particles to form a dense layer close to the wall is more pronounced at the points HIABC, on the upper arc of the contour, than at points GFED on the lower arc. This is not surprising, since the latter points lie closer to the broken contour representing "laminar flow", for which there is no segregation. The value of K is also found to be larger, indicating that the flow is more "turbulent" at points on the upper arc of the contour than at adjacent points on the lower arc.

The part of the contour in the first quadrant represents riser flow and at points A and E Figure 6.16(a) shows that there is downflow of particles in the dense layer adjacent to the pipe wall. These points correspond to zero net solids flux, but for points on the contour close to them, with $\bar{V}_s > 0$, the flow in the dense layer must also be directed downward. This agrees with the usual

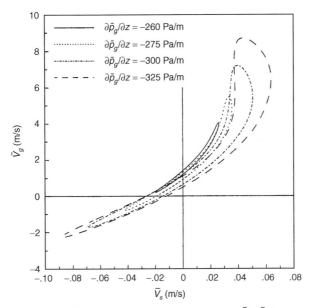

Figure 6.17. Contours of constant pressure gradient in the $(\bar{V}_s, \bar{V}_g)$-plane for various values of $d\bar{p}^f/dz$. (Reprinted from Dasgupta et al., 1998. Copyright 1998, with permission from Elsevier Science.)

observation of downflow at the wall for risers with a low net solids flux. For high values of the solids flux, in contrast, at points such as B, C, and D, there is little or no downflow even near the wall. Recently reported experimental results of Issangya et al. (1998) confirm the absence of any region with downflux of solids near the wall when the solids flux is large. At points I and F, where there is zero net gas flux, the solids are circulating in the pipe and carrying equal flows of gas up in the central region and down in an annulus adjacent to the wall. Even at points G and H where $\bar{V}_s$ and $\bar{V}_g$ are both negative, representing cocurrent downflow, there is a small upward flux of particles close to the axis. Though a standpipe operating in these conditions would effect a transfer of particles from a region of lower gas pressure above to a region of higher gas pressure below, there would nevertheless be some back leakage of gas up the pipe near its axis. The ability to predict this could be important in practice.

Figure 6.17 reproduces the contour for $d\bar{p}^f/dz = -300$ Pa/m and also shows corresponding contours for other negative values of $d\bar{p}^f/dz$. These form an envelope in the second quadrant that defines the condition of flooding in countercurrent flow.

In Figure 6.11 we saw plots of the pressure gradient as a function of the gas flow rate, for riser flow with various fixed values of the solids flux, as

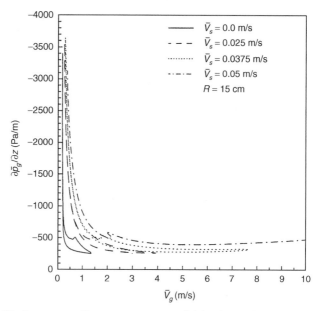

Figure 6.18. Pressure gradient versus gas superficial velocity, for various small values of the solids superficial velocity. (Reprinted from Dasgupta et al., 1998. Copyright 1998, with permission from Elsevier Science.)

predicted by Sinclair & Jackson (1989). Figures 6.18 and 6.19 show comparable predictions from the model of Dasgupta et al. (1998), for low and high flow ranges, respectively. At the lower values of $\bar{V}_s$ the curves form closed loops, but apparently they do not for values of $\bar{V}_s$ greater than about 0.05 m/s. However, even at these higher solids flow rates, there are kinks in the curves that indicate the possibility of multiple solutions, with more than one value of the gas flow rate for given values of the pressure gradient and the solids flow rate, or more than one pressure gradient for given values of the solids and gas flow rates. Apart from these complications the overall shape of the curves is similar to that found by Sinclair & Jackson (Figure 6.11), with a minimum in the pressure gradient as a function of gas flow, corresponding to the choking condition.

Dasgupta et al. also provide some quantitative comparisons of their theoretical predictions with available experimental measurements. These are of two sorts, namely, measurements of the pressure gradient and of the radial solids fraction profile as functions of the flow rates of the gas and solid phases. Figure 6.20 compares theoretical solids fraction profiles with measurements of Weinstein et al. (1986). The agreement is fairly good when the ratio of solids flow to gas flow is not too large but in Figure 6.20(c), where the ratio of solids flow to gas flow is largest, the observed solids fraction greatly exceeds

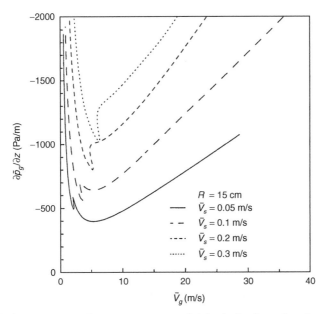

Figure 6.19. Pressure gradient versus gas superficial velocity for various larger values of the solids superficial velocity. (Reprinted from Dasgupta et al., 1998. Copyright 1998, with permission from Elsevier Science.)

the predictions. Figure 6.21 shows the predicted relation between pressure gradient and solids flow rate, for a fixed value of the gas flow, compared with experimental data of Monceaux et al. (1986) and of Yerushalmi (1986). Again the agreement is fair, except for Figure 6.21(c), where the ratio of solids flow to gas flow is large. The poor agreement with experiment at high solids/gas flow ratio may be an indication that the flow is not fully developed at the location of the measurements since the entry length, over which the profiles reach their fully developed forms, increases with this ratio.

In a third paper Dasgupta et al. (1997) have supplemented their work on fully developed flow by exploring steady, but spatially developing flow. For a riser this presents a difficulty of principle since, as we have seen, it is common to find upflow near the axis and downflow near the wall when the flow is fully developed. To study the spatial evolution of such a flow pattern it would be necessary to specify conditions at both the top and the bottom of the pipe, leading to a two-point boundary value problem. For situations in which the velocities of particles and gas are everywhere in the same direction, however, conditions need to be specified only at the inlet, yielding an initial value problem. The two-point boundary value problem can be avoided in two ways, by limiting attention either to cocurrent downflow or to flow in the absence of gravity. Here we shall

272

6. *Riser flow*

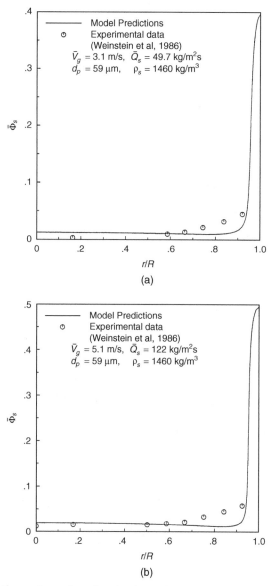

Figure 6.20. Comparison of predicted volume fraction profiles with measurements. (Reprinted from Dasgupta et al., 1998. Copyright 1998, with permission from Elsevier Science.)

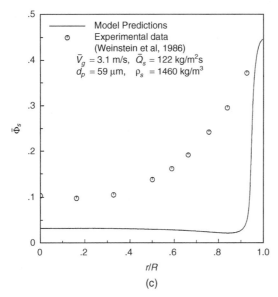

Figure 6.20. (*Cont.*)

discuss only the gravity-free case; computations for downflow, together with a limited number of results for upflow, can be found in the original publication.

Dasgupta et al. (1997) study developing flow in a channel confined between parallel plane walls, rather than a cylindrical channel as in their work on fully developed flow. The equations of motion are essentially the same, except that the solution now depends on the axial coordinate. At the inlet the particles are uniformly distributed between the plates and they enter moving slowly, so they are accelerated by a faster moving gas stream. Since the gas velocity is largest at the midpoint of the channel the particles there are accelerated most, and consequently their concentration decreases relative to the concentration near the walls. This is a purely kinematic effect and has nothing to do with the pseudoturbulent mechanism responsible for the increase in concentration near the wall in fully developed flow; it occurs over a small fraction of the distance needed to establish the fully developed flow pattern. This kinematically generated concentration gradient is accompanied by a corresponding gradient in the particle-phase pressure, which becomes larger near the wall than in the centre of the channel and thus drives the particles back towards the centre, flattening the concentration profile once more. Indeed, if the pseudoturbulent fluctuations are suppressed by setting $K = 0$, sufficiently far downstream the particle concentration profile again becomes flat, and this represents the fully developed flow pattern. However, if K is not constrained to vanish, the tendency of the

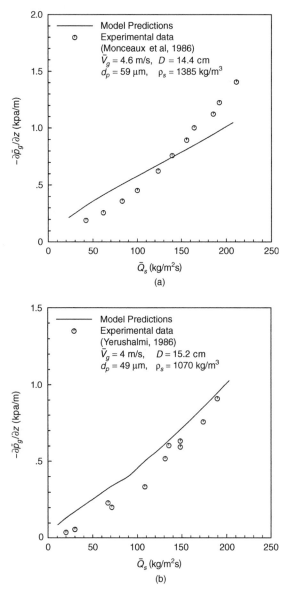

Figure 6.21. Pressure gradient versus solids flow rate, for a fixed value of the gas flow rate. Comparison of predictions and measurements. (Reprinted from Dasgupta et al., 1998. Copyright 1998, with permission from Elsevier Science.)

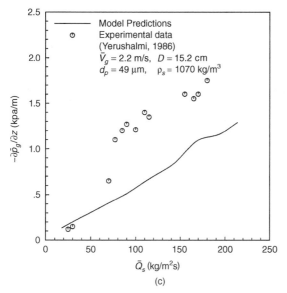

Figure 6.21. (*Cont.*)

particle-phase pressure to flatten the concentration profile is eventually overcome by the effect of the "Reynolds stress", which drives the particles towards
the wall. The fully developed concentration and velocity profiles are reached
when the opposing effects of the particle-phase pressure and the pseudoturbulent Reynolds stress come into balance, as we have seen already from the work
on fully developed flow.

This description of the process is illustrated by the sequence of concentration
profiles shown in Figure 6.22. Corresponding particle-phase velocity profiles
are shown in Figure 6.23. The transverse coordinate is expressed as a fraction
y/a of the half-width of the channel, while the axial coordinate is quoted in
dimensionless terms as $z^* = z/L$, where the characteristic length L is defined
by $L = a(Re)$ in terms of a Reynolds number

$$Re = \frac{\rho_s \bar{V}_s a}{\mu^{ep}}.$$

Here μ^{ep} is evaluated using (6.48b) with $\bar{\phi} = \bar{V}_s/\bar{V}$, where $\bar{V} = \bar{V}_s + \bar{V}_g$. The
channel half-width is $a = 0.1$ m, the particle concentration at inlet is $\bar{\phi} = 0.2$,
independent of y, and the solids and gas velocities at inlet are given by

$$\bar{v}_z = 1 - (y/a)^5 \text{ m/s}, \quad \bar{u}_z = 5[1 - (y/a)^5] + 0.5 \text{ m/s}.$$

Correspondingly, $\bar{V}_s = (1/6)$ m/s and $\bar{V}_g = 0.4 + (20/6)$ m/s, so $\bar{V} = 3.9$ m/s.
Under these conditions the characteristic length L is 35.26 m. In these figures

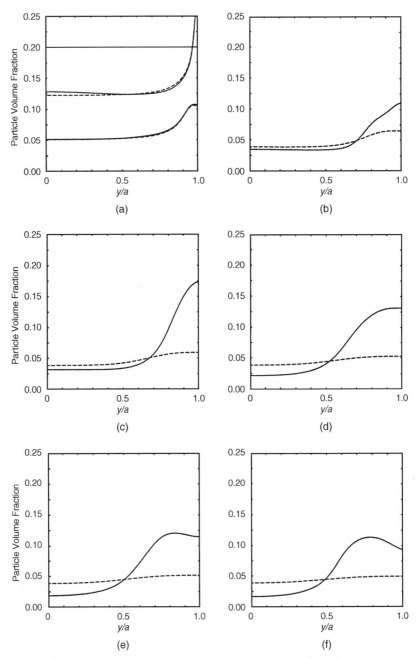

Figure 6.22. Profiles of $\bar{\phi}$ at different axial location. —— turbulent flow model. - - - laminar flow limit. (a) $z^* = 0$, 0.000057 and 0.0012; (b) $z^* = 0.0117$; (c) $z^* = 0.0302$; (d) $z^* = 0.0756$; (e) $z^* = 0.0974$; (f) $z^* = 0.134$. (Reprinted by permission from Dasgupta et al., 1997. Copyright 1997 American Chemical Society.)

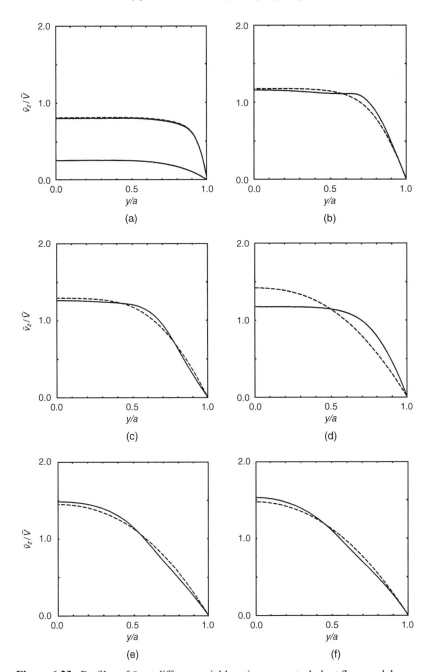

Figure 6.23. Profiles of $\bar{v}_z$ at different axial locations. —— turbulent flow model. - - - laminar flow limit. (a) $z^* = 0$ and 0.0012; (b) $z^* = 0.0117$; (c) $z^* = 0.0302$; (d) $z^* = 0.0756$; (e) $z^* = 0.0974$; (f) $z^* = 0.134$. (Reprinted by permission from Dasgupta et al., 1997. Copyright 1997 American Chemical Society.)

continuous lines indicate the profiles predicted by the pseudoturbulent model, while broken lines show the corresponding profiles when fluctuations are eliminated by setting $K = 0$.

In Figure 6.22(a) the continuous and broken profiles almost coincide, so the Reynolds stresses associated with pseudoturbulence have not yet had any significant influence on the motion. The concentration profile at $z^* = 0.000057$ already exhibits the marked segregation of the particles towards the wall that occurs as a purely kinematic consequence of the gas velocity profile, and the profile at $z^* = 0.0012$ shows how the particles are subsequently pushed back towards the centre of the channel by the particle-phase pressure. In physical terms these two axial distances are only 0.2 cm and 4.23 cm, respectively, so the acceleration length is very short indeed, and the particle pressure gradient also exerts its influence within a short distance. The resegregation towards the wall as a result of the Reynolds stresses is much slower. The effect of this is greatest at $z^* = 0.03$, or at an axial distance of 106 cm, after which the particles gradually redeploy towards the centre of the channel. In the "laminar" case we see from the broken curves that, following the initial "kinematic" segregation towards the wall, the concentration profile flattens out gradually on moving up and clearly approaches a uniform concentration distribution at a large value of the axial distance. Figure 6.23 shows that the particle-phase velocity profiles are almost uninfluenced by pseudoturbulence in the lower part of the channel. They separate, with the pseudoturbulent profile flatter than the "laminar" profile at intermediate distances, but by $z^* = 0.1$ they are again quite close together. Thus, profiles of particle concentration for the pseudoturbulent and the "laminar" cases approach each other at large distances, and the same is true for the profiles of particle velocity. Both these effects result from a gradual decay in the intensity of velocity fluctuations, represented by K, on moving along the channel, but it is not clear whether this captures a real physical effect or is a defect of the model.

Dasgupta et al. also present the results of computations for both downflow and upflow in the presence of gravity though, as noted above, the upflow calculations must be limited to cases where there is no reversal in the direction of particle flow near the wall.

6.4 The Model of Hrenya and Sinclair

The equations of Sinclair & Jackson (1989), presented in Section 6.2, neglect pseudoturbulent fluctuations altogether and consequently yield results with an unrealistically strong dependence on the values of the particle–particle and particle–wall coefficients of restitution. The model of Dasgupta et al. (1994,

1997, 1998) in contrast introduces the pseudoturbulence and, at the same time, removes the unwelcome dependence on the coefficients of restitution by the simple expedient of adopting particle-phase pressure and viscosity closures of the form (6.48), which do not depend on these measures of inelasticity. However, this does not resolve the physical question of whether the presence of pseudoturbulence is, in itself, sufficient to mitigate the unwelcome sensitivity to the values of the coefficients of restitution. This matter has recently been addressed directly by Hrenya & Sinclair (1997) who introduce pseudoturbulence while also retaining expressions for the particle-phase pressure and viscosity that are essentially those used by Sinclair & Jackson, (6.7)–(6.9), which contain the particle–particle coefficient of restitution and also depend on T. What follows is an account of the Hrenya–Sinclair model, slightly simplified to bring out its relation to the work described in the last two sections but retaining its essential features.

For fully developed flow the time-averaged axial and radial components of the particle-phase momentum equation have the form of (6.42) and (6.43), simplified by neglecting the last term in (6.43). Thus

$$\frac{1}{r}\frac{\partial}{\partial r}\left[r(\mu^p + \rho_s\bar{\phi}\nu_T)\frac{\partial\bar{v}_z}{\partial r}\right] = \rho_s\bar{\phi}g + \frac{\partial\bar{p}^f}{\partial z} \tag{6.53}$$

and

$$\frac{\partial}{\partial r}\left[\bar{p}^p + \frac{2}{3}\rho_s\bar{\phi}K\right] = 0. \tag{6.54}$$

With neglect of inertial and viscous terms for the gas, the time-averaged form of the gas-phase axial momentum balance is

$$\frac{\partial\bar{p}^f}{\partial z} + \bar{\beta}(\bar{u}_z - \bar{v}_z) = 0. \tag{6.55}$$

The closures for $\bar{p}^p$ and μ^p are essentially those used by Sinclair & Jackson and given by (6.7)–(6.9), with ϕ and T replaced by their time-averaged values. (These relations are actually modified slightly, in a manner suggested by Louge et al. (1991), to take account of the presence of the bounding wall.) Since $\bar{T}$ now appears in these closures it must be determined by a time-averaged form of the pseudothermal energy balance. Decomposing variables as sums of time averages and fluctuations introduces a number of correlations for which there are no plausible closures. Hrenya & Sinclair therefore neglected many of these, simply replacing variables in the pseudothermal energy equation by their time averages. The resulting time-averaged energy balance is

$$\frac{1}{r}\frac{\partial}{\partial r}\left[r\left(-\bar{q}_r + \frac{3}{2}\rho_s\bar{\phi}\nu_T\frac{\partial\bar{T}}{\partial r}\right)\right] - \bar{Q}_c + \mu^p\left(\frac{\partial\bar{v}_z}{\partial r}\right)^2 + \rho_s\bar{\phi}\varepsilon^p = 0. \tag{6.56}$$

The closure for the pseudothermal energy flux $\bar{q}_r$ has the form (6.11), with the term in $\nabla \bar{\phi}$ omitted and the term in $\nabla \bar{T}$ modified slightly (as in the case of μ^p) to account for the presence of the walls. The simplest expression for $\bar{Q}_c$ takes the form of (6.10), with ϕ and T replaced by their time averages.

To complete the closure of the differential equations it then remains only to specify the three quantities ν_T, K, and ε^p associated with the pseudoturbulent fluctuations. For this purpose the K–ε model, used by Dasgupta et al., is replaced by a simpler mixing length approach, giving

$$\nu_T = l_m^2 \left| \frac{\partial \bar{v}_z}{\partial r} \right|, \quad K = 3.54 \left(l_m \frac{\partial \bar{v}_z}{\partial r} \right)^2, \quad \varepsilon^p = \frac{0.15 K^{3/2}}{l_m}, \tag{6.57}$$

with the mixing length l_m specified, as a function of radial position, by

$$l_m = a \left[0.14 - 0.08 \left(\frac{r}{a} \right)^2 - 0.06 \left(\frac{r}{a} \right)^4 \right] \left(1 - \frac{\bar{\phi}}{\phi_m} \right). \tag{6.58}$$

The last factor in parentheses is introduced to ensure that the mixing length tends to zero as the particles approach random close packing.

To determine a solution, boundary conditions are required at the pipe wall, and a simplified version of those used by Sinclair & Jackson is found to be adequate. For small particles (e.g., fluidized cracking catalyst) Pita & Sundaresan (1991) showed that the boundary condition (6.17) for the particle-phase velocity reduces approximately to the no-slip condition $\bar{v}_z = 0$. Similarly, (6.19) can be replaced by a no-slip condition for the gas velocity, $\bar{u}_z = 0$. Finally, with the above value for $\bar{v}_z$, the energy flux boundary condition (6.18) reduces to

$$\bar{q}_r = \frac{\sqrt{3}\pi (\bar{\phi}/\phi_m) \rho_s \bar{T}^{3/2} (1 - e_w^2)}{4 \left[1 - (\bar{\phi}/\phi_m)^{1/3} \right]} \quad \text{at } r = a.$$

The degree of success of this Hrenya–Sinclair model in eliminating the dependence of predicted performance on the coefficients of restitution can be judged by comparing its results with those from the model of Sinclair & Jackson (1989). Figure 6.24 shows solids concentration and granular temperature profiles from both models. Here $\dot{Q}_s$ denotes solids mass flow rate per unit area, V_{sg} denotes gas superficial velocity, and the system parameters are listed in Table 6.3.

Figure 6.24(a) gives the results from the Sinclair–Jackson model, which is seen to predict strong segregation of the particles towards the pipe wall when the particle–particle coefficient of restitution is unity but a reversal of this, with a maximum concentration on the axis, when e is reduced to 0.99!

Table 6.3. *Parameter values for the example used to compare*
the Sinclair–Jackson and Hrenya–Sinclair models

Particle diameter	$d = 75\,\mu\mathrm{m}$
Solid material density	$\rho_s = 1{,}654\,\mathrm{kg/m^3}$
Pipe radius	$a = 0.0375\,\mathrm{m}$
Gas viscosity	$\mu_g = 1.81 \times 10^{-5}\,\mathrm{kg/(m \cdot s)}$
Maximum solids volume fraction	$\phi_m = 0.65$

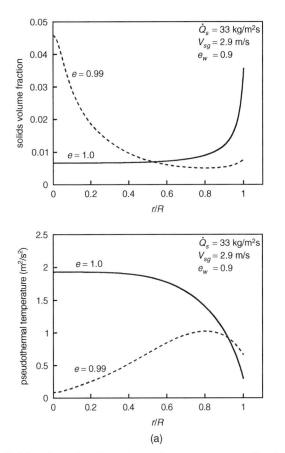

(a)

Figure 6.24. Solids volume fraction and granular temperature profiles for the system specified in Table 6.3. Comparison of (a) Sinclair–Jackson model with (b) Hrenya–Sinclair Model 1. (From Hrenya & Sinclair 1997. Reproduced with permission of the American Institute of Chemical Engineers. Copyright 1997 AIChE. All rights reserved.)

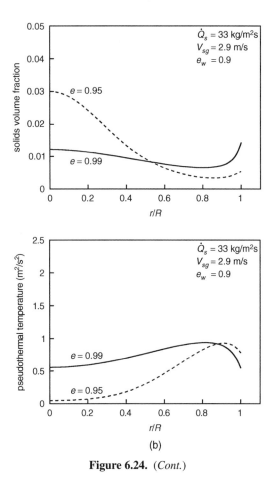

Figure 6.24. (*Cont.*)

The results of the Hrenya–Sinclair model, shown in Figure 6.24(b), are a little less sensitive to the value of e. The particles are still concentrated near the wall when $e = 0.99$, but the maximum concentration shifts to the axis if the coefficient of restitution is reduced further to 0.95. Thus, though the undesirable sensitivity is reduced by the introduction of the pseudoturbulence, the Hrenya–Sinclair model, as just described, still leaves an unacceptably strong dependence of the particle concentration profile on the inelasticity of collisions. More specifically, with a realistic value for e (namely 0.95) the particle concentration at the axis is far larger than it should be. The reason for this becomes clear on examining the relative values of the laminar and pseudoturbulent contributions to the effective particle-phase pressure, represented by the two terms within the square brackets in (6.54). Despite the introduction

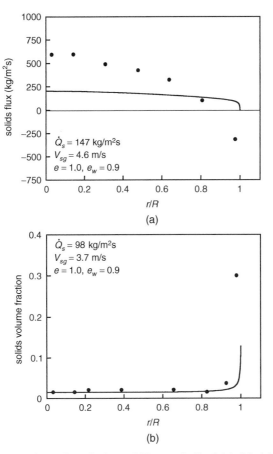

Figure 6.25. Comparison of predictions of Hrenya & Sinclair's Model 1 with measurements of Bader et al. (1988). (From Hrenya & Sinclair, 1997. Reproduced with permission of the American Institute of Chemical Engineers. Copyright 1997 AIChE. All rights reserved.)

of pseudoturbulence it is found that the laminar contribution still dominates, so the undesirable behaviour of the Sinclair–Jackson model with inelastic particle–particle collisions still survives, though to a reduced extent.

According to (6.53) the effective particle-phase viscosity is also the sum of laminar and turbulent contributions, represented by the two terms in parentheses, but in this case it is found that the turbulent contribution dominates. This has the consequence of flattening unduly the radial profiles of the solid-phase flux, as seen in Figure 6.25, which compares the predictions of the model with data of Bader et al. (1988). The model predicts a flux almost independent of position, except for a sharp decrease to zero at the wall, while the measured flux decreases smoothly throughout the interval of r/R, with downflow in an annulus adjacent to the wall.

Both the above shortcomings of the model can be traced to the term $\bar{Q}_c$ in the pseudothermal energy balance (6.56), which appears to be overestimating the dissipation due to collisions when it is evaluated as suggested above, simply using the expression (6.10) with ϕ and T replaced by their time-averaged values. Now (6.10) has the form

$$Q_c = \frac{\rho_s}{d} f_1(e, \phi) \, f_2(T), \qquad (6.59)$$

with

$$f_1(e, \phi) = \frac{12}{\pi^{1/2}}(1 - e^2)\phi^2 g_0(\phi), \quad f_2(T) = T^{3/2}, \qquad (6.60)$$

and the Hrenya–Sinclair model as described above, designated "Model 1" by the authors, makes the crude approximation

$$\bar{Q}_c \approx \frac{\rho_s}{d} f_1(e, \bar{\phi}) \, f_2(\bar{T}),$$

which neglects all correlations involving fluctuations in ϕ and T. As a better approximation Hrenya and Sinclair suggest that the fluctuations might be introduced through linearised forms of f_1 and f_2, writing

$$Q_c = \frac{\rho_s}{d}\left[f_1(\bar{\phi}) + \left(\frac{\partial f_1}{\partial \phi}\right)_{\bar{\phi}} \phi' \right]\left[f_2(\bar{T}) + \left(\frac{\partial f_2}{\partial T}\right)_{\bar{T}} T' \right].$$

Then, on time-averaging, we get

$$\bar{Q}_c = \frac{\rho_s}{d}\left\{ f_1(\bar{\phi}) f_2(\bar{T}) + \left(\frac{\partial f_1}{\partial \phi}\right)_{\bar{\phi}} \left(\frac{\partial f_2}{\partial T}\right)_{\bar{T}} \overline{(\phi'T')} \right\}. \qquad (6.61)$$

The first term in this is positive and is the estimate of $\bar{Q}_c$ used in Model 1. Each of the partial derivatives is positive, but the correlation $\overline{(\phi'T')}$ would be expected to be negative, since the particle temperature tends to be depressed in local regions of increased concentration. Thus the second term of (6.61) should have the desired effect of reducing the dissipation term associated with collisions.

To complete this modified form of their model, designated "Model 2", the authors propose a closure for $\overline{(\phi'T')}$ of the following form:

$$\overline{(\phi'T')} = \alpha \bar{\phi}_{\mathrm{av}} K_{\mathrm{av}} [1 + (\phi_m - \bar{\phi}_w - 1)(r/a)^n], \qquad (6.62)$$

where α and n are parameters available for fitting, $\bar{\phi}_w$ indicates the value of $\bar{\phi}$ at the pipe wall, and the suffix "av" means that an average value must be taken over a radius of the pipe. The values assigned to the fitting parameters are $\alpha = -5, n = 8$. With this choice the predictions of Model 2 are compared with

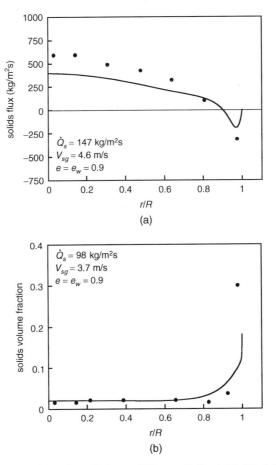

Figure 6.26. Comparison of predictions of Hrenya & Sinclair's Model 2 with measurements of Bader et al. (1988). (From Hrenya & Sinclair, 1997. Reproduced with permission of the American Institute of Chemical Engineers. Copyright 1997 AIChE. All rights reserved.)

the measurements of Bader et al. (1988) in Figure 6.26. The shortcomings of Model 1, seen in Figure 6.25, are now largely eliminated, and other comparisons with experiment can be found in the original publication.

It is also interesting to compare the predictions of Model 2 with those of Dasgupta et al. (1998). For this purpose Hrenya & Sinclair computed a contour of constant gas pressure gradient, $\partial \bar{p}^f / \partial z = -300\,\text{Pa/m}$, in the $(\bar{V}_s, \bar{V}_g)$-plane for particles with the same physical properties and a pipe of the same diameter as specified in Table 6.2 above. The result, with $e = e_w = 0.9$, which is shown in Figure 6.27, should be compared with the contour for the same pressure gradient

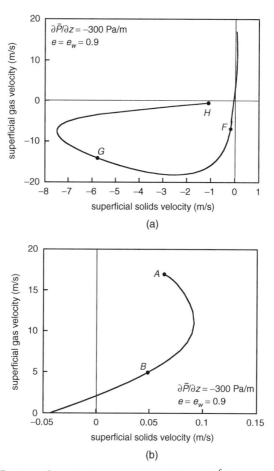

Figure 6.27. Contour of constant gas pressure gradient, $\partial \bar{p}^f / \partial z = -300$ Pa/m, pre-dicted by Hrenya & Sinclair's Model 2 for particles and pipe specified in Table 6.2. $e = e_w = 0.9$. (From Hrenya & Sinclair, 1997. Reproduced with permission of the American Institute of Chemical Engineers. Copyright 1997 AIChE. All rights reserved.)

in Figure 6.17, as found from the model of Dasgupta et al. The Hrenya–Sinclair contour does not form a closed loop like that of Dasgupta et al. However, both pass through a maximum of $\bar{V}_s$ in the first quadrant, though the maximum value is at least twice as large in Figure 6.27 as in Figure 6.17. It also occurs at about twice the value of the superficial gas velocity. The contours found by both investigators pass through the second quadrant and Hrenya & Sinclair demonstrate that their contours for different values of $\partial \bar{p}^f / \partial z$, like those of Dasgupta et al., form an envelope. However, the part of the second quadrant cut off by this envelope is found by Hrenya & Sinclair to be about twice as large as that found by Dasgupta et al. Thus, in the first and second quadrants, the contours from both models are of similar shape, but that predicted by Hrenya & Sinclair is, roughly speaking, twice as large. In the third quadrant the contours behave quite differently. That of Hrenya & Sinclair forms a large loop resembling the loops found in the earlier work of Sinclair & Jackson, seen in Figure 6.2. The contour of Dasgupta et al. is different in shape and is confined to values of $\bar{V}_s$ and $\bar{V}_g$ much smaller in magnitude.

Clearly, much more experimental information would be needed to conduct a thorough test of both models, and information on cocurrent downflow, where the disparity is greatest, would be of particular value. Both models fix parameter values in their respective closures for the pseudoturbulence to match those values used in the corresponding closures for turbulence in a single-phase fluid, so these are not regarded as adjustable parameters. This leaves only two adjustable parameters in each model. For Dasgupta et al. these are the proportionality constants A and B in Equations (6.48), which determine the magnitudes of the particle-phase pressure and viscosity, respectively. For Hrenya & Sinclair, they are the parameters α and n in (6.62). It seems likely that these are less related to the properties of specific particles than the parameters A and B so one might, perhaps, anticipate that the Hrenya–Sinclair model would have better predictive capability.

Finally, mention should be made of one other model of the type discussed in Sections 6.2 to 6.4. Nieuwland et al. (1996) describe work on fully developed riser flow based on equations that are, for the most part, minor variants of those used by Sinclair & Jackson (1989). Expressions for β and g_0 somewhat more elaborate than (6.3) and (6.9) are used, but the main difference is the replacement of the fluid-phase viscosity μ^f in (6.13) by a sum $\mu^f + \mu^{ft}$, where μ^{ft} is an "eddy viscosity to account for turbulent momentum transport". An expression for μ^{ft} is proposed that vanishes when the particle concentration approaches random close packing and degenerates into the Prandtl mixing length form for a single-phase fluid when the concentration approaches zero. Curiously, there is no corresponding supplement to the fluid-phase pressure, and it is not clear whether μ^{ft} is intended to take account of velocity fluctuations at the scale of the "thermal" motion of the particles, or at the mesoscale of the "pseudoturbulent" fluctuations discussed at length above. If the latter, then it is difficult to justify the absence from the particle-phase momentum equation of corresponding terms representing the effect of fluctuations in particle velocity on this scale.

Like Sinclair & Jackson (1989) and Pita & Sundaresan (1991), Nieuwland et al. comment on the extreme sensitivity of their results to the value of the coefficient of restitution for collisions between particles, noting that: "calculations with e values which differ slightly from unity (e.g., $e = 0.999$) result in extinguishment of the granular temperature and, as a consequence, fail to produce the experimentally observed lateral solids segregation". The mere introduction of a turbulent supplement to the gas-phase viscosity therefore fails to save the day.

6.5 Computational Fluid Dynamic Modelling of Riser Flow

In Chapter 5 we saw that direct numerical solution of the time-dependent equations of motion of the fluid–particle system was practicable, and it proved capable of resolving some long-standing puzzles about bubble and cluster formation. The problems were formulated carefully to minimize demands on the numerical methods; for example, spatially periodic systems replaced real beds of finite depth, and the frame of reference was set in motion to immobilize steep concentration gradients, as far as possible, in the computational grid. If one contemplates a corresponding approach to motion in an entire riser such devices are no longer available, so the computational difficulties are clearly much greater. Nevertheless, there is now a rapidly increasing interest in this type of approach stimulated, in large part, by pioneering attempts due to Gidaspow and coworkers.

The first significant contribution of this type is due to Tsuo & Gidaspow (1990), who based their computations on equations of motion with a level of sophistication comparable to those used in the work described in Chapter 5. More specifically, Newtonian closures were assumed for the gas- and particle-phase stress tensors, with no dependence on granular temperature, so that there was no need for a balance equation of pseudothermal energy. Specific details of the equations can be found in the original publication. The simulations started with an empty pipe and specified flow rates of particles and gas were initiated at time zero, at the bottom of the pipe. The inlet volume fraction of solids was also specified and this, in turn, determined the velocities of the two phases at inlet. All these were taken to be uniform over the cross section, ensuring plug flow at the inlet. A no-slip condition was imposed on the gas velocity at the pipe wall, while the particle velocity was assumed to satisfy a boundary condition of the form

$$v_z = -L_p \frac{\partial v_z}{\partial r}.$$

The inlet conditions were then held constant and the resulting initial-boundary value problem for the particle concentration and the velocities of both phases within the pipe was solved numerically, using an extension of the K-FIX computer code due to Syamlal (1985).

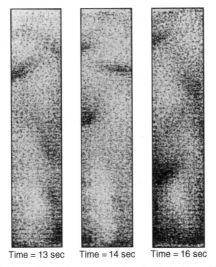

Time = 13 sec   Time = 14 sec   Time = 16 sec

Figure 6.28. Distribution of particle concentration in the 7.62 cm diameter riser at three successive times. (From Tsuo & Gidaspow, 1990. Reproduced with permission of the American Institute of Chemical Engineers. Copyright 1990 AIChE. All rights reserved.)

Two systems were studied, for each of which experimental measurements of velocities and particle concentrations were available. The first (Gidaspow et al., 1989) used a pipe of length 5.5 m and diameter 7.62 cm, with glass beads of diameter 520 μm and density 2.62 g/cm^3. The inlet conditions were $\bar{V}_s = 0.0095$ m/s, $\bar{V}_g = 4.86$ m/s, and $\phi = 0.0246$. Correspondingly, $\bar{V}_s/\bar{V}_g = 0.002$, $v_z = 0.386$ m/s, and $u_z = 4.98$ m/s. The second pipe (Bader et al., 1988) was of length 11.2 m and diameter 0.305 m, with cracking catalyst particles of mean diameter 76 μm and density 1.714 g/cm^3. Its inlet conditions were $\bar{V}_s = 0.057$ m/s, $\bar{V}_g = 2.78$ m/s, and $\phi = 0.25$. Correspondingly, $\bar{V}_s/\bar{V}_g = 0.02$, $v_z = 0.228$ m/s, and $u_z = 3.7$ m/s. Note that the ratio of solids flow to gas flow in the second pipe is ten times larger than that in the first.

In each case there was a "start-up" period, during which the particles were carried up from the inlet to the top of the pipe, followed by a period in which conditions appeared either to oscillate or to remain approximately steady. Figure 6.28 shows the distribution of particle concentration in the pipe of 7.62 cm diameter at three instants, all after the end of the start-up period. Clustering of the particles is apparent, as is the tendency of the clusters to congregate near the wall. The dense regions are transient phenomena, forming and subsequently dissolving, but while they survive they drift downward, and examination of the velocity fields reveals that the particles within them also move downward. The

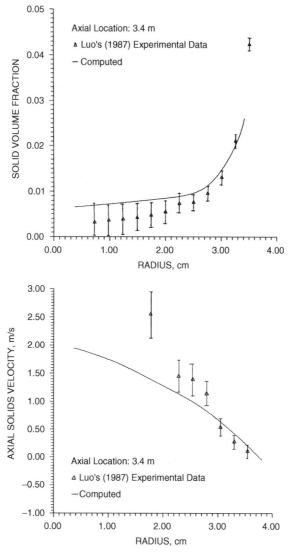

Figure 6.29. Comparison of predicted and measured profiles of ϕ, v_z, and u_z at a height of 3.4 m in the 7.62 cm diameter riser. (From Tsuo & Gidaspow, 1990. Reproduced with permission of the American Institute of Chemical Engineers. Copyright 1990 AIChE. All rights reserved.)

computed flux of particles from the top of the pipe is found to oscillate, more or less periodically, with a period of a few seconds. This is comparable with the time taken by a typical cluster to drift downward through the length of the pipe. Figure 6.29 compares experimental measurements with computed radial

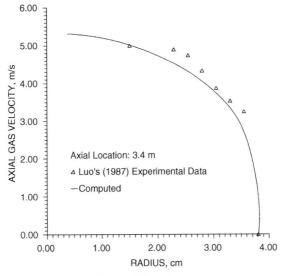

Figure 6.29. (*Cont.*)

profiles of ϕ, v_z, and u_z, averaged over time (after the start-up period) at a height of 3.4 m from the bottom of the pipe. There appears to be fair agreement.

Figure 6.30 reveals that the instantaneous distributions of particle concentration in the 0.305 m diameter pipe are quite different in nature. There is no longer much evidence of clustering but the segregation of particles to the wall survives in the form of a core–annular pattern. The particles move up rapidly in the core region of low concentration and drift downward more slowly in a dense annulus adjacent to the wall. Computed radial profiles of the time-averaged particle concentration and v_z, at a height of 9.1 m, are compared with measurements of Bader et al. in Figure 6.31.

These results are encouraging. The radial displacement of the particles towards the wall is predicted, for both low and high particle loadings. The clusters are seen only when the average concentration of the particles is low, as would be expected from the work of Glasser et al. (1998), described in Chapter 5. Low-frequency oscillatory behaviour is observed in the clustering system and, finally, quantitative agreement with available measurements is fairly good.

Further computations of riser flow, based on a mathematical description at the same level of sophistication, have recently been reported by Benyahia et al. (1998). These used a different code for the numerical work and again predicted a core–annular type of behaviour. From the rather sparse results presented they appear to relate to experiments with about the same success as the work of Tsuo & Gidaspow.

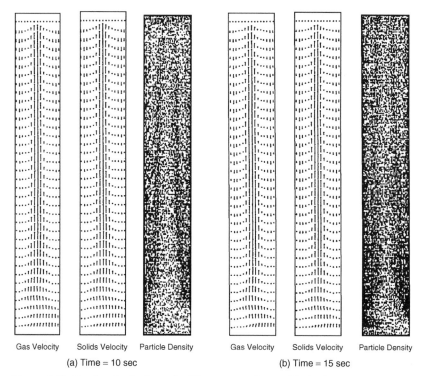

Gas Velocity   Solids Velocity   Particle Density   Gas Velocity   Solids Velocity   Particle Density

(a) Time = 10 sec   (b) Time = 15 sec

Figure 6.30. Predicted velocity and solids concentration profiles in the 0.305 m diameter riser. (From Tsuo & Gidaspow, 1990. Reproduced with permission of the American Institute of Chemical Engineers. Copyright 1990 AIChE. All rights reserved.)

A more ambitious attempt at direct simulation by numerical solution of the equations of motion has been reported by Gidaspow et al. (1992). These authors address the behaviour of a complete circulating fluidized bed loop, consisting of a riser, a standpipe or downcomer, and the upper and lower connections between them. A more elaborate closure for the equations of motion, based on the kinetic theory of granular materials, is invoked. Thus, their equations of motion match the level of sophistication of those used in the Hrenya–Sinclair model, described in Section 6.4, and an equation of balance of pseudothermal energy must be solved in addition to the continuity and momentum equations for each of the two phases. The system modelled is a large industrial CFB and a full discussion of the results can be found in Gidaspow (1994). Figure 6.32 is a density plot showing the instantaneous distribution of particle concentration 5 s after the initiation of integration. It gives a general impression of conditions in both the riser and downcomer sections and also shows the quite coarse scale of spatial discretization needed to make the computations practicable. The authors

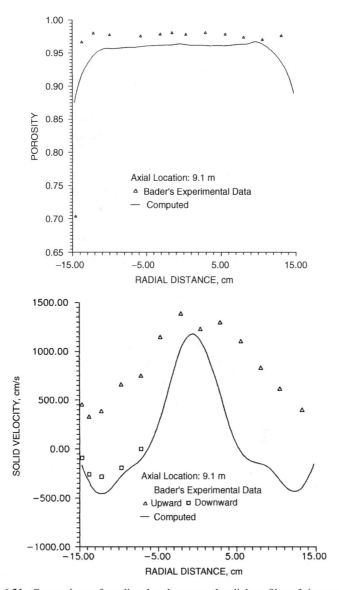

Figure 6.31. Comparison of predicted and measured radial profiles of time averaged void fraction and particle axial velocity at a height of 9.1 m in the 0.305 m diameter riser. (From Tsuo & Gidaspow, 1990. Reproduced with permission of the American Institute of Chemical Engineers. Copyright 1990 AIChE. All rights reserved.)

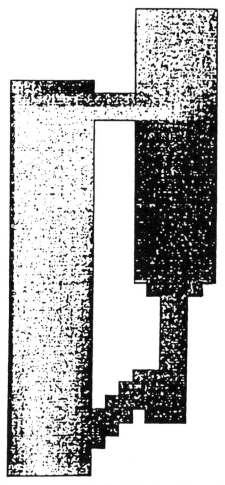

Figure 6.32. Particle concentration 5 s after initiation of the motion in the CFB loop simulated by Gidaspow et al. (1992). (From Gidaspow, 1994.)

indicate that the choice of the value for the coefficient of restitution in particle–particle collisions has a significant effect on the solutions obtained, which may spur some concern about the detailed form of the equations of motion used.

Clearly, much more work is needed before the computational fluid dynamics approach can be established as a reliable and practicable tool for design. The limited number of solutions found seem to have the right qualitative features, but evidence is lacking that they are quantitatively accurate solutions of the equations of motion. Extensive testing of their sensitivity to changes in scale of both spatial and temporal discretization is needed, and it is also desirable to demonstrate that the solutions can be reproduced, in quantitative detail, by at least two completely disparate numerical techniques. As in the case of turbulent motion of a single-phase fluid, it may be necessary to devise some explicit subgrid scale closure to be able to address the performance of a complete apparatus, such as a riser, with an adequate resolution of the important mesoscale phenomena such as clusters and streamers, while maintaining acceptable quantitative accuracy over long enough time intervals. Despite these challenges it seem likely that this will be the direction of the future. If successful it offers the possibility of finding reliable answers to the sort of detailed questions practicing engineers must pose, for example, the effect on riser performance of proposed modifications to the geometry of the junction between the downcomer and the riser, changes in the arrangements for introducing gas at its foot, or restructuring the device for separating particles from gas at the top of the riser.

References

Alam, M. & Nott, P. R. 1998. Stability of plane Couette flow of a granular material. *J. Fluid Mech.* **377**, 99–136.

Arastoopour, H. & Gidaspow, D. 1979. Analysis of IGT pneumatic conveying data and fast fluidization using a thermohydrodynamic model. *Powder Technol.* **22**, 77–87.

Arastoopour, H., Lin, S-C., & Weil, S. A. 1982. Analysis of vertical pneumatic conveying of solids using multiphase flow models. *AIChE J.* **28**, 467–473.

Bader, R., Findlay, J., & Knowlton, T. M. 1988. Gas/solid flow patterns in a 30.5-cm diameter circulating fluidized bed. *Int. Circulating Fluidized Bed Conf.*, Compiègne, France.

Bartholomew, R. & Casagrande, R. 1957. Measuring solids concentration in fluidized systems by gamma ray absorption. *Ind. Eng. Chem.* **49**, 428–431.

Basu, P. & Nag, P. K. 1987. An investigation into heat transfer in circulating fluidized beds. *Int. J. Heat & Mass Transfer* **30**, 2399–2408.

Benyahia, S., Arastoopour, H., & Knowlton, T. 1998. Prediction of solid and gas flow behaviour in a riser using a computational multiphase flow approach. In *Fluidization IX*, ed. L-S Fan & T. M. Knowlton, p. 493. Engineering Foundation.

Buyevich, Yu. A. 1994. Fluid dynamics of coarse dispersions. *Chem. Eng. Sci.* **49**, 1217–1228.

Dasgupta, S., Jackson, R., & Sundaresan, S. 1994. Turbulent gas–particle flow in vertical risers. *AIChE J.* **40**, 215–228.

Dasgupta, S., Jackson, R., & Sundaresan, S. 1997. Developing flow of gas–particle mixtures in vertical ducts. *Ind. Eng. Chem. Res.* **36**, 3375–3390.

Dasgupta, S., Jackson, R., & Sundaresan, S. 1998. Gas–particle flow in vertical pipes with high mass loading of particles. *Powder Technol.* **96**, 6–23.

Gidaspow, D. 1994. *Multiphase Flow and Fluidization*, Chapter 10. Academic Press.

Gidaspow, D., Tsuo, Y. P., & Luo, K. M. 1989. Computed and experimental cluster formation in velocity profiles in circulating fluidized beds. *Int. Fluidization Conf.*, Banff, Canada.

Gidaspow, D., Bezbaruah, R., & Ding, J. 1992. Hydrodynamics of circulating fluidized beds: kinetic theory approach. In *Fluidization VII*, ed. O. E. Potter & D. J. Nicklin, p. 75. Engineering Foundation.

Glasser, B. J., Sundaresan, S., & Kevrekidis, I. G. 1998. From bubbles to clusters in fluidized beds. *Phys. Rev. Lett.* **81**, 1849–1852.

Goldhirsch, I., Tan, M-L., & Zanetti, G. 1993. A molecular dynamical study of granular fluids. I: the unforced granular gas in two dimensions. *J. Sci. Comput.* **8**, 1–40.

Grace, J. R. & Tuot, J. 1979. A theory for cluster formation in vertically conveyed suspensions of intermediate density. *Trans. Inst. Chem. Eng.* **57**, 49–54.

Hopkins, M. A. & Louge, M. Y. 1991. Inelastic microstructure in rapid granular flows of smooth disks. *Phys. Fluids* **A3**, 47–57.

Hrenya, C. M. & Sinclair, J. L. 1997. Effects of particle phase turbulence in gas–solid flows. *AIChE J.* **43**, 853–869.

Issangya, A. S., Bai, D., & Grace, J. R. 1998. Solids flux profiles in a high density circulating fluidized bed riser. In *Fluidization IX*, ed. L-S. Fan & T. M. Knowlton, p. 197. Engineering Foundation.

Johnson, P. C. & Jackson, R. 1987. Frictional–collisional constitutive relations for granular materials, with application to plane shearing. *J. Fluid Mech.* **176**, 67–93.

Koch, D. L. 1990. Kinetic theory for a monodisperse gas–solid suspension. *Phys. Fluids* **A2**, 1711–1723.

Lam, C. K. G. & Bremhorst, K. A. 1981. Modified form of the K–ε model for predicting wall turbulence. *J. Fluids Eng.* **103**, 456.

Louge, M. Y., Mastorakos, E., & Jenkins, J. T. 1991. The role of particle collisions in pneumatic transport. *J. Fluid Mech.* **231**, 345–359.

Lun, C. K. K., Savage, S. B., Jeffrey, D. J., & Chepurniy, N. 1984. Kinetic theories for granular flow: inelastic particles in Couette flow and slightly inelastic particles in a general flow field. *J. Fluid Mech.* **140**, 223–256.

Matsen, J. M. 1982. Mechanisms of choking and entrainment. *Powder Technol.* **32**, 21–33.

Monceaux, L., Aziz, M., Molodtsov, Y., & Large, J. 1986. Particle mass flux profiles and flow regime characterization in a pilot scale fast fluidized bed. In *Fluidization V*, ed. M. A. Sorensen & J. Ostergaard, p. 337. Engineering Foundation.

Nieuwland, J. J., van Sint Annaland, M., Kuipers, J. A. M., & van Swaaij, W. P. M. 1996. Hydrodynamic modelling of gas/particle flows in riser reactors. *AIChE J.* **42**, 1569–1582.

Patel, V. C., Rodi, W., & Scheurer, G. 1985. Turbulence models for near-wall and low Reynolds number flows: a review. *AIAA J.* **23**, 1308–1319.

Pita, J. A. & Sundaresan, S. 1991. Gas–solid flow in vertical tubes. *AIChE J.* **37**, 1009–1018.

Savage, S. B. 1992. Instabilities of unbounded uniform granular shear flow. *J. Fluid Mech.* **241**, 109–123.

Saxton, A. & Worley, A. 1970. Modern catalytic cracking design. *Oil & Gas J.* **68**, 82.

Schmid, P. J. & Kytomaa, H. K. 1994. Transient and asymptotic stability of granular shear flow. *J. Fluid Mech.* **264**, 255–275.

Sinclair, J. L. & Jackson, R. 1989. Gas–particle flow in a vertical pipe with particle–particle interactions. *AIChE J.* **35**, 1473–1486.

Syamlal, M. 1985. Multiphase hydrodynamics of gas–solids flow. Ph.D. thesis, Illinois Institute of Technology.

Tsuo, Y. P. & Gidaspow, D. 1990. Computation of flow patterns in circulating fluidized beds. *AIChE J.* **36**, 885–896.

Wang, C-H., Jackson, R., & Sundaresan, S. 1996. Stability of bounded rapid shear flows of a granular material. *J. Fluid Mech.* **308**, 31–62.

Wang, C-H., Jackson, R., & Sundaresan, S. 1997. Instabilities of fully developed rapid flow of a granular material in a channel. *J. Fluid Mech.* **342**, 179–197.

Weinstein, H., Shao, M., & Wasserzug, L. 1984. Radial solid density variation in a fast fluidized bed. *AIChE Symp. Ser.* **80**, 117–120.

Weinstein, H., Shao, M., & Schnitzlein, M. 1986. Radial variation in solid density in high velocity fluidization. In *Circulating Fluidized Bed Technology I*, ed. P. Basu, p. 201. Pergamon Press.

Yerushalmi, J. 1986. *Gas Fluidization Technology*, p. 174. Wiley.

Yerushalmi, J., Cankurt, N., Geldart, D., & Liss, B. 1978. Flow regimes in vertical gas–solid contact systems. *AIChE Symp. Ser.* **74**, 1–13.

Youchou, L. & Kwauk, M. 1980. The dynamics of fast fluidization. *In Fluidization*, ed. J. Grace & J. Matsen. Plenum Press.

Zenz, F. & Othmer, D. 1960. *Fluidization and Fluid–Particle Systems.* Reinhold Publishing Co.

7

Standpipe flow

7.1 Introduction

Standpipes are devices that enable particles to be transferred from a region of lower gas pressure to a region of higher pressure and, as such, they are vital elements in particle circulation loops. For example, in a catalytic cracker the pressure drops as the gas and suspended catalyst move up the riser reactor. At the top of the riser, where the gas pressure is lowest, the particles are separated from the bulk of the gas in cyclones. Subsequently they must pass through the regenerator where coke is burnt off and be reintroduced into the riser at its foot, where the pressure is highest. This is accomplished by taking advantage of an effect that is very simple, in principle.

Consider a vertical pipe containing particles maintained as a fluidized bed by an upflow of gas. A force balance shows that the pressure drop in the gas, over the height of the fluidized bed, is equal to the buoyant weight of the suspended particles plus the hydrostatic pressure drop associated with the density of the gas. Since this is usually much smaller than that of the solid material both buoyancy and hydrostatic contributions are small, and it is a good approximation to equate the pressure drop to the weight of the suspended particles. Now if the whole assembly of particles is in motion down the pipe, the same fluidized condition is maintained if the *relative* velocity of particles and gas is the same, so we should see the same value for the gas pressure drop. But then the particles are in motion from a region of lower gas pressure to a region of higher pressure, which is what was required. The rate of increase of pressure on moving down the pipe will match the bulk density of the suspension, so this should be as large as possible, and the overall rise in pressure will then be proportional to the length of the pipe.

This simple argument relies on the assumption that the particles are supported in a state of uniform motion entirely by the drag force exerted on them by the gas

298

as a result of the relative motion. This is not quite correct. When the suspension is flowing down the pipe there are also frictional forces between the particles and the bounding wall and these provide part of the support for the suspension, but this is a relatively small effect unless the bulk density is low and the velocity of descent is high. However, the argument is seriously erroneous once the bulk density becomes high enough that the particles are in continuous mutual contact, forming a moving packed bed rather than a suspension. Then they can be supported by a gradient in the axial component of a stress that represents the collective effect of the contact forces; indeed, it is this stress gradient that supports a heap of particles resting on a horizontal plane in the absence of any gas flow. Thus, once the relative velocity of gas and particles drops below the minimum fluidization velocity, the particle assembly collapses into a packed bed, no longer supported entirely by the drag forces, and the pressure buildup in the pipe no longer matches the weight of the particle assembly.

These arguments suggest that, in principle, the pipe should be operated under conditions where the relative velocity of gas and particles is just above the minimum fluidization velocity and that the pressure rise should then match the weight of the suspension at the concentration corresponding to incipient fluidization. In practice, however, a pressure buildup as large as this is rarely, if ever, attained. The difficulty, as we shall see, lies in maintaining a uniform suspension at incipient fluidization throughout the pipe. As in the case of risers, treated in the last chapter, there are local inhomogeneities such as streamers and clusters, when the concentration of particles is low, or bubbles, when the concentration is high, as well as radial variations in the time-averaged concentration. In addition there may be significant variations in bulk density in the axial direction; for example, the upper part of the pipe may be occupied by a fluidized suspension while the lower part is filled by a sliding packed bed. The significance of bubbles and the possibility that different regimes of flow might exist at different axial positions were both recognized quite early (Matsen, 1973; Leung & Wilson, 1973), and it was also understood that the arrangements for feeding particles to the top of the pipe and removing them from the bottom could have a marked influence on the performance of the system (Leung, 1977; Leung & Jones, 1978a,b; Judd & Rowe, 1978).

The earliest reference to standpipes as a means of closing particle circulation loops appears to be the patent of Campbell et al. (1948) in connection with the FCC process for gasoline production, and, shortly after this, Kojabashian (1958) made a first attempt to classify possible regimes of standpipe flow. The idea that all possible modes of flow could be classified into a small number of distinct and precisely definable regimes was further exploited by Leung & Jones (1978a), who identified four such regimes, which they called LEANFLO,

DENFLO, TRANSPACFLO, and PACFLO. The first two correspond to low and high density fluidized suspensions, respectively, and the last to a sliding packed bed. The third most likely represents the condition of a Geldart "group A" material in an expanded state where, as we saw in Chapter 3, it is still capable of transmitting stress through forces at sustained contacts between particles. With our present understanding of fluidization and granular materials, and our enhanced ability to compute their behaviour, it is perhaps best to avoid such hard-and-fast definitions and simply to examine the predicted behaviour of each standpipe, together with its upper and lower appendages, on its own merits.

There is no single, standard configuration for a standpipe system. (See Knowlton (1997) for a good overview.) The pipe itself may be vertical or inclined, or it may consist of sections with different inclinations. Its upper end may form a weir at the surface of a dense fluidized bed, so that particles spill over the rim into the pipe, or it may be immersed at any level within the bed. Alternatively, the particles may be fed to the pipe from a hopper. The lower end of the pipe may again be immersed in a fluidized bed, or discharge of the particles may be controlled by a constriction imposed by some type of mechanical valve, often a slide valve. Alternatively, the discharge may be regulated pneumatically by gas injection in a J-valve or L-valve. Finally, in long standpipes it is common to inject supplementary "aeration" gas at one or more points along the pipe to compensate for the reduction in volume of the interstitial gas as its pressure increases.

As one example of a standpipe system we may cite the dipleg used to return entrained particles from the freeboard of a fluidized bed after they have been separated from gas in a cyclone. The feed device in this case is the hopper section at the lower end of the cyclone, which connects directly to the top of the standpipe, and the lower end of the dipleg is immersed in the fluidized bed itself, often without any restriction. A second example is provided by the now classic case of the standpipe used to return particles to the foot of the riser in an FCC unit. Here the feed device is a dense fluidized bed in the regenerator, with the upper end of the pipe opening into the lower part of this bed, while the discharge device is the slide valve, J-valve, or L-valve, which regulates the rate of return of particles from the foot of the standpipe to the riser.

In spite of the diversity of standpipe arrangements it is possible to construct predictive models for their behaviour by developing separate mathematical descriptions of flow in the feed device, the discharge device, and the pipe itself, then linking these together using matching conditions for the gas pressure and the flows of the two phases. The mathematical model of each separate device may be as simple, or as sophisticated, as seems possible or appropriate. As we shall see, it turns out that the overall picture is so dominated by effects due to interactions among the feed device, the pipe proper, and the discharge device, that quite primitive mathematical descriptions of each of these separately are capable of predicting the dominant qualitative features of the system as a whole with some realism. Some of these features are, perhaps, unexpected and they point to circumstances under which the system may malfunction in disastrous ways not unfamiliar in commercial operation.

The role of interaction between the pipe and its feed device can be understood through a simple example. Consider a conical hopper to which a vertical pipe, with the same diameter as the hopper exit, may be attached. With its exit hole plugged the hopper is charged with a known mass of a particulate material; then the plug is removed and the time required for complete discharge of the material is recorded. If this experiment is repeated, both with and without the pipe attached to the exit, it is found that the discharge time is shorter with the pipe than without. Furthermore, this is not a small effect. As a lecture demonstration the author uses a hopper with a wall angle of approximately 25° and an exit hole of diameter 0.5 in, to which a pipe of internal diameter 0.5 in and length 4 ft can be attached. The particulate material is poppy seed and it is found that the discharge time with the pipe attached is less than half that for the bare hopper. In a larger laboratory apparatus, with a pipe of diameter 1 in and length 10 ft, the bare hopper discharge rate is increased by a factor of about eight when the pipe is attached! It is also found that this flow enhancement depends on the existence of a good, airtight seal at the junction of the pipe and the hopper. If this seal is broken the discharge time with the pipe increases to a value somewhat longer than that for the bare hopper. Clearly then, the phenomenon must depend on interaction between the particles and the interstitial air, and this realization points immediately to its explanation. The particles falling down the pipe entrain air, so the downward flow of particles is accompanied by a downward flow of air. Nevertheless, since the downward velocity of the entrained air is less than that of the entraining particles, this air is flowing upward *relative to the particles*. Thus, if we view the situation from the point of view of an observer moving with the particles, we see a stationary assembly of particles with air flowing up through them. Consequently, the air pressure must decrease with increasing height. Hence, since the pressure is atmospheric at the bottom of the pipe, it must be subatmospheric at the top where the pipe joins the hopper. But the pressure above the fill of particles in the hopper is atmospheric, so there is a pressure drop between the surface of the fill and the hopper exit, and this drives a flow of air *downward* relative to the slow-moving particles in the hopper. (This is just the air that becomes the entrained gas lower in the pipe.) As a result drag forces are exerted downward on the particles in the hopper and these supplement the force of gravity, thus increasing the hopper discharge rate.

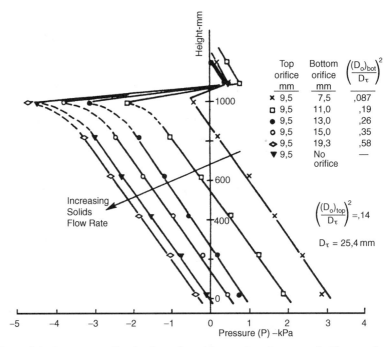

Figure 7.1. Pressure profiles for flow of cracking catalyst from a conical hopper down a standpipe of diameter 25.4 mm and length 1 m. Top orifice diameter is 9.3 mm. (From Judd & Rowe, 1978.)

This experiment can be repeated with the air pressure where the particles exit, at the bottom of the pipe, maintained at a value higher than atmospheric pressure and, of course, this reduces the flow of the particulate material. The higher the value of the back pressure, the smaller the flow until, if the back pressure is large enough, downflow can no longer occur. The maximum value of the pressure rise, against which particles can be delivered, increases with the length of the pipe. Thus, the arrangement of hopper and tailpipe can be used either to provide a large flow of particulate material against a small or vanishing back pressure of air or a more modest flow against a significant back pressure. This is just the function of a standpipe.

The above ideas are illustrated clearly by results of Judd & Rowe (1978) shown in Figure 7.1. These authors worked with a pipe fed by a hopper and provided with orifices top and bottom, whose diameters could be changed. By decreasing the diameter of the bottom orifice it was then possible to generate increasing back pressures at the bottom of the pipe. The figure shows pressure profiles, through the hopper and the pipe, for various bottom orifice diameters and a fixed size of top orifice. With the largest bottom orifice, of diameter 19.3 mm, the pressure at the bottom of the pipe is approximately atmospheric and a vacuum of about 5 kPa exists below the top orifice. As the diameter of the bottom orifice is decreased the pressure at the bottom of the pipe increases and the vacuum below the top orifice softens by about the same amount. Thus, the pressure rise in the pipe between the two orifices remains approximately the same. With the smallest orifice used, of diameter 7.5 mm, the pressure at the bottom of the pipe has risen to about 3 kPa, while the vacuum below the top orifice is reduced to less than 0.5 kPa. The overall effect is then to transport particles from a region at atmospheric pressure to a point at the bottom of the pipe where there is a gauge pressure of 3 kPa.

We shall now translate these qualitative considerations into mathematical terms. The treatment of flow in the standpipe will be one dimensional. Consequently, as in the case of early work on riser flow discussed in the last chapter, all effects due to clusters, bubbles, streamers, or radial maldistribution of the particles will be subsumed in a modified expression for an "effective" slip velocity between the two phases.

7.2 Analysis of a Simple Standpipe System

7.2.1 Definition of the Problem

As noted above, standpipes and their associated feed and discharge devices are found in a number of different configurations, each of which requires a somewhat different mathematical model. Here we will focus on one particular configuration that has been studied rather extensively, but similar problems must be faced in modelling any other type of system. Thus, this example should provide a strategy applicable to most other cases, even though the detailed models of the feed and discharge arrangements may be quite different.

The system considered here, shown in Figure 7.2, has been studied, both theoretically and experimentally, by Ginestra et al. (1980), Chen et al. (1984), and Mountziaris & Jackson (1991). It consists of a vertical standpipe of length H, fed with particles at its upper end from a conical hopper, and terminated at its lower end by a circular orifice coaxial with the pipe. The angle α, between the walls of the feed hopper and the vertical, is sufficiently small that mass flow can be assumed; in other words all the particles within the hopper are in motion. Supplementary aeration is introduced at an intermediate level in the pipe, and the flow of aeration gas, expressed as a superficial velocity based on the cross-sectional area of the pipe, is $\bar{u}_a$. In the pipe a coordinate z is measured downward from the top, as indicated, while in the feed hopper radial distance r is measured from the vertex of the cone defined by its walls. The position of

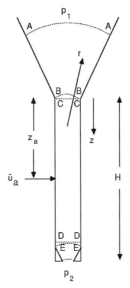

Figure 7.2. Standpipe system.

the aeration point is denoted by z_a. The gas pressure above the fill of particles in the hopper is denoted by p_1 and the pressure below the bottom orifice by p_2.

Throughout the system the motions of both particles and gas satisfy continuity equations

$$\frac{\partial}{\partial t}(\rho_g \varepsilon) + \nabla \cdot (\rho_g \varepsilon \mathbf{u}) = 0 = \frac{\partial}{\partial t}(\rho_s \phi) + \nabla \cdot (\rho_s \phi \mathbf{v})$$

and momentum equations of the form (2.30) and (2.31), namely

$$\rho_g \varepsilon \left[\frac{\partial \mathbf{u}}{\partial t} + \mathbf{u} \cdot \nabla \mathbf{u} \right] = \varepsilon \nabla \cdot \mathbf{S}^g - n\mathbf{f}_1 + \rho_g \varepsilon \mathbf{g} \qquad (7.1)$$

and

$$\rho_s \phi \left[\frac{\partial \mathbf{v}}{\partial t} + \mathbf{v} \cdot \nabla \mathbf{v} \right] = \nabla \cdot \mathbf{S}^p + \phi \nabla \cdot \mathbf{S}^g + n\mathbf{f}_1 + \rho_s \phi \mathbf{g}. \qquad (7.2)$$

For the present purpose these can be simplified in several ways. First, attention will be limited to steady flows so that time derivatives vanish. Second, because of the small density of the gas, inertial and gravitational terms can be omitted from the gas-phase momentum equation. Third, viscous stress transmission in the gas phase can be neglected, so we can write $\mathbf{S}^g = -p^g \mathbf{I}$. Fourth, the density of the solid material can be regarded as constant, so it cancels from the particle-phase continuity equation. Finally, though we shall see that the compressibility of the gas cannot be neglected, it will be assumed that heat transfer between particles and gas is sufficiently rapid for the large thermal inertia of the solid material to maintain essentially isothermal conditions in the gas during its passage through the pipe. Then the density of the gas can be assumed to be proportional to its pressure. With these simplifications the equations of motion become

$$\nabla \cdot (p^g \varepsilon \mathbf{u}) = 0 = \nabla \cdot (\phi \mathbf{v}), \tag{7.3}$$

$$0 = \varepsilon \nabla p^g + n\mathbf{f}_1, \tag{7.4}$$

$$\rho_s \phi \mathbf{v} \cdot \nabla \mathbf{v} = \nabla \cdot \mathbf{S}^p - \nabla p^g + \rho_s \phi g \mathbf{i}, \tag{7.5}$$

where $\mathbf{i}$ is the unit vector in the downward vertical direction.

In a standpipe system the volume fraction of solids can vary widely, between quite small values and ϕ_m, representing a bed of particles at random close packing. Consequently, the closure adopted for the fluid–particle interaction force $n\mathbf{f}_1$ should be one that is valid throughout this wide interval of concentration. Mountziaris & Jackson (1991) used the following form:

$$n\mathbf{f}_1 = [w\beta_{RZ}(\phi) + (1 - w)\beta_{ER}(\phi, |\mathbf{u} - \mathbf{v}|)](\mathbf{u} - \mathbf{v}), \tag{7.6}$$

where β_{RZ} and β_{ER} are the familiar Richardson–Zaki and Ergun expressions, valid respectively at the limits of high and low concentration

$$\beta_{RZ} = \frac{\rho_s \phi g}{v_t \varepsilon^{n-2}}, \quad \beta_{ER} = \frac{\phi}{d}\left[\frac{150\mu_g \phi}{d\varepsilon} + 1.75\rho_g |\mathbf{u} - \mathbf{v}|\right], \tag{7.7}$$

and w is a function of ϕ whose value varies smoothly between one, when $\phi \to 0$, and zero, when $\phi \to \phi_m$. Mountziaris & Jackson chose

$$w = \exp\left[-C_w \frac{\phi}{\phi_m - \phi}\right]. \tag{7.8}$$

Then the transition between the Richardson–Zaki and Ergun forms can be made abrupt or gradual, depending on the value selected for C_w.

Equations (7.3) to (7.5) can be rendered dimensionless in various ways. Mountziaris & Jackson chose the following scaling variables:

length: H,
density: ρ_s,
velocity: $(gr_p)^{1/2}$,
pressure and stress: $\rho_s g H \phi_m$,
viscosity: $\rho_s H^2 (g/r_p)^{1/2}$,

where r_p is the radius of the standpipe. In terms of dimensionless variables, so defined, the continuity equations (7.3) retain the same form, while the momentum balances (7.4) and (7.5) become

$$0 = \frac{\phi_m}{\phi} \nabla p^g + (\mathbf{u} - \mathbf{v}) \left[\frac{w}{v_t \varepsilon^{n-1}} + \frac{(1-w)}{d\varepsilon} \right.$$
$$\left. \times \left\{ \frac{150 \mu_g \phi}{d\varepsilon} + 1.75 \rho_g r_p |\mathbf{u} - \mathbf{v}| \right\} \right] \qquad (7.9)$$

and

$$r_p \mathbf{v} \cdot \nabla \mathbf{v} = \frac{\phi_m}{\phi} [\nabla \cdot \mathbf{S}^p - \nabla p^g] + \mathbf{i}. \qquad (7.10)$$

Here, and in everything that follows, *all* the symbols stand for dimensionless variables, defined as above.

Before applying these equations to each section of the system shown in Figure 7.2 we introduce one more simplifying assumption. For materials such as glass beads or sand the transition from a suspension to a packed bed is quite sharp. For particle concentrations smaller than ϕ_m the suspension behaves like an easily compressible fluid; in other words, its resistance to deformation is relatively small. Once ϕ reaches the value ϕ_m, however, the assembly of particles becomes a packed bed with a very much higher resistance to deformation. In particular, the particle concentration remains very close to ϕ_m. We shall idealize this distinction, classifying the material at any point in the system as either a suspension or a packed bed. For a suspension we shall set $\mathbf{S}^p = 0$ and regard ϕ as a variable quantity to be determined. For a packed bed, we shall set $\phi = \phi_m = $ constant and leave the particle-phase stress to be determined, treating the bed of particles as a Coulomb material. (This mirrors the well-known case of an incompressible fluid, where the pressure must be determined dynamically.)

We shall now proceed to apply the above equations to each section of the apparatus depicted in Figure 7.2. Then we shall show how the resulting models of the separate sections can be linked by matching conditions at their junctions to provide a means of computing the performance of the complete standpipe system.

7.2.2 Motion of Material in the Feed Hopper

The motion in the feed hopper is simplified by assuming that the velocities of both gas and particles are everywhere directed radially towards the vertex of the cone defined by the bounding walls. If the hopper angle is small, so that the walls are steeply inclined, the line of action of the gravity force nowhere deviates much from the radial direction and this assumption should not be greatly in error. Then the continuity equations for both phases can be integrated immediately to give

$$u_r = -\frac{\bar{u}r_p^2 p_1}{2(1 - \cos \alpha)\varepsilon_m r^2 p^g} \tag{7.11}$$

and

$$v_r = -\frac{\bar{v}r_p^2}{2(1 - \cos \alpha)\phi_m r^2}. \tag{7.12}$$

Here $\bar{u}$ and $\bar{v}$ are the dimensionless volumetric flow rates of gas and particles, respectively, expressed as superficial velocities based on the cross-sectional area of the pipe, α is the inclination of the hopper wall to the vertical, and $\varepsilon_m = 1 - \phi_m$. The flow rate $\bar{u}$ is evaluated at pressure p_1 and the ratio p_1/p^g appears in (7.11) because of the compressibility of the gas.

The radial component of the gas-phase momentum balance (7.9) reduces to

$$\frac{dp^g}{dr} = -\frac{(u_r - v_r)}{\varepsilon_m d}\left[\frac{150\mu_g \phi_m}{\varepsilon_m d} + 1.75\rho_g r_p |u_r - v_r|\right], \tag{7.13}$$

and, after substituting from (7.11) and (7.12) for u_r and v_r, this can be integrated in the direction of decreasing r starting from $p^g = p_1$ at the surface of the fill, thus generating the gas pressure profile in the hopper.

The momentum balance for the particle phase is less easily dealt with since it contains the stress tensor $\mathbf{S}^p$, for which some constitutive assumptions must be made. Mountziaris & Jackson (1991) treated the particle assembly as a Coulomb material and expanded the variables in powers of θ, the polar angle of a spherical coordinate system with origin at the vertex of the hopper. The procedure, which is a generalization of that due to Nguyen et al. (1979), is complicated and a full account can be found in Mountziaris (1989); only the results will be quoted here. It is found that the radial component S_{rr}^p of the particle-phase stress must satisfy the following differential equation:

$$\frac{dS_{rr}^p}{dr} - \frac{2(\tilde{K} - 1)S_{rr}^p}{r} = \frac{\bar{v}^2 r_p^5}{2(1 - \cos\alpha)^2 \phi_m^2 r^5} - 1 + \frac{(u_r - v_r)}{\varepsilon_m d}$$
$$\times \left[\frac{150\mu_g \phi_m}{\varepsilon_m d} + 1.75\rho_g r_p |u_r - v_r|\right]. \tag{7.14}$$

The parameter $\tilde{K}$ depends on the angle of internal friction ψ of the particle assembly and the angle of friction ψ_w between the particle bed and the wall in the following way:

$$\tilde{K} = 1 + \frac{2\sin\psi}{1 - \sin\psi}\left[1 + \frac{\gamma}{\alpha}\right],$$

where γ is that root of the equation

$$\sin \psi_w = \sin \psi \sin(2\gamma - \psi_w)$$

lying in the interval $[0, \pi/4 + \psi/2]$.

Substituting from (7.11) and (7.12) for u_r and v_r we can express the right-hand side of (7.14) as a function of r and p^g. But p^g is known as a function of r from the solution of (7.13), so (7.14) can be integrated from an initial condition $S_{rr}^p = 0$ at the surface of the fill. Having solved both (7.13) and (7.14) we can now find the values of the gas pressure p^g and the particle-phase stress S_{rr}^p at the spherical cap BB (Figure 7.2) spanning the top of the standpipe by setting $r = r_p/\sin \alpha$. This completes the model for flow in the feed hopper.

7.2.3 Motion of Material in the Standpipe

In the standpipe the problem will be simplified by neglecting any dependence of the velocities, gas pressure, and particle-phase stress components on the radial coordinate. Then it follows immediately from the continuity equations that

$$u_z = \frac{\bar{u}\, p_1}{\varepsilon p^g}, \qquad v_z = \frac{\bar{v}}{\phi}. \tag{7.15}$$

However, we must recognize that the particle assembly may be moving down the pipe either as a packed bed or as a suspension. In the former case the problem is analogous to that discussed in Chapter 3 and, as there, we follow the classical analysis of Janssen (1895), modified by the effect of the drag force exerted between the two phases and simplified by our assumption that $\phi = \phi_m = $ constant for this mode of flow. Then the axial components of the momentum equations for the gas and particle phases take the following forms:

$$\frac{dp^g}{dz} = -\frac{1}{\varepsilon_m d}\left(\frac{\bar{u}\, p_1}{\varepsilon_m p^g} - \frac{\bar{v}}{\phi_m}\right)$$
$$\times \left[\frac{150\mu_g \phi_m}{\varepsilon_m d} + 1.75\rho_g r_p \left|\frac{\bar{u}\, p_1}{\varepsilon_m p^g} - \frac{\bar{v}}{\phi_m}\right|\right] \tag{7.16}$$

and

$$\frac{dS_{zz}^p}{dz} + \frac{2\tan\psi_w S_{zz}^p}{K r_p} = 1 + \frac{1}{\varepsilon_m d}\left(\frac{\bar{u}\, p_1}{\varepsilon_m p^g} - \frac{\bar{v}}{\phi_m}\right)$$
$$\times \left[\frac{150\mu_g \phi_m}{\varepsilon_m d} + 1.75\rho_g r_p \left|\frac{\bar{u}\, p_1}{\varepsilon_m p^g} - \frac{\bar{v}}{\phi_m}\right|\right]. \tag{7.17}$$

Here the second term on the left-hand side of (7.17) represents support of the material by frictional forces at the pipe wall, where the angle of friction is ψ_w.

The factor $\tan \psi_w / K$ is the Janssen coefficient, defined as the ratio of the shear stress at the wall to the axial stress S_{zz}^p. If it is further assumed that the material moving down the pipe is in a state of active failure; in other words, if it is on the point of compacting everywhere, then the constant K is related to the angle of internal friction ψ by

$$K = \frac{1 + \sin \psi}{1 - \sin \psi}.$$

Integration of (7.16) and (7.17) permits p^g and S_{zz}^p to be found, as functions of z, throughout the packed section of the pipe given their values at its upper end.

If, however, the particle assembly is moving down the pipe as a suspension then, according to our simplifying assumptions, $S_{zz}^p = 0$. But since the particle concentration no longer takes the constant value ϕ_m, we need equations to determine both p^g and ϕ. The differential equation for the gas pressure is (7.9), which now takes the form

$$\frac{\phi_m}{\phi} \frac{dp^g}{dz} = -\left(\frac{\bar{u} p_1}{\varepsilon p^g} - \frac{\bar{v}}{\phi} \right) \left[\frac{w}{v_t \varepsilon^{n-1}} + \frac{(1 - w)}{\varepsilon d} \right.$$
$$\left. \times \left\{ \frac{150 \mu_g \phi}{\varepsilon d} + 1.75 \rho_g r_p \left| \frac{\bar{u} p_1}{\varepsilon p^g} - \frac{\bar{v}}{\phi} \right| \right\} \right]. \qquad (7.18)$$

Also, using the second of (7.15), the particle-phase momentum equation (7.10) reduces to

$$\frac{r_p \bar{v}^2}{\phi^3} \frac{d\phi}{dz} = \frac{\phi_m}{\phi} \frac{dp^g}{dz} - 1, \qquad (7.19)$$

and integration of (7.18) and (7.19) then determines p^g and ϕ as functions of z throughout the section of the pipe occupied by a suspension.

7.2.4 Motion of Material in the Discharge Region

The pipe is terminated below by a circular orifice, of diameter less than or equal to the pipe diameter, above which shoulders of immobile material lodge, so that the material discharges through a roughly conical region bounded by these shoulders. In Figure 7.2 this region extends from the spherical cap EE down to a second spherical cap spanning the terminal orifice. Motion in this region can be approximated in the same way as flow in the feed hopper, using equations similar to (7.11–7.14). The angle α must now be replaced by the half-angle β of the cone formed by the shoulders of immobile material and, according to Wieghardt (1975), this is related to the angle of internal friction of the particulate material by $\beta = \pi/4 - \psi/2$. The wall angle of friction ψ_w is also replaced by $\tan^{-1}(\sin \psi)$. The equations can then be integrated to yield

p^g and S_{rr}^p throughout the discharge region, and at the exit orifice these must satisfy the conditions $p^g = p_2$ and $S_{rr}^p = 0$.

The above is actually a significant simplification of the complete treatment of the discharge region, since it is possible that this region is not entirely filled by a moving bed with $\phi = \phi_m$. The subregions occupied by a suspension and a moving bed must then be dealt with separately. Details can be found in Mountziaris & Jackson (1991).

7.2.5 Matching Conditions

Though we have now found how to propagate values of p^g, and either $\mathbf{S}^p$ or ϕ, in the direction of motion through each separate section of the system the picture is not yet complete, since we have not specified how the values of these variables at the bottom of one section are to be related to those at the top of the next section below. If we had not limited attention to approximate, one-dimensional solutions in each section this would have presented no problem of principle. However, because of the changes from conical to cylindrical geometry, and back again, pointwise matching of all variables is not possible at these transitions using only the one-dimensional solutions in the adjacent regions. Instead we must be satisfied with matches between quantities integrated over the cross sections.

Consider first the transition from the spherical cap BB at the feed hopper exit to the plane CC at the top of the standpipe. The particulate material enclosed between these surfaces is in steady motion so the net flux of z momentum carried out of this region by the particles must balance the net force in the z direction exerted on them by the surface tractions across BB and CC. (Since the thickness of the region between BB and CC is small, assuming that the angle α is small, we can neglect body forces on the enclosed material due to gravity and drag.) It then follows that

$$S_{zz}^p = S_{rr}^p - \frac{\bar{v}^2 r_p}{\phi_m^2} \left[\frac{\phi_m}{\phi} - \frac{1 + \cos\alpha}{2} \right], \tag{7.20}$$

where ϕ denotes the solids volume fraction on CC, the left-hand side of the equation is the axial component of the particle-phase stress at the top of the standpipe, and the first term on the right-hand side is the radial component of the particle-phase stress on the spherical cap at the exit from the feed hopper.

There are now two cases to consider. If the top of the pipe contains a packed bed of particles, then $\phi = \phi_m$, and the above reduces to

$$S_{zz}^p = S_{rr}^p - \frac{\bar{v}^2 r_p}{\phi_m^2} \left[\frac{1 - \cos\alpha}{2} \right], \tag{7.21}$$

which simply allows the value of the particle-phase stress to be translated down from the hopper exit to the top of the pipe. If, however, the top of the pipe is occupied by a suspension, then $S_{zz}^p = 0$ and (7.20) becomes a condition on the radial component of stress at the hopper exit, namely

$$S_{rr}^p = \frac{\bar{v}^2 r_p}{\phi_m^2} \left[\frac{\phi_m}{\phi} - \frac{1 + \cos \alpha}{2} \right]. \tag{7.22}$$

For the gas phase the matching condition is simple. Since the inertia of the gas is regarded as negligible the gas pressure at the top of the pipe is the same as that at the exit from the hopper.

Similar matching conditions apply at an interface between a suspension and a packed bed within the standpipe itself. If S_{zz}^p is the particle-phase stress in the packed bed and ϕ the solids volume fraction in the suspension, at points adjacent to the interface, then

$$S_{zz}^p = \frac{r_p \bar{v}^2}{\phi_m} \left[\frac{1}{\phi} - \frac{1}{\phi_m} \right], \tag{7.23}$$

and again there is no change in the gas pressure.

At an aeration point the only thing that changes is the total gas flow. If this is $\bar{u}$ above the aeration point it must be replaced by $\bar{u} + \bar{u}_a$ everywhere below.

Finally, a matching condition is needed at the transition in geometry where the flow enters the conical discharge region at the lower end of the standpipe. This is analogous to (7.21) above, but there are certain complications and these are explained fully in the original publications (Chen et al., 1984; Mountziaris & Jackson, 1991).

7.2.6 Sketch of the Solution Procedure

The procedure for generating a solution in the complete standpipe system will now be sketched briefly. This is not intended as a detailed prescription for the computations but as an illustration of the nature of the boundary value problems that must be addressed. Similar considerations would be expected to arise in standpipe configurations other than this particular one.

The situation is simplest in the absence of aeration. Then the pattern of motion takes one of the three forms sketched in Figure 7.3. For given values of p_1 and p_2 these appear in succession as the diameter of the discharge orifice at the foot of the pipe is increased. In regime 1 the whole system is filled with a moving bed; in effect it is just a hopper of rather unconventional shape. As the orifice diameter is increased the particle-phase stress at the top of the pipe decreases and eventually drops to zero, at which point a short section occupied

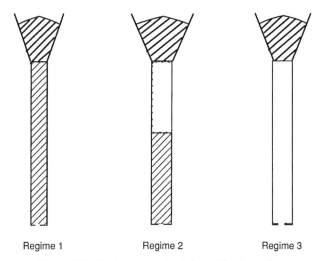

Figure 7.3. Regimes in a standpipe without aeration.

by a suspension appears below the feed hopper, and regime 2 begins. In regime 2 the feed hopper is terminated below by a free-fall surface spanning the top of the standpipe. Above this the hopper is occupied by a moving bed but below it the particles fall as a suspension. However, there is a second interface at a lower level below which flow again occurs as a moving bed. This interface moves down the pipe as the diameter of the discharge orifice is increased until finally it reaches the exit, and in regime 3 flow occurs as a suspension everywhere below the top of the standpipe. (The original paper divides regime 2 into two subregimes depending on whether the transition back to moving bed flow is located in the pipe proper or in the short conical discharge region above the exit orifice. This distinction is not important in principle but it affects the details of the computational procedure.)

The main objective of the computations is to determine how the flows $\bar{u}$ and $\bar{v}$ of gas and particles are related to the pressures p_1 and p_2 and the diameter of the discharge orifice. However, for computational convenience it is best to start with a given orifice diameter and a specified value for $\bar{v}$. The calculations are then initiated by assuming a value for $\bar{u}$. The subsequent course of the calculations then differs for the different flow regimes. For regime 1 Equations (7.13) and (7.14) are integrated down the feed hopper from the starting conditions $p^g = p_1$, $S_{rr}^p = 0$ at the surface of the fill. Thus the values of these variables are found on the spherical cap BB in Figure 7.2. On the plane CC at the top of the pipe p^g remains the same, and S_{zz}^p is found from (7.21). Equations (7.16) and (7.17) can then be integrated down the pipe to give p^g and S_{zz}^p on the plane

DD at the bottom, and their values are finally propagated to the exit orifice using matching conditions between surfaces DD and EE, then using differential equations analogous to (7.13) and (7.14) in the discharge region. The condition of vanishing particle-phase stress at the discharge orifice will not, in general, be met so it is necessary to repeat the calculation, iterating on the assumed value of $\bar{u}$, until this stress vanishes. The value of p^g at the exit orifice then determines p_2 and the calculation is complete.

For regime 3 the calculation is started in the same way, but we now require that S_{zz}^p should vanish on the plane CC. The condition for this is (7.22) and if there is to be no jump in the value of ϕ we can set $\phi = \phi_m$ in this equation. The initially assumed value of $\bar{u}$ must then be adjusted until (7.22) is satisfied. Equations (7.18) and (7.19) can then be integrated down the standpipe to determine $\phi(z)$ and $p^g(z)$ and, provided ϕ is nowhere found to exceed ϕ_m, this completes the solution. Once again the exit value of p^g gives p_2.

This procedure fails if the value of $\phi(z)$ increases beyond ϕ_m anywhere in the pipe or the discharge region, in which case we must seek a solution in regime 2. The calculation then starts in the same way as for regime 3, but at some point before ϕ reaches the value ϕ_m we must postulate the presence of an interface separating the suspension from a moving bed below. The particle-phase stress S_{zz}^p at the top of this moving bed is determined by (7.23). Then p^g and S_{zz}^p are propagated down from this interface to the exit orifice in the same way as described for regime 1. In general the value of S_{zz}^p at the exit will not vanish, as it should, so it is necessary to iterate on the assumed position of the interface separating the suspension from the moving bed in the pipe until this condition is satisfied.

7.3 Predicted Behaviour of the Unaerated Standpipe

For given values of the pipe length and the upper pressure p_1 the predictions of the theory sketched above can be presented as curves relating the dimensionless particle flow rate, represented by $\bar{v}$, to the dimensionless pressure rise $\Delta p = p_2 - p_1$, for various fixed values of the exit orifice diameter, represented by $w_v = r_o/r_p$, the ratio of the orifice radius to the pipe radius. Figure 7.4 shows a set of these curves computed by Chen (1983) for a fine sand. The geometry of the system and the physical properties needed by the theory are specified in Table 7.1.

Curve 1 in Figure 7.4 corresponds to the pipe with no restriction at its lower end, while curves 2, 3, ... show the relation between $\bar{v}$ and Δp for successively smaller discharge orifices. The parts of each curve corresponding to the different flow regimes are distinguished by the type of line used, as indicated on the

Table 7.1. *Parameter values for the standpipe studied*

Pipe length H	3.27 m
Pipe diameter r_p	25.4 mm
Feed hopper angle α	9.5°
Particle mean diameter d	154 μm
Solids density ρ_s	2,620 kg/m^3
Maximum solids volume fraction ϕ_m	0.64
Discharge cone angle β	28°
Angle of internal friction ψ	34°
Angle of wall friction ψ_w	16.5°
Richardson–Zaki exponent n	5
Richardson–Zaki parameter v_t	3.0 m/s
Coefficient in drag force expression C_w	0.01

diagram, and each point where there is a change in regime is marked by a black circle. For the bare feed hopper, with no pipe attached, the flow rate would correspond to $\bar{v} \approx 0.7$.

Focussing first on curve 1, when $\Delta p = 0$ we see that the flow is in regime 3 and the discharge rate is between seven and eight times as large as for the bare feed hopper. As Δp is increased from zero $\bar{v}$ decreases, quite slowly at first, but then rather more rapidly. However, when $\Delta p \approx 0.55$ there is a change in flow regime. A moving bed begins to form at the bottom of the pipe and the slope of the curve reverses as regime 2 becomes established. Now $\bar{v}$ and Δp decrease together on moving along the curve in the same sense, and the interface separating the suspension from the moving bed moves up the pipe. When the interface reaches the top of the pipe regime 1 is established, and at this point the flow rate is a little smaller than the value for the bare hopper. The whole system is now occupied by a moving bed and the slope of the curve once more reverses, so $\bar{v}$ now decreases as Δp is increased, and it falls to zero when the back pressure becomes large enough to prevent downflow of particles.

Rangachari & Jackson (1982) have presented theoretical arguments indicating that segments of the curves in Figure 7.4 with positive slope correspond to unstable steady states, so these states should not be realized in practice. As a result there should be a large hysteresis in the flow rate as Δp is first increased, then subsequently decreased. As Δp increases through 0.55 the flow rate should drop discontinuously by about an order of magnitude as the pipe fills up with a moving bed. This sudden drop is associated with a sudden transition from regime 3 to regime 1. If Δp is now reduced progressively the system continues to operate in regime 1, with flow rates less than that of the bare feed hopper, until $\Delta p \approx 0.25$. At that point there is a discontinuous increase in the discharge

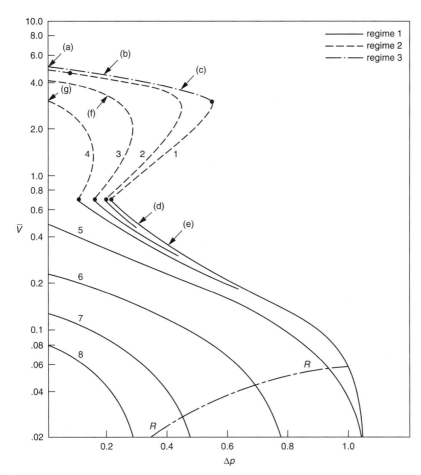

Figure 7.4. Particle flow rate $\bar{v}$ as a function of Δp, with various values of w_v, for the system of Table 7.1. 1: $w_v = 1.0$, 2: $w_v = .94$, 3: $w_v = .90$, 4: $w_v = .87$, 5: $w_v = .80$, 6: $w_v = .70$, 7: $w_v = .60$, 8: $w_v = .50$. (From Chen, 1983.)

rate by a factor between six and seven as the moving bed disperses and flow in regime 3 is reestablished.

It is interesting that this prominent hysteresis phenomenon is entirely a consequence of the compressibility of the gas. Results of calculations essentially identical to those described here, but with the factor p^g omitted from the gas-phase continuity equation (7.3), have been reported by Ginestra et al. (1980), who found $\bar{v}$ to be a continuous, monotone decreasing function of Δp in all cases. On reflection it is not surprising that the gas compressibility should have a large influence on the performance of the system. The change in pressure on moving down the pipe induces a corresponding change in the volume flow rate of the gas, and hence in its velocity. The fractional change in the gas velocity should be comparable with the fractional change in pressure in the pipe, and typically this is not large. However, the *relative* velocity of particles and gas is usually much smaller than the velocity of the gas down the pipe, so a small fractional change in the latter quantity may represent a large fractional change in the former, and it is this relative velocity that is important in determining the behaviour of the system.

Now let us turn our attention to cases in which the pipe is restricted below by a discharge orifice. For curve 2 in Figure 7.4 the diameter of the discharge orifice is 0.94 times the pipe diameter, so it presents only a mild resistance to flow. The overall shape resembles that of curve 1, but now regime 3 is found only for quite small values of Δp. High discharge rates persist to much larger values of the pressure rise, but with a flow pattern corresponding to regime 2. This regime continues along the segment of the curve with positive slope, representing unstable states of operation, with a transition to regime 1 only when the slope again becomes negative. Thus, in contrast with curve 1, we now find an extended interval of *stable* operation in regime 2, with high discharge rates. Curves 3 and 4, for successively smaller values of the discharge orifice diameter, are similar in shape to curve 2 but there is no longer any interval of operation in regime 3. As the orifice diameter is decreased the interval of Δp preceding the discontinuous drop in flow rate shrinks, and for curves 5 to 8 there is no such interval. In other words, for these restrictive discharge conditions, the system always operates in regime 1, completely filled with a moving bed. Correspondingly, the discharge rate is always smaller than that for a bare feed hopper. Clearly these are undesirable operating conditions. Finally, note the broken curve labelled RR in the diagram. Above this curve the flow of gas is downward relative to the pipe, while below it is upward. This distinction may be important in practice, since standpipes may be required to seal against a backflow of gas.

The predictions of Figure 7.4 should be compared with experimental results of Chen (1983) shown in Figure 7.5. The overall pattern of behaviour seems to be the same in both figures, except that the unstable states predicted theoretically are, of course, not seen in the experiments. For large orifice diameters ($w_v \geq 0.94$) there is an interval of operation at high discharge rate, terminated by a catastrophic drop in rate if the back pressure is increased too much. To recover the high flow rate it is then necessary to decrease Δp well below the value that induced the catastrophic drop, so we see the characteristic hysteresis loop predicted theoretically. For smaller orifice diameters (curves 5 and 6) the upper branch is missing, just as the theory predicts. Despite this encouraging similarity between experiment and theory there are two points of difference.

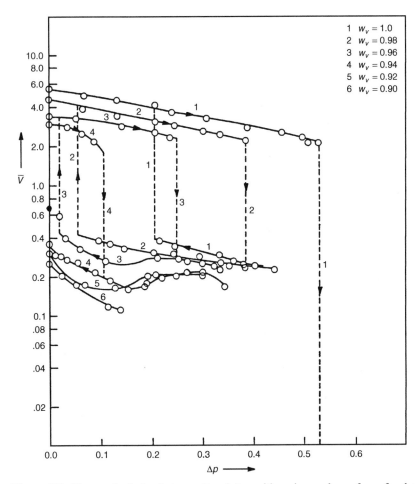

Figure 7.5. Measured relation between $\bar{v}$ and Δp, with various values of w_v, for the system of Table 7.1. (From Chen, 1983.)

First, the experiments did not identify any conditions of operation in regime 1 for the unrestricted pipe with $w_v = 1$. Second, the minimum and maximum seen in the experimental curve 5 and the lower branches of curves 3 and 4 have no analogue in the theoretical predictions.

At first glance the overall quantitative agreement between the predicted (Figure 7.4) and measured (Figure 7.5) curves appears to be remarkably good. However, closer scrutiny shows that the experimental system is much more sensitive to restriction by the discharge orifice than is predicted theoretically. For example, when the value of w_v is reduced to 0.92 the experiments can find no upper branch with high discharge rate, while the theory predicts quite an extensive upper branch for this value of w_v. Indeed, according to the theory w_v must be reduced to 0.8 before the upper branch is suppressed. Clearly, the model for flow through the discharge region, while predicting the correct trends, is far from quantitative accuracy. It should also be noted that the good agreement between theoretical and measured values of $\bar{v}$ on the upper branch, for small values of Δp and with $w_v = 1$, was *imposed* by adjusting the value of v_t used in the Richardson–Zaki term of the drag force. Indeed, the value of this parameter reported in Table 7.1 is about 50% larger than the true terminal velocity of fall of an isolated particle. The need for an adjustment of this sort is not surprising. From the results in Chapter 6 we expect that the spatial distribution of the particle concentration will be nonuniform in suspension flow down the pipe, so the effective slip velocity between the phases, averaged over the cross section, will be larger than would be predicted for a uniform suspension. This effect can be accommodated crudely in a one-dimensional flow model, such as that used here, only by reducing the effective drag force (i.e., increasing the value of v_t).

Figures 7.6, 7.7, and 7.8 show gas pressure profiles under various operating conditions. Each curve is labelled with a letter and the corresponding operating conditions are identified in Figure 7.4. Profiles (a), (b), and (c) in Figure 7.6 all correspond to points on the upper stable segment of curve 1 in Figure 7.4, so they represent rapid flows with no restriction at the bottom of the pipe. For profile (a) there is no pressure rise across the system, while the pressure rise is larger for (b) and (c), in that order. The pressure profiles in the pipe are almost parallel for all three cases. Thus the vacuum at the top of the pipe becomes successively weaker in the order (a)–(b)–(c). This shows clearly how the vacuum at the feed hopper exit controls the flow rate of particles, which decreases in the same order. The pressure profiles within the pipe are not quite linear; the rate of increase of pressure with depth becomes somewhat larger on moving down the pipe. This reflects the decrease in volume flow of gas as the pressure increases, with a consequent increase in the bulk density of the suspension. Figure 7.6 should be compared with the set of pressure profiles measured by Judd & Rowe (1978) and shown in Figure 7.1.

The pressure profiles (d) and (e) of Figure 7.7 again belong to points on curve 1 in Figure 7.4, but now they lie on the lower stable segment of this curve. The pressure now increases monotonically on moving down throughout the system. The vacuum at the feed hopper exit is thus replaced by a small positive pressure, and this is why the solids flow rates at points (d) and (e) in Figure 7.4 are somewhat smaller than the bare hopper discharge rate. Despite the low flow rates achieved both points (d) and (e) lie above the curve RR, so the pipe still provides a seal against upflow of gas.

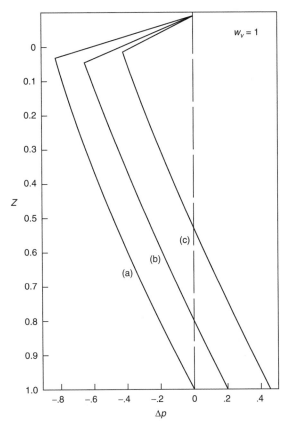

Figure 7.6. Predicted pressure profiles for the system of Table 7.1, with $w_v = 1.0$. Curves labelled (a), (b), and (c) refer to correspondingly labelled operating points in Figure 7.4. (After Chen, 1983.)

Figure 7.8 shows two pressure profiles, (f) and (g), for operation with restrictive discharge orifices. The corresponding points in Figure 7.4 both lie on the upper stable segments of their respective curves, but in each case the system operates in regime 2. In the upper part of the pipe, where the particles descend as a suspension, they are moving downward relative to the gas. Consequently, the gas pressure increases on moving down. For each pressure profile the interface between the suspension above and the moving bed below is easily identified by the point at which the slope of the curve reverses. At the transition to a moving bed the particle velocity decreases suddenly, and below the interface the gas moves down relative to the particles. Consequently, the gas pressure decreases on moving down. In each case there is a vacuum at the exit from the feed hopper, so the flow of particles exceeds the discharge rate

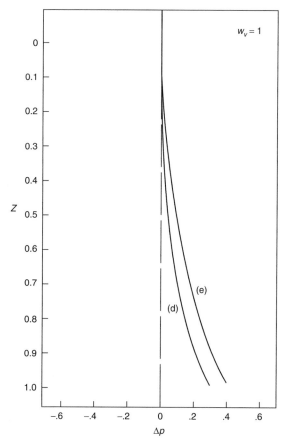

Figure 7.7. Predicted pressure profiles for the system of Table 7.1, with $w_v = 1.0$.
Curves labelled (d) and (e) refer to correspondingly labelled operating points in
Figure 7.4. (After Chen, 1983.)

of the bare hopper. The vacuum is stronger for curve (f) than for curve (g) and
correspondingly, as seen from Figure 7.4, the flow rate is larger at point (f). The
discontinuity in the pressure gradient that occurs at the suspension–moving
bed interface has also been observed in experimental measurements, providing
clear evidence of the coexistence of suspension and moving bed flow within
the pipe.

    In summary, it appears that the simple, one-dimensional model described
here is capable of giving a good account of the qualitative features of flow in
an unaerated standpipe, although for quantitatively realistic prediction of the
relations among discharge rate, orifice size, and pressure rise it is necessary
to regard the parameter v_t as adjustable, choosing a value significantly larger

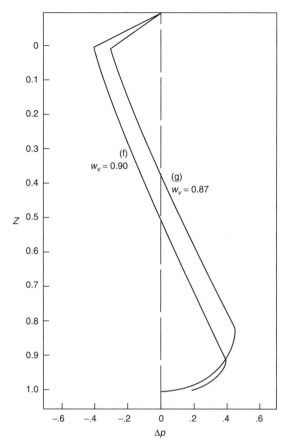

Figure 7.8. Predicted pressure profiles for the system of Table 7.1. Curves labelled (f) and (g) refer to correspondingly labelled operating points in Figure 7.4. Curve (f): $w_v = 0.90$. Curve (g): $w_v = 0.87$. (After Chen, 1983.)

than the true terminal velocity to simulate the effect of nonuniform spatial distribution of the particles.

7.4 Predicted Behaviour of the Aerated Standpipe

The potential advantage to be gained from aeration is clear from the pressure profiles of Figure 7.8. Since the buildup of pressure down the pipe is reversed when the suspension gives way to a moving bed, the overall pressure buildup would be greater if the moving bed in the lower section of the pipe could be eliminated. Then flow could be sustained against a larger back pressure, or the discharge rate could be increased for the same value of the back pressure.

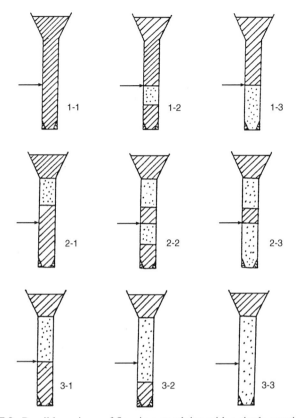

Figure 7.9. Possible regimes of flow in a standpipe with a single aeration point.

The transition from suspension flow to moving bed flow within the pipe is a consequence of the decrease in the volume flow of gas resulting from its increase in pressure, so this transition can be delayed or eliminated by supplementing the gas flow with injected aeration gas. The theoretical model described above can be adapted to take account of this merely by increasing the gas flow by an increment $\bar{u}_a$ on moving down past the aeration point. However, this simple change has profound consequences for the performance of the system.

If too large an aeration flow is introduced at a point in the pipe where the particles are originally flowing as a suspension, a moving bed may form above the aeration point. Consequently, each of the three flow regimes identified in Figure 7.3 may be found, both in the section of the pipe above the aeration point and in the section below. This gives a total of nine possible flow regimes in the system as a whole, as sketched in Figure 7.9. Each of these can be identified by a pair of numbers, the first of which describes the flow pattern above the aeration point and the second the pattern below. Thus, in a 2-2 regime each section of the pipe contains a suspension above a moving bed, whereas in a 1-3 regime the upper section is entirely filled with a moving bed and the lower section with a suspension.

There are now three operating variables whose values determine $\bar{u}$ and $\bar{v}$, namely w_v, Δp, and $\bar{u}_a$, and the task of identifying the correct flow regime and generating a complete solution for given values of these parameters is much more complicated than was the case for the unaerated pipe. It is also more difficult to form an overall picture of the influence of the operating variables on the particle flow. For the unaerated pipe $\bar{v}$ depended only on Δp and w_v, so a set of curves, such as the sections of constant w_v shown in Figure 7.4, provided all the necessary information. For the aerated pipe, in contrast, each section of constant w_v is a *surface* in the three-dimensional space of $\bar{v}$, Δp, and $\bar{u}_a$.

Despite these problems Mountziaris & Jackson (1991) obtained results for a selection of operating conditions in a standpipe system identical with that of Section 7.3, except for the addition of aeration. The advantages and dangers of aeration are illustrated well by Figure 7.10, which shows four computed plots of $\bar{v}$ versus aeration rate for a fixed discharge orifice diameter $w_v = 0.8$, and for four values of the pressure rise Δp, corresponding to the four panels. In each case the single aeration point is located at $z = z_a = 0.54$. Figure 7.4 shows that, without aeration, this system operates in regime 1 for all values of Δp, and consequently the discharge rate is always low.

Panel (i) of Figure 7.10, corresponding to $\Delta p = 0$, illustrates the effect of aeration in its simplest manifestation. Starting at a small value of $\bar{u}_a$ the system operates in regime 1-1, completely filled with a moving bed, and $\bar{v}$ is quite small. As $\bar{u}_a$ is increased a regime transition (indicated by an open circle) occurs. Suspensions appear in the sections of the pipe both above and below the aeration point, so the system now operates in regime 2-2. At the same time the particle flow rate begins to increase significantly. With further increase in the aeration rate the parts of the pipe occupied by suspensions grow in length at the expense of the parts occupied by moving beds. First, the length of the moving bed above the aeration point shrinks to zero, causing a transition to regime 3-2. Then the moving bed below the aeration point follows suit, leaving the pipe occupied entirely by a suspension. This is the transition to regime 3-3. Immediately following this transition the discharge rate reaches its largest value, which is about ten times the value without aeration. However, increasing the aeration rate beyond this point leads to a quite rapid decrease in the particle flow and eventually there is a transition to regime 2-3, with a section of moving bed appearing above the aeration point. The aeration rate cannot be increased much beyond this regime transition before the flow becomes unstable and there is a

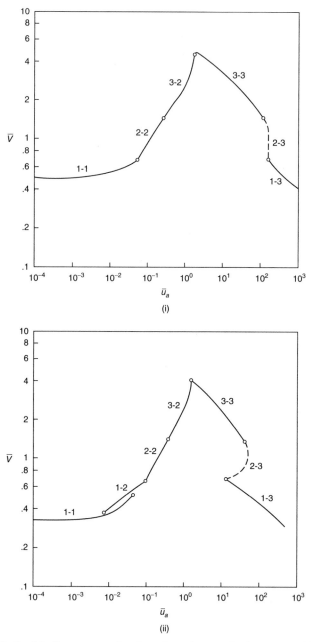

Figure 7.10. Particle flow rate as a function of aeration rate for the system of Table 7.1, with $w_v = 0.8$, $z_a = 0.54$, and four values of Δp. (i) $\Delta p = 0$; (ii) $\Delta p = 0.2$; (iii) $\Delta p = 0.4$; (iv) $\Delta p = 0.7$. (Reprinted from Mountziaris & Jackson, 1991. Copyright 1991, with permission from Elsevier Science.)

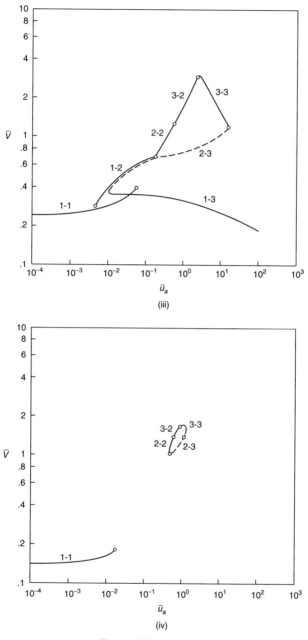

Figure 7.10. (*Cont.*)

discontinuous drop in the discharge rate, accompanied by an abrupt transition to regime 1-3, with the whole region above the aeration point occupied by a moving bed. Though this sequence of flow patterns could not have been anticipated without solving the problem formally, it is not surprising that the largest particle flow occurs when the aeration rate is just sufficient to ensure that the pipe is filled with a suspension. Increasing the aeration rate beyond this value merely reduces the particle concentration, thereby reducing the pressure difference needed to support the suspension.

Panel (ii) of Figure 7.10, corresponding to a pressure rise $\Delta p = 0.2$, has much the same structure as panel (i). The only significant difference is the appearance of a segment on which the system operates in regime 1-2. This overlaps the 1-1 segment, with the result that there is a small hysteresis loop. As the aeration flow is increased from a low value the operating point moves to the right along the 1-1 segment until this segment ends, at which point it jumps discontinuously to the point on the 1-2 segment immediately above it. If the aeration flow is thereupon decreased the operating point then moves back to the left, but now along the 1-2 segment until it reaches its end, at which point it drops back onto the 1-1 segment. For larger values of the aeration rate the sequence of operating regimes is the same as in panel (i).

Panel (iii), for $\Delta p = 0.4$, retains the same sequence of operating regimes as panel (ii), but the right-hand part of the diagram is now pushed back so far that it overlies part of the 1-1 segment, and this changes the sequence of regimes encountered when the aeration flow is first increased, then decreased. Starting from a low aeration rate the operating point moves out along the 1-1 segment to its end, before jumping onto the 1-2 segment, as in panel (ii). With further increase in the aeration flow the discharge rate increases more rapidly and the mode of operation changes from 1-2 to 2-2, then to 3-2, and finally to 3-3, at which point the discharge rate reaches its largest value. As before, further increase in the aeration rate causes a rapid decrease in the particle flow, followed by a catastrophic discontinuous drop by a factor of about six as the system relapses into the 1-3 mode. If the aeration rate is then decreased the operating point moves to the left along the 1-3 segment until this becomes unstable, when the system jumps to the 1-2 segment, follows this down to its end, then jumps back onto the 1-1 segment from which it started.

For each of the three cases discussed above there is an optimum value of the aeration rate. Higher aeration rates are counterproductive and can lead to a sudden and disastrous drop in the particle flow rate. There have been reports of just such an effect of overaeration in commercial operation of standpipes.

The situation for $\Delta p = 0.7$, represented in panel (iv), is particularly interesting. Some features of the sequence of regimes seen in the earlier panels can still be recognized, but now the 2-3 and 2-2 segments have met to form an isolated closed loop. The uppermost point of this loop represents the maximum possible rate of particle discharge and the regime there is of type 3-3, as before. However, operating states on the isolated loop cannot be reached by increasing the aeration rate from an initially small value. To reach these states it is necessary to manipulate at least two of the operating variables. For example, by decreasing Δp or by increasing the diameter of the discharge orifice the two disjoint parts of the diagram can be linked together again. Then it is possible to increase the aeration rate from a low initial value to somewhere near its optimum value. The original values of Δp and the orifice diameter can then be restored, reestablishing the isolated loop, but now the operating point is stranded on this loop and the aeration rate can be adjusted to maximize the particle flow.

The predicted effects of varying the aeration rate are seen to be complicated, but a limited number of experimental tests (Mountziaris & Jackson, 1991) suggest that these effects are real. For example, Figure 7.11, which shows the results of measurements of $\bar{v}$ as a function of $\bar{u}_a$, should be compared to the predicted curve in Figure 7.10(ii). The two are of the same form, except that the experiments could not be extended to an aeration rate large enough to precipitate the catastrophic drop in particle flow at the right-hand end of the 3-3 segment. Even the small predicted hysteresis loop between the 1-1 and 1-2 segments is found in the experiments. Other experimental evidence can be found in Mountziaris & Jackson (1991).

In long standpipes the practice is to introduce aeration progressively through a sequence of aeration ports spaced down the pipe, and this further increases the number of conceivable flow regimes within the system. However, one might anticipate that many of these would occur only with unreasonable strategies for distributing the aeration among the ports. When there is just one aeration point, as in the calculations described above, it is interesting to vary its position and calculate how the optimum aeration rate (i.e., the value of $\bar{u}_a$ that maximizes the particle flow rate) and the corresponding value of the particle flow respond. Figure 7.12 shows this for a system with $w_v = 0.9$ and $\Delta p = 0$. The length of the pipe has been doubled from 3.27 to 6.54 m to reveal the features of greatest interest. As the aeration point is moved down from the top of the pipe both the maximum particle discharge rate, and the aeration rate needed to achieve this, increase monotonically. The flow pattern, which starts as regime 3-2, switches to regime 3-3 at $z_a \approx 0.23$. This regime persists until $z_a \approx 0.62$, but if the aeration point is moved any further down the pipe there is a sudden transition to regime 1-1, with the whole system occupied by a moving bed, and a corresponding discontinuous drop in the particle flow rate to a much smaller value. Thus, with a single aeration port, the best way to run the standpipe is

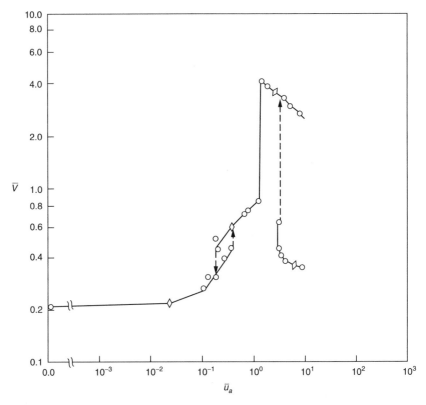

Figure 7.11. Measured relation between particle flow rate and aeration rate for the system of Table 7.1, with $w_v = 0.92$, $z_a = 0.54$, and $\Delta p = 0$. (Reprinted from Mountziaris & Jackson, 1991. Copyright 1991, with permission from Elsevier Science.)

to place this as low as possible without precipitating a catastrophic drop in the discharge rate. Physically this strategy makes sense. To achieve the largest pressure rise as much as possible of the pipe should be occupied by a suspension whose concentration is as high as it can be without inducing collapse to form a moving bed. The addition of aeration should therefore be delayed as long as possible; extra gas is introduced only when it is essential to prevent the particles from coming together as a moving bed.

7.5 General Comments

Detailed dynamic modelling of standpipes has received much less attention than that devoted to risers, though both components are vital to the operation of any circulating fluidized bed system. Comparison of Chapters 6 and 7 underlines

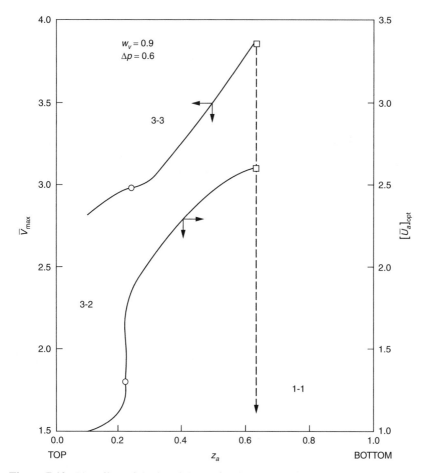

Figure 7.12. The effect of the location z_a of a single aeration point on the maximum particle flow rate and the aeration rate needed to achieve this, for the system of Table 7.1, but with pipe length increased to 6.54 m. Operating conditions: $w_v = 0.9$, $\Delta p = 0.6$. (After Mountziaris, 1989.)

this difference. The models of riser flow described in Chapter 6 take account of radial as well as axial variations in velocities and concentration, and they also attempt to incorporate the effect of "pseudoturbulent" fluctuations in these quantities. The treatment of standpipe flow, in contrast, has been limited to a one-dimensional model comparable to those used for riser flow fifteen to twenty years earlier. Nevertheless, because of the dominant role played by interactions among the feed system, the pipe proper, and the discharge arrangements at its foot, this simple model appears to give a remarkably good qualitative account of the rather complicated behaviour of the complete system. To advance to a quantitatively predictive model it will be necessary to replace the one-dimensional treatment of suspension flow in the pipe by a treatment of comparable sophistication to that used in riser models. Since these models are available there would appear to be no insuperable difficulty in doing this.

An important feature of the model described here is the sharp distinction between a suspension and a moving bed. For materials such as sand, or glass beads, this is a valid simplification, but some of the most important industrial standpipes handle fluidized cracking catalyst. As seen in Chapter 3 this is a material that can support some degree of compressive stress over an extended range of bulk densities, so the transition from a suspension to an essentially constant-density moving bed is no longer a sharp one. Though a model of the type described in this chapter has not yet been extended to encompass this sort of material there is no reason why this should present any great difficulty. The existence of expanded states that retain a yield stress is the most likely reason for the success of FCC standpipes in building pressure efficiently. The residual yield stress in the particle assembly can restrain the instabilities that break up a true suspension into bubbles, clusters, and streamers, so the effective slip velocity between the gas and the particles will remain small.

References

Campbell, D. L., Martin, H. Z., & Tyson, C. W. 1948. U. S. Patent 2,451,803.

Chen, Y-M. 1983. Flow of particulate materials in a vertical standpipe. Ph.D. thesis, University of Houston.

Chen, Y-M., Rangachari, S., & Jackson, R. 1984. Theoretical and experimental investigation of fluid and particle flow in a vertical standpipe. *Ind. Eng. Chem. Fundam.* **23**, 354–370.

Ginestra, J. C., Rangachari, S., & Jackson, R. 1980. A one-dimensional theory of flow in a vertical standpipe. *Powder Technol.* **27**, 69–84.

Janssen, H. A. 1895. Versuche uber getriededruck in silozellen. *Z. Ver. Deutsche Ing.* **39**, 1045–1049.

Judd, M. R. & Rowe, P. N. 1978. Dense phase flow of powder down a standpipe. In *Fluidization*, ed. J. F. Davidson & D. L. Keairns, pp. 110–115. Cambridge University Press.

Knowlton, T. M. 1997. Standpipes and return systems. In *Circulating Fluidized Beds*, ed. J. R. Grace & A. Avidan, pp. 214–260. Blackie.

Kojabashian, C. 1958. Properties of dense-phase fluidized solids in vertical downflow. Ph.D. thesis, Massachusetts Institute of Technology.

Leung, L. S. 1977. Design of fluidized gas–solids flow in standpipes. *Powder Technol.* **16**, 1–6.

Leung, L. S & Jones, P. J. 1978a. Flow of gas–solid mixtures in standpipes. A review. *Powder Technol.* **20**, 145–160.

Leung, L. S. & Jones, P. J. 1978b. Coexistence of fluidized solids flow and packed bed flow in standpipes. In *Fluidization* ed. J. F. Davidson & D. L. Keairns, pp. 116–121. Cambridge University Press.

Leung, L. S. & Wilson, L. A. 1973. Downflow of solids in standpipes. *Powder Technol.* **7**, 343–349.

Matsen, J. M. 1973. Flow of fluidized solids and bubbles in standpipes and risers. *Powder Technol.* **7**, 93–96.

Mountziaris, T. J. 1989. The effects of aeration on the gravity flow of particulate materials in vertical standpipes. Ph.D. thesis, Princeton University.

Mountziaris, T. J. & Jackson, R. 1991. The effects of aeration on the gravity flow of particles and gas in vertical standpipes. *Chem. Eng. Sci.* **46**, 381–407.

Nguyen, T. V., Brennen, C., & Sabersky, R. H. 1979. Gravity flow of granular materials in conical hoppers. *J. Appl. Mech.* **46**, 529–535.

Rangachari, S. & Jackson, R. 1982. The stability of steady states in a one-dimensional model of standpipe flow. *Powder Technol.* **31**, 185–196.

Wieghardt, K. 1975. Experiments in granular flow. *Annu. Rev. Fluid Mech.* **7**, 89–114.

Author Index

Where page numbers are given in *italics*, they refer to publications cited in lists of references.

Subject Index

aggregative fluidization, 99, 153
angle of repose, 144
angular momentum equation
  particle phase average, 24, 25
  single particle, 24
averages, ensemble, 18
averages, local spatial
  comparison with ensemble averages, 18,
    60–2
  dependence on separation of scales, 19
  fluid-particle force, 23, 27, 35
  fluid phase; definition, 20
  fluid phase; differentiation, 20
  hard, 19
  mass weighted; definition, 21
  overall; definition, 19
  particle-particle force, 24
  particle phase; definition, 21
  particle phase; differentiation, 22
  radius of, 19
  soft, 19
  solid phase; definition, 21
  weighting functions for, 19

bed height
  measurements of, 91–3, 95–6
  prediction of, 83–9
bifurcations
  one-dimensional structures, 189–91, 221–2
  theory of Göz, 227
  two-dimensional structures, 203–10, 223–4
boundary conditions
  momentum, for gas phase, 241–3
  momentum, for particle phase, 239–40
  pseudo-thermal energy, 240–1
bubbles
  Collins' model of, 174–5

Davidson cloud accompanying, 159, 160–2,
    176–8, 206, 211
  Davidson's model of, 157–63
  fluid pressure field near, 179
  fluid streamlines near, 160, 161, 168, 169,
    172, 173, 175, 214
  growth of, computed, 208–16, 227–29
  Jackson's model of, 163–7
  Murray's model of, 167–73
  photographs of, 154, 155
  pressure, effect on, 156–7
  relation to clusters, 221–4
  rise velocity, 165, 172
  tracer injection study of, 160–3
  void fraction near, 167, 179–82
  X-ray studies of, 154–5, 182
bulk density, 18
buoyancy, 26–32

catalytic cracking, 15, 233, 298
closures for averaged quantities
  effective fluid phase stress, 34, 38, 54–5
  effective particle phase stress, 38, 39–40,
    41, 54–5
  fluid-particle interaction force, 35, 38, 39,
    41, 48–53
  fluid phase stress, 33
  pseudothermal energy flux, 38, 39–40
compressor characteristics, 8
computational fluid dynamics (CFD), 2,
    288–95
consolidation loci, 68–9
continuity equations, 22
  linearized form, 104
continuity waves, 110
  and stability criterion, 120
cyclone, 300

336